Kinematics and Robot Design IV, KaRD2021

Kinematics and Robot Design IV, KaRD2021

Editor

Raffaele Di Gregorio

MDPI • Basel • Beijing • Wuhan • Barcelona • Belgrade • Manchester • Tokyo • Cluj • Tianjin

Editor
Raffaele Di Gregorio
Engineering Department
University of Ferrara
Ferrara
Italy

Editorial Office
MDPI
St. Alban-Anlage 66
4052 Basel, Switzerland

This is a reprint of articles from the Special Issue published online in the open access journal *Robotics* (ISSN 2218-6581) (available at: https://www.mdpi.com/journal/robotics/special_issues/KaRD2021).

For citation purposes, cite each article independently as indicated on the article page online and as indicated below:

LastName, A.A.; LastName, B.B.; LastName, C.C. Article Title. *Journal Name* **Year**, *Volume Number*, Page Range.

ISBN 978-3-0365-3615-6 (Hbk)
ISBN 978-3-0365-3616-3 (PDF)

Cover image courtesy of Alba Perez-Gracia

Contents

About the Editor

Raffaele Di Gregorio is currently a Full Professor of machine mechanics at the Engineering Department of the University of Ferrara, Italy. He received an M.Sc. degree in nuclear engineering, an M.Sc. degree in mechanical engineering and an M.Sc. degree in automotive engineering from the Polytechnic University of Turin in 1982, 1985 and 1988, respectively, and a Ph.D. degree in applied mechanics from the University of Bologna, Italy, in 1992.

In 1983, he received a grant from the Italian Automotive Technical Association (ATA) to spend one year as a research fellow at the FIAT Research Center, Orbassano (Italy)

Between 1984 and 1992, he served as an officer of the Technical Corp in the Italian Army, first at the Military School of Turin and, successively, at the STAVECO (Military Vehicle Factory) of Bologna. In 1993, he taught at ITIS O. Belluzzi of Bologna, Italy. Since 1994, he has been part of the Engineering Department of the University of Ferrara, Italy.

His research interests include kinematics and dynamics of mechanisms and machines, biomechanics, robotics, vibration mechanics and vehicle mechanics.

He has been an author and co-author of about 85 technical papers published in refereed international journals and is the author of as many papers published in conference proceedings.

He has been an observer/member of the IFToMM Technical Committee "Computational Kinematics" since 2007 and a member of the IFToMM Permanent Commission "Standardization of Terminology" since 2009. Moreover, he is an ASME member and served as a general member of the ASME Mechanisms and Robotics Committee between 2007 and 2012; since 2013, he has continued to serve this committee as a past member. In these roles, he has actively participated in the organization of International Conferences on Mechanisms and/or Robotics and served as the chair of the "ASME—Mechanisms & Robotics' Honors and Awards Sub-Committee" from 2014 to 2016.

In addition to being an editorial board member of many international journals, he serves as an associate editor of ACTAPRESS International Journal of Robotics and Automation, ROBOTICA and Mechanism and Machine Theory and served as associate editor of the ASME Journal of Mechanisms and Robotics from July 2014 to October 2020. Furthermore, he has been a Guest Editor of the Special Issue series "Kinematics and Robot Design" published by the journal Robotics since 2018 and has served as "Editor in Chief" of the Journal of Mechanical Engineering and Automation since 2012.

An updated list of his publications together with more details can be found at the following websites: UNIFE: http://docente.unife.it/raffaele.digregorio-en

ResearchGate: http://www.researchgate.net/profile/Raffaele_Di_Gregorio

Scopus Author ID: 7003465861

ORCID: https://orcid.org/0000-0003-3925-3016

ResearchID: http://www.researcherid.com/rid/F-9316-2014

Google Scholar: http://scholar.google.it/citations?user=n_Ml9FEAAAAJ&hl=it

A list of his professional links can be found at http://www.linkedin.com/profile/view?id=128075898&trk=tab_pro

Preface to "Kinematics and Robot Design IV, KaRD2021"

Kinematics is intimately related to nearly all the design aspects of robotic/automatic systems. Topics such as the analysis and synthesis of mechanisms, robot modelling and simulation, robot control, mobility and singularity analysis, performance measures, accuracy analysis, path planning and obstacle avoidance, collaborative robotics, novel manipulator architectures, metamorphic mechanisms, compliant mechanism analysis and synthesis, micro/nano-manipulator design, origami-based robotics, medical and rehabilitation robotics, bioinspired robotics, etc., deal with kinematics. All of these topics have a deep social impact and somehow delineate future perspectives of human welfare, which attracts much economic interest.

Kinematics and Robot Design IV (KaRD2021) is the 4th edition of the KaRD series, hosted by MDPI's *Robotics*. The KaRD series of open-access Special Issues began in 2018 and is now an open environment where researchers can present their work and discuss all topics focused on the many aspects that involve kinematics in the design of robotic/automatic systems by also using supplementary multimedia materials uploadable during the submission. All of the papers are peer reviewed as soon as they are submitted and, if accepted, they are immediately published in MDPI's *Robotics* and appear on the website of the KaRD issue. All of the papers in each KaRD edition are also collected into freely downloadable e-books. In short, KaRD series is an "agora", where researchers efficiently exchange their experiences.

In the KaRD series, as in all the well-established serial international conferences/publications, the activity of the guest editor is supervised/supported by a scientific committee that is composed of the following members from all over the world, whose service is gratefully acknowledged: Massimo Callegari (Polytechnic University of Marche, Italy), Juan Antonio Carretero (University of New Brunswick, Canada), Yan Chen (Tianjin University, China), Daniel Condurache ("Gheorghe Asachi" Technical University of Iași, Romania), Xilun Ding (Beijing University of Aeronautics & Astronautics, China), Mary Frecker (Penn State—College of Engineering, USA), Clement Gosselin (Laval University, Canada), Just Herder (TU Deft, Netherlands), Larry Howell (Brigham Young University, USA), Xianwen Kong (Heriot-Watt University, UK), Pierre Larochelle (South Dakota School of Mines & Technology, USA), Giovanni Legnani (University of Brescia, Italy), Haitao Liu (Tianjin University, China), Daniel Martins (Universidade Federal de Santa Catarina, Brazil), Andreas Mueller (Johannes Kepler Universität, Austria), Andrew Murray (University of Dayton, USA), Leila Notash (Queen's University, Canada), Matteo Palpacelli (Polytechnic University of Marche, Italy), Alba Perez (Remy Robotics, Barcelona, Spain), Victor Petuya (University of the Basque Country, Spain), José Maria Rico Martinez (Universidad de Guanajuato, Mexico), Nina Robson (California State University, Fullerton, USA), Jon M. Selig (London South Bank University, UK), Bruno Siciliano (University of Naples Federico II, Italy), Tao Sun (Tianjin University, China), Yukio Takeda (Tokyo Institute of Technology, Japan), Federico Thomas (Institute of Industrial Robotics, Spain), and Volkert Van Der Wijk (TU Deft, Netherlands).

All of those involved in the KaRD series editorial project believe in and aim to realize a peaceful world where nobody is left behind and where people and nations respect each other and cooperate to make the world a better place. Unfortunately, these goals are still far from being reached, and disseminating open science is our contribution on the road toward them.

KaRD2021 (https://www.mdpi.com/journal/robotics/special_issues/KaRD2021), after a rigorous peer-review process, accepted 12 papers. The accepted papers cover some theoretical and many design/applicative aspects. This book collects the twelve papers published on KaRD2021. The book is organized as follows. The first five papers [1–5] propose novel methodologies useful for the machine design. The next four papers [6–9] solve problems related to the design of healthcare/rehabilitation robots. Then, the successive paper [10] addresses design issues of a bio-inspired robot. Finally, the last two papers [11,12] deal with two applicative issues of industrial interest.

References

1. Hassanzadeh, N.; Perez-Gracia, A. Mixed Position and Twist Space Synthesis of 3R Chains. Robotics 2022, 11(1), 13; https://doi.org/10.3390/robotics11010013.
2. Valderrama-Rodríguez, J.; Rico, J.; Cervantes-Sánchez, J.; García-García, R. A Screw Theory Approach to Computing the Instantaneous Rotation Centers of Indeterminate Planar Linkages. Robotics 2022, 11(1), 6; https://doi.org/10.3390/robotics11010006.
3. Di Gregorio, R. Dimensional Synthesis of a Novel 3-URU Translational Manipulator Implemented through a Novel Method. Robotics 2022, 11(1), 10; https://doi.org/10.3390/robotics11010010.
4. Gamper, H.; Luthi, A.; Gattringer, H.; Mueller, A.; Di Castro, M. Kinematic Model Pruning: A Design Optimization Technique for Simultaneous Optimization of Topology and Geometry. Robotics 2022, 11(2), 31; https://doi.org/10.3390/robotics11020031.
5. Baigunchekov, Z.; Laribi, M.; Mustafa, A.; Kassinov, A. Kinematic Synthesis and Analysis of the RoboMech Class Parallel Manipulator with Two Grippers. Robotics 2021, 10(3), 99; https://doi.org/10.3390/robotics10030099.
6. Zakaryan, N.; Harutyunyan, M.; Sargsyan, Y. Bio-Inspired Conceptual Mechanical Design and Control of a New Human Upper Limb Exoskeleton. Robotics 2021, 10(4), 123; https://doi.org/10.3390/robotics10040123.
7. Metcalf, A.; Gallagher, J.; Jackson, A.; Levesley, M. Multi-Domain Dynamic Modelling of a Low-Cost Upper Limb Rehabilitation Robot. Robotics 2021, 10(4), 134; https://doi.org/10.3390/robotics10040134.
8. Tan, D.; Ng, P.; Noor, E.; Saptari, A.; Hue, C.; Ng, Y. Development and Usability Testing of a Finger Grip Enhancer for the Elderly. Robotics 2022, 11(1), 5; https://doi.org/10.3390/robotics11010005.
9. Lanese, N.; Myszka, D.; Bazler, A.; Murray, A. Six-Bar Linkage Models of a Recumbent Tricycle Mechanism to Increase Power Throughput in FES Cycling. Robotics 2022, 11(1), 26; https://doi.org/10.3390/robotics11010026.
10. Costa, D.; Scoccia, C.; Palpacelli, M.; Callegari, M.; Scaradozzi, D. Design of a Labriform-Steering Underwater Robot Using a Multiphysics Simulation Environment. Robotics 2022, 11(1), 11; https://doi.org/10.3390/robotics11010011.
11. Righettini, P.; Strada, R.; Cortinovis, F. Modal Kinematic Analysis of a Parallel Kinematic Robot with Low-Stiffness Transmissions. Robotics 2021, 10(4), 132; https://doi.org/10.3390/robotics10040132.
12. Giberti, H.; Abbattista, T.; Carnevale, M.; Giagu, L.; Cristini, F. A Methodology for Flexible Implementation of Collaborative Robots in Smart Manufacturing Systems. Robotics 2022, 11(1), 9; https://doi.org/10.3390/robotics11010009.

Raffaele Di Gregorio
Editor

robotics

MDPI

Article

Mixed Position and Twist Space Synthesis of 3R Chains

Neda Hassanzadeh [1] and Alba Perez-Gracia [1,2,*]

[1] Department of Mechanical Engineering, Idaho State University, Pocatello, ID 83209, USA; hassneda@isu.edu
[2] Department of Mechanical Engineering, Universitat Politecnica de Catalunya, 08028 Barcelona, Spain
* Correspondence: alba.perez.gracia@upc.edu

Abstract: Mixed-position kinematic synthesis is used to not only reach a certain number of precision positions, but also impose certain instantaneous motion conditions at those positions. In the traditional approach, one end-effector twist is defined at each precision position in order to achieve better guidance of the end-effector along a desired trajectory. For one-degree-of-freedom linkages, that suffices to fully specify the trajectory locally. However, for systems with a higher number of degrees of freedom, such as robotic systems, it is possible to specify a complete higher-dimensional subspace of potential twists at particular positions. In this work, we focus on the 3R serial chain. We study the three-dimensional subspaces of twists that can be defined and set the mixed-position equations to synthesize the chain. The number and type of twist systems that a chain can generate depend on the topology of the chain; we find that the spatial 3R chain can generate seven different fully defined twist systems. Finally, examples of synthesis with several fully defined and partially defined twist spaces are presented. We show that it is possible to synthesize 3R chains for feasible subspaces of different types. This allows a complete definition of potential motions at particular positions, which could be used for the design of precise interaction with contact surfaces.

Keywords: kinematic synthesis of robots; mixed-position synthesis; twist systems

Citation: Hassanzadeh, N.; Perez-Gracia, A. Mixed Position and Twist Space Synthesis of 3R Chains. *Robotics* **2022**, *11*, 13. https://doi.org/10.3390/robotics11010013

Academic Editor: Raffaele Di Gregorio

Received: 15 November 2021
Accepted: 6 January 2022
Published: 10 January 2022

1. Introduction

Mixed-position synthesis, in which a combination of finite and infinitesimal positions are defined for the dimensional synthesis, appears as an extension of Burmester's theory for planar motion [1]. Mixed-position approximate synthesis for planar linkages can be found in [2,3]. In [4], exact synthesis for planar linkages includes higher-order derivatives applied to envelope a planar curve with a tangent line.

The problem was studied for spatial dyads and their combined linkages by Chen and Roth [5,6] to unify the formulation for finite and infinitesimal positions, and later by Tsai and Roth [7], who used the spatial triangle. In all cases, it is applied to define one rigid-body velocity per position at a set of desired positions of the end-effector. The objectives are, among others, to achieve better guidance of the end-effector along a trajectory or to obtain a more accurate approximation to a desired motion curve. Second-order derivatives were included in the synthesis of spatial mechanisms in [8], and related to the curvature of contacts for planar mechanisms in [9].

First- and second-order derivatives can describe contact conditions in the synthesis problem [10], associating first- and second-order kinematics to curvatures and forces at the points of contact [11]. So far, this identification has mainly been used for the design of robotic hands. In [12], multiple rigid-body velocities were used at finite positions in order to synthesize hands for local manipulation conditions. The numerical synthesis allowed to fully or partially define the set of potential directions of motion for the fingertips at specified points. Optimized fingers for grasping objects with a minimally articulated hand were designed in [13] using these type of constraints.

On the other hand, subspaces of twists and wrenches have been used as a design tool for the structural or type synthesis of parallel robots. The potential motions of the platform are found by intersecting the twist or wrench subspace generated by each of the supporting legs; see, for instance, [14].

In this work, we explore the relation between the twist subspaces that can be defined in combination with finite positions, and the topology and geometry of the kinematic chain that imposes the restrictions on the motion. The interest is on defining the twist systems that a given chain can generate, and on stating the set of synthesis equations that are created when the subspace of twists is fully defined at a position. This study is applied to one simple but significant spatial linkage, the 3R spatial chain. The finite-position synthesis of the spatial 3R chain was partially solved in [15,16]. In [17], interval analysis was applied to obtain a set of solutions for the exact synthesis problem, and in [18], numerical algebraic geometry was used to obtain the number of solutions of the problem and to numerically solve the set of equations, which are based on a vector formulation. While algebraic solutions for the finite-position synthesis of spatial chains tends to be a very complex problem, it is possible to state a simpler set of equations for the mixed-position synthesis. The resulting set of equations can be easily solved using the numerical Groebner basis and eigensystem methods.

2. The 3R Serial Chain

Let $S_i = s_i + \epsilon s_i^0$ be the Plücker coordinates of the ith joint axis of the chain shown in Figure 1, with $i = 1,2,3$, at the reference configuration. Let θ_i be the joint variables measured from the reference configuration and $\dot{\theta}_i$ be the joint rates at each of the axes $i = 1,2,3$.

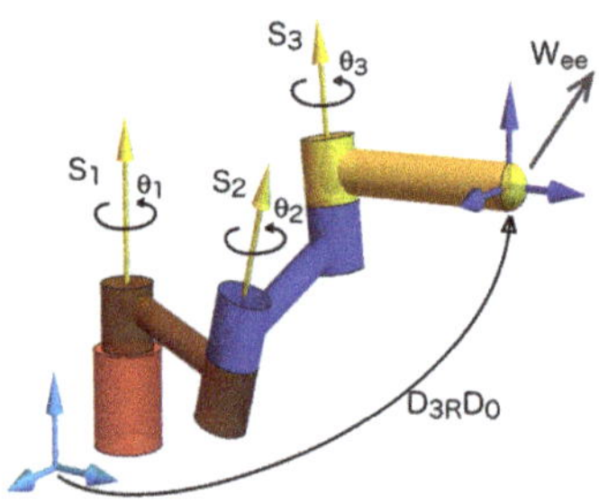

Figure 1. The spatial 3R chain. Forward kinematics is the product of the relative displacement D_{3R} and the displacement to the reference configuration D_0. Twist of the end-effector expressed in the fixed frame is W_{ee}.

2.1. Forward and Inverse Kinematics of the 3R Chain

The forward and inverse kinematics of the 3R chain can be computed using different methods and mathematical tools. Here, we present the approach that is most suited to use them as a first step in the creation of the synthesis equations.

Dual quaternion algebra is used in the development of these equations because it allows for a direct identification of the joint axes. We denote a dual quaternion as $\hat{Q} = q_w + \epsilon q_{w0} + Q$, where Q is the six-dimensional dual vector.

The relative displacements of the 3R serial chain can be expressed using the product of exponentials,

$$\hat{D}_{3R} = e^{S_1 \frac{\theta_1}{2}} e^{S_2 \frac{\theta_2}{2}} e^{S_3 \frac{\theta_3}{2}} \tag{1}$$

and arranged as a linear system on the product of joint angles,

$$\hat{D}_{3R} = [\hat{I} \quad \hat{S}_1 \quad \hat{S}_2 \quad \hat{S}_3 \quad \hat{S}_{12} \quad \hat{S}_{13} \quad \hat{S}_{23} \quad \hat{S}_{123}] \begin{Bmatrix} c_{\theta_1}c_{\theta_2}c_{\theta_3} \\ s_{\theta_1}c_{\theta_2}c_{\theta_3} \\ c_{\theta_1}s_{\theta_2}c_{\theta_3} \\ c_{\theta_1}c_{\theta_2}s_{\theta_3} \\ s_{\theta_1}s_{\theta_2}c_{\theta_3} \\ s_{\theta_1}c_{\theta_2}s_{\theta_3} \\ c_{\theta_1}s_{\theta_2}s_{\theta_3} \\ s_{\theta_1}s_{\theta_2}s_{\theta_3} \end{Bmatrix}, \tag{2}$$

where $\hat{S}_i$ are the pure dual quaternions corresponding to the joint axes, $\hat{S}_{ij}$ is the product $\hat{S}_i\hat{S}_j$ using the dual quaternion product, and s_{θ_i}, c_{θ_i} are the sin and cosine of the half angle θ_i.

Equating to a target position $\hat{P} = p_w + \epsilon p_{w0} + P$ of the end-effector, we obtain a system with $2^3 = 8$ unknowns on the joint variables. We use the notation $x_1 = c_{\theta_1}c_{\theta_2}c_{\theta_3}$ to $x_8 = s_{\theta_1}s_{\theta_2}s_{\theta_3}$, where these variables are related by 3 independent bilinear conditions. The system, with a degree of over-determination of 3, is shown below:

$$\begin{aligned} &\hat{I}x_1 + \hat{S}_1x_2 + \hat{S}_2x_3 + \hat{S}_3x_4 + \hat{S}_1\hat{S}_2x_5 + \hat{S}_1\hat{S}_3x_6 + \hat{S}_2\hat{S}_3x_7 + \hat{S}_1\hat{S}_2\hat{S}_3x_8 = \hat{P}, \\ &x_1x_5 = x_2x_3, \\ &x_1x_8 = x_4x_5, \\ &x_2x_7 = x_3x_6, \end{aligned} \tag{3}$$

It is possible to solve the 8 linear equations directly for the unknowns; the resulting unique solution will comply with the quadratic conditions only if the position belongs to the workspace.

A smaller system of equations is obtained if we post-multiply by the inverse of the last rotation in order to split in 2R and 1R sides,

$$[\hat{I} \quad \hat{S}_1 \quad \hat{S}_2 \quad \hat{S}_1\hat{S}_2] \begin{Bmatrix} c_{\theta_1}c_{\theta_2} \\ s_{\theta_1}c_{\theta_2} \\ c_{\theta_1}s_{\theta_2} \\ s_{\theta_1}s_{\theta_2} \end{Bmatrix} = [\hat{P} \quad -\hat{P}\hat{S}_3] \begin{Bmatrix} c_{\theta_3} \\ s_{\theta_3} \end{Bmatrix}, \tag{4}$$

so that the total of unknowns is 6. Denote $x_1 = c_{\theta_1}c_{\theta_2}$, $x_2 = s_{\theta_1}c_{\theta_2}$, $x_3 = c_{\theta_1}s_{\theta_2}$, $x_4 = s_{\theta_1}s_{\theta_2}$, and $x_5 = c_{\theta_3}$, $x_6 = s_{\theta_3}$, then we have the system

$$\begin{aligned} &\hat{I}x_1 + \hat{S}_1x_2 + \hat{S}_2x_3 + \hat{S}_1\hat{S}_2x_4 = \hat{P}x_5 - \hat{P}\hat{S}_3x_6, \\ &x_1x_4 = x_2x_3. \end{aligned} \tag{5}$$

This system of equations can be used to solve the inverse kinematics of the chain, as well as to state the kinematic synthesis equations.

2.2. Space of Potential Twists of the End-Effector

Let us consider the twist as the six-dimensional vector $W = (w, v)$, where w is the angular velocity of the body and v is the velocity of a point belonging to the body. The set of twists of the end-effector at position i can be defined as

$$\mathsf{W}^i_{ee} = S^i_1\dot{\theta}^i_1 + S^i_2\dot{\theta}^i_2 + S^i_3\dot{\theta}^i_3, \tag{6}$$

where S^i_j is the jth joint axis at a configuration defined by i, and $\dot{\theta}^i_j$ is its angular joint rate.

This is, in the most general case, a three-dimensional subspace of the six-dimensional space of twists. For a given configuration of the robot, defining three end-effector twists, W_1, W_2, and W_3, fully specifies the subspace of potential twists of the end-effector at position i,

$$\mathsf{W}^i_{ee} = \lambda_1 W_1 + \lambda_2 W_2 + \lambda_3 W_3, \tag{7}$$

as the linear combination of the three independent twists. However, for a generic twist subspace such as Equation (7) to coincide with the subspace of twists of the end-effector of the robot, some additional conditions are needed. The dimension of the specified twist space must be equal to the dimension of the space spanned by the joint axes at that configuration, and also some conditions on the principal pitches of the screw system must be satisfied, as explained in Sections 3 and 4.

3. Twist Spaces of Dimension 3 for the Serial 3R Chain

The classification used for the screw systems of dimension 3 follows [19,20]. There are 12 different types of screw systems of dimension 3, counting special sub-types. Each type comprises screw systems with different pitch values that define equivalence classes up to rigid motion. Twists belonging to the end-effector of a kinematic chain such as those shown in Equation (6) form a twist system.

3.1. *Finite and Infinitesimal Motion*

The completion group of a screw system is defined by the exponential map of the smallest Lie subalgebra containing the screw system; see, for instance, [20] for the completion subgroups of the different types of screw systems.

For any twist system generated by a kinematic chain, the finite motion of the chain must belong to the completion group of the system, which is a subgroup of $SE(3)$.

Some twist subspaces are *invariant*. Invariant twist systems are closed under the Lie bracket, that is, they are subalgebras of $se(3)$. These systems are symmetric under the action of a rigid motion by a screw belonging to the system. They guarantee full-cycle mobility for closed chains, and exhibit remarkable invariance properties [21].

For those twists systems generated by a kinematic chain, this means that the screw system will remain constant along the motion of the chain. For the case in which the dimension of the twist space coincides with the global degrees of freedom of the chain, this implies that its finite motion must belong to the corresponding subgroup of $SE(3)$.

If the twist system remains of the same type and also keeps the pitches of the principal screws, but it does not remain constant along the motion, then it is a *persistent* screw system [21,22]. Those undergo a displacement when changing the configuration of the kinematic chain that generates them, that is, remain within the equivalence class. Invariant systems can be considered as particular cases of persistent systems [23].

While the composition of two revolute joints creates persistent twist systems, in general, for three revolute joints in series this is not the case, and the twist system may change when passing from one configuration of the chain to another. For this most general case, we say that the chain has a *variable* screw system corresponding to the subspace of twists at different configurations.

Along this work, we will refer to the *twist system* when denoting the screw system corresponding to a given twist subspace.

3.2. *Kinematic Generators of a Twist System*

A set of screws $S_1, \ldots, S_n$ corresponding to the twists of kinematic pairs connected in series span a screw system, and the chain is the *instantaneous kinematic generator* of the twist subspace corresponding to that system. Every twist system of dimension 3 can be obtained by specializing the instantaneous kinematic generator formed by three cylindrical joints assembled in series. In any case, the individual twist of each of the joints belongs to the twist system at a given configuration.

In the case of the 3R serial robot, the pitch is zero for each screw corresponding to a joint, and this limits the twist systems that can be generated and correspondingly, the combination of finite and infinitesimal positions that we can define.

The concept of instantaneous kinematic generators has its equivalence in finite motion using the kinematic bond, which is the set of allowed relative displacements between two rigid bodies, belonging to a given kinematic chain. The mechanical generator of the bond is the kinematic chain (that is, the set of kinematic pairs and their connectivity) that generates the bond [24].

3.3. Feasible Screw Systems for the 3R Serial Chain

We consider a screw system as feasible by a given mechanical generator when the instantaneous kinematic generator can span the screw system at a given configuration of the chain.

The identification of those systems of twists of dimension 3 that are feasible for a spatial 3R chain is presented in Table 1. In this table, we present all screw systems of dimension 3 according to the classification in [20], with their pitches and normal form in a particular reference frame, the identification of their completion group, and their invariance properties (whether they are invariant, persistent or variable by the set of instantaneous kinematic generators presented in the normal form). For each system, we include in the last column its feasibility as generated by a 3R mechanical generator.

Table 1. Twist systems of dimension 3 [20] and their synthesis properties.

Completion Group	Gibson Hunt Type	Normal Form	h_1	h_2	h_3	Invariance	Feasible
$\mathbb{R}^3$	IID	$(0,0,0,1,0,0)$ $(0,0,0,0,1,0)$ $(0,0,0,0,0,1)$	∞	∞	∞	Invariant	NO
$SO(3)$	$IIA\ (h=0)$	$(1,0,0,0,0,0)$ $(0,1,0,0,0,0)$ $(0,0,1,0,0,0)$	0	0	0	Invariant	YES
$SE(2)$	$IIC\ (h=0)$	$(1,0,0,0,0,0)$ $(0,0,0,0,1,0)$ $(0,0,0,0,0,1)$	0	∞	∞	Invariant	YES
$\mathbb{H}_p \ltimes \mathbb{R}^2$	$IIC\ (h \neq 0)$	$(1,0,0,h,0,0)$ $(0,0,0,0,1,0)$ $(0,0,0,0,0,1)$	h	∞	∞	Invariant	NO
$SE(2) \ltimes \mathbb{R}$	$IC\ (h \neq 0)$	$(1,0,0,0,0,0)$ $(0,0,0,0,1,0)$ $(0,0,0,1,0,h)$	0	∞	∞	Variable	NO
$SE(2) \ltimes \mathbb{R}$	IC^0	$(1,0,0,0,0,0)$ $(0,0,0,0,1,0)$ $(0,0,0,1,0,0)$	0	∞	∞	Persistent	NO
$SE(3)$	IIB	$(1,0,0,h,0,0)$ $(0,1,0,0,h,0)$ $(0,0,0,0,0,1)$	h	h	∞	Variable	NO
$SE(3)$	IIB^0	$(1,0,0,0,0,0)$ $(0,1,0,0,0,0)$ $(0,0,0,0,0,1)$	0	0	∞	Variable	YES
$SE(3)$	$IB_0\ (h_1 \neq h_2)$	$(1,0,0,h_1,0,0)$ $(0,1,0,0,h_1,0)$ $(0,0,0,1,0,h_2)$	h_1	h_1	∞	Variable	YES
$SE(3)$	$IB_3\ (h_1 \neq h_2)$	$(1,0,0,h_1,0,0)$ $(0,1,0,0,h_2,0)$ $(0,0,0,0,0,1)$	h_1	h_2	∞	Variable	YES $Sign(h_1) \neq Sign(h_2)$
$SE(3)$	$IIA\ (h \neq 0)$	$(1,0,0,h,0,0)$ $(0,1,0,0,p,0)$ $(0,0,1,0,0,h)$	h	h	h	Variable	NO

Table 1. *Cont.*

Completion Group	Gibson Hunt Type	Normal Form	h_1	h_2	h_3	Invariance	Feasible
$SE(3)$	IA_2 $(h_1 \neq h_2)$	$(1,0,0,h_1,0,0)$ $(0,1,0,0,h_2,0)$ $(0,0,1,0,0,h_2)$	h_1	h_2	h_2	Variable	YES $Sign(h_1) \neq Sign(h_2)$
$SE(3)$	IA_1 $(h_1 \neq h_2 \neq h_3)$	$(1,0,0,h_1,0,0)$ $(0,1,0,0,h_2,0)$ $(0,0,1,0,0,h_3)$	h_1	h_3	h_2	Variable	YES $Sign(h_i) \neq Sign(h_j)$

In the table, we can see that a 3R chain can be designed with specific geometry to have some invariant twist systems, with the corresponding subgroup of motion, and also can be designed for several variable twist systems that allow general motion in $SE(3)$. However, the 3R chain cannot be designed to generate a persistent twist system of dimension 3 that is not invariant. This agrees with the results in [23].

The methodology to prove the feasibility of the 3R chain to generate these twist systems, at any or some particular configuration, is presented in Section 4, and it is used in the subsection below, in particular the matrix formulation presented in Equations (18)–(22).

3.4. Proofs of Feasibility

The principal screws [25] for a 3-twist system can be computed and written, after a coordinate transformation, as

$$W_{p1} = \begin{Bmatrix} \theta_1 \\ 0 \\ 0 \\ t_1 \\ 0 \\ 0 \end{Bmatrix}, W_{p2} = \begin{Bmatrix} 0 \\ \theta_2 \\ 0 \\ 0 \\ t_2 \\ 0 \end{Bmatrix}, W_{p3} = \begin{Bmatrix} 0 \\ 0 \\ \theta_3 \\ 0 \\ 0 \\ t_3 \end{Bmatrix}. \tag{8}$$

We use this form as well as the equivalent normal form [26] used in classifications [19] and presented in Table 1 to derive the reachability conditions for some special cases. For this, we consider the pitch $h_i = t_i/\theta_i$.

The minors of matrices used to prove linear dependence, described in detail in Section 4, can be obtained for the principal screws of the system and the joint axes S_j^i,

$$[M_{pi}] = \begin{bmatrix} S_1^i & W_{p1} & W_{p2} & W_{p3} \end{bmatrix}, \tag{9}$$

The computation of the corresponding 4x4 minors provides us with simplified equations in terms of the pitches of principal screws,

$$\begin{aligned} s_x t_1 - s_x^0 \theta_1 &= 0, \\ s_y t_2 - s_y^0 \theta_2 &= 0, \\ s_z t_3 - s_z^0 \theta_3 &= 0. \end{aligned} \tag{10}$$

For twist systems with finite pitch, the Plücker conditions on the moment of the lines, $ss^0 = 0$, yield the equations,

$$h_1 s_x^2 + h_2 s_y^2 + h_3 s_z^2 = 0. \tag{11}$$

For the case of one or more pitches being infinite or zero, the minors' conditions are modified as detailed below.

Table 1 also shows the pitches of the generator screws as h_1, h_2, h_3. Using Equations (10) and (11), we can conclude which of the screw systems are reachable for a 3R system.

Screw System IID**:** From the ∞ condition for the three pitches, we conclude that θ_i, $i = 1, 2, 3$ is 0, and from Equation (10) we conclude that

$$\begin{Bmatrix} s_x \\ s_y \\ s_z \end{Bmatrix} = \begin{Bmatrix} 0 \\ 0 \\ 0 \end{Bmatrix}, \tag{12}$$

for every joint axis. That proves that this system is not feasible for a 3R system.

Screw System $IIA(h = 0)$**:** The 0 condition on all three pitches implies $\mathbf{s}_\mathrm{i}^0 = 0$ for each joint S_i, which means that all three axes must intersect at the center of the twist system. They must also be independent in order for the dimension to be three. The system is feasible.

Screw System $IIC(h = 0)$**:** A similar strategy for this system reveals that, for the normal form of the table, s_y, s_z and s_{x0} must be zero, while s_x is not restricted. The joint axes have direction x and are parallel, located anywhere in the perpendicular plane. The system is feasible.

Screw System $IIC(h \neq 0)$**:** In this system also using Equation (10), we conclude that s_y and s_z must be zero, so the joint axes must be directed along the x axis. Imposing Equation (11) gives

$$h_1 s_x^2 = 0, \tag{13}$$

and thus, s_x is also zero, which proves that the system is not feasible.

Screw System $IC(h \neq 0)$**:** In this system, since the screw system is not in the principal form, we use Equation (9) to conclude $s_y = 0$, $s_z = 0$ $ps_x^0 = s_z^0$. Imposing from the Plücker coordinate condition of $s_i s_i^0 = 0$, we have $s_x s_x^0 = 0$ and since $s_x \neq 0$, we get $s_x^0 = s_z^0 = 0$. This will result in the S_i being

$$S_i = \begin{Bmatrix} 1 \\ 0 \\ 0 \end{Bmatrix} + \epsilon \begin{Bmatrix} 0 \\ s_y^0 \\ 0 \end{Bmatrix}. \tag{14}$$

The resulting axes are clearly linearly dependent and create a 2-system, which will not be able to generate linear velocity, for instance, in the x direction, although they are contained in this screw system. We can conclude that the system is not completely feasible, only a subspace of twists is.

Screw System IC^0**:** Similar to system IC, the conditions for this system result in three linearly dependent axes of the same form as the previous one, forming a subspace of dimension two. The system cannot be fully generated by a 3R chain.

Screw System IIB**:** In this system, Equation (10) provides that $s_z = 0$, and Equation (11) turns to

$$h(s_x^2 + s_y^2) = 0, \tag{15}$$

for which s_x and s_y must be zero and makes the system not feasible by the 3R chain.

Screw System IIB^0**:** Applying linear dependence, we obtain that $s_z = 0$, $s_{x0} = s_{y0} = 0$. A 3R chain with axes on the $x - y$ plane can generate this system.

Screw System $IB_0(h_1 \neq h_2)$**:** In this case, imposing linear dependence yields the equations

$$\begin{aligned} &s_z = 0, \\ &s_{y0} - h_1 s_y = 0, \\ &h_2(h_1 s_x - s_{x0}) + s_{z0} = 0, \end{aligned} \tag{16}$$

which means that we have three free parameters to define the three independent joint axes in order to generate the twist system. From the equations, we can see that $s_x \neq 0$.

Screw System $IB_3(h_1 \neq h_2)$**:** A similar approach provides $s_z = 0$ and $h_1 s_x^2 + h_2 s_y^2 = 0$. This makes the twist feasible with the condition that the pitches of the imposed screw system have opposite signs.

Screw System $IIA((h \neq 0)$**:** For this system, the condition from Equation (11) yields $h(s_x^2 + s_y^2 + s_z^2) = 0$, that requires $s = 0$, which cannot span a revolute joint, or $h = 0$, which is a different screw system. The system is not feasible.

Screw System $IA_2(h_1 \neq h_2)$**:** Equation (11) for this system gives $h_1 s_x^2 + h_2(s_y^2 + s_z^2) = 0$. It is possible to create three revolute joints to obtain this twist system, with $s_x \neq 0$ and as long as the pitches have opposite signs.

Screw System $IA_1(h_1 \neq h_2 \neq h_3)$**:** This is the most general three-dimensional screw system. In this case, we obtain $h_1 s_x^2 + h_2 s_y^2 + h_3 s_z^2 = 0$, which is feasible when one of the pitches has an opposite sign to the other two.

3.5. The 3R Chain as a Kinematic Generator of Twists

A 3R chain can generate those twists systems that are feasible, for at least one particular configuration of the chain. The interesting information of Table 1 is that it shows the twist systems of dimension 3 (with the corresponding completion subgroup of $SE(3)$) and whether or not they can be realized with a spatial 3R chain.

In general, the relative finite motion defined by D_{3R} in Equation (1) must be obtained from the twist of the end-effector, W_{ee} in Equation (6) at the initial configuration. When the complete twist space is defined at that position, the subsequent finite motion must belong to the completion group of the twist system.

4. Synthesis with a Specified Twist Space

Once the feasible twist spaces that can be generated by a 3R chain are identified, we can state a synthesis problem in which finite precision positions, as well as the full subspace of potential velocities at those positions, are imposed to the chain. The type of twist system (whether it is invariant, persistent or variable), as well as the completion group of the targeted system, are important inputs in this process.

One important point to consider is that we can define arbitrary finite motions of the chain for synthesis purposes only in the cases for which the completion group of the imposed twist system is the whole $SE(3)$. This consideration is not necessary if we only partially specify the twist system.

Another important point is that if we specify an invariant or persistent screw system, it is enough to specify it at the reference configuration, which simplifies the equations.

In this section, we derive the synthesis equations that can be imposed for the twist system and the finite solutions, and how to solve the overall system of equations.

Let us consider a 3R chain with joint axes S_j^i, $j = 1, 2, 3$. The superscript i indicates the configuration of the chain. Each revolute joint axis S_j can apply an angular velocity of magnitude $\dot{\theta}_j$.

Assume that, at a given position i, we impose a twist system given by twists W_k, $k = 1, 2, 3$ if we fully specify the 3D twist system. The matching of the twists at position i,

$$S_1^i \dot{\theta}_{1k}^i + S_2^i \dot{\theta}_{2k}^i + S_3^i \dot{\theta}_{3k}^i = W_k, \quad k = 1, 2, 3, \tag{17}$$

is equivalent to the condition on the rank of the space spanned by the axes and the twists,

$$rank\begin{bmatrix} S_1^i & S_2^i & S_3^i & W_1 & W_2 & W_3 \end{bmatrix} = 3. \tag{18}$$

This condition can be simplified. In order for the 3R chain to be able to perform the twists defined in Equation (7), each of the joint rotations $S_j^i \dot{\theta}_j^i$ must belong to the subspace W^i, in particular with angular velocity equal to one. To study the solutions of the problem, note that the specified twists can be defined at any configuration of the robot, in particular at the reference configuration; it is always possible to switch the reference configuration as needed.

This allows us to write the three conditions for the minors of matrices M_j, with

$$[M_j] = \begin{bmatrix} S_j & W_1 & W_2 & W_3 \end{bmatrix}, \tag{19}$$

such that

$$rank([M_j]) = 3, \quad j = 1,2,3. \tag{20}$$

Notice that this condition implies only that the twist system of the three axes is contained in the specified twist system,

$$< S_1, S_2, S_3 > \subseteq < W_1, W_2, W_3 > . \tag{21}$$

In order to impose equality, we need the condition that the three twists of the screw axes are independent,

$$rank(\begin{bmatrix} S_1 & S_2 & S_3 \end{bmatrix}) = 3. \tag{22}$$

However, this condition is implied in the additional finite-position synthesis equations.

The 3 independent minors per axis are linear in the Plücker coordinates of the axes, for a total of 9 linear equations.

To this condition for the twist system at the reference configuration, we add the condition that the chain must pass through two exact positions. The finite-position synthesis equations are written using Equation (5) and considering the six-dimensional dual vector part only, for one relative position of dual vector P,

$$Ic_{\theta_1}c_{\theta_2} + S_1 s_{\theta_1} c_{\theta_2} + S_2 c_{\theta_1} s_{\theta_2} + S_1 S_2 s_{\theta_1} s_{\theta_2} = Pc_{\theta_3} - PS_3 s_{\theta_3}, \tag{23}$$

where c_{θ_i} and s_{θ_i} denote the cosine and sine of the half angle for each joint variable.

In summary, the exact synthesis for a specified twist space of dimension 3 at a position can be stated for the 3R serial chain if the chain must reach two exact positions, having a fully defined compatible twist space in one of them, in particular in position 1, which will be taken as the reference configuration.

Solving the System of Equations

The nine linear equations obtained for the twist space from Equation (20) can be used to express some of the Plücker coordinates of the joints as a function of the others. For instance, we can keep as independent the directions of the joint axes s_i, with

$$s_i^0 = \frac{1}{w_1 \cdot w_2 \times w_3}(v_1 s_i \cdot w_2 \times w_3 + v_2 s_i \cdot w_3 \times w_1 + v_3 s_i \cdot w_1 \times w_2), \quad i = 1,2,3. \tag{24}$$

The selection of variables in this step depends on the twist system that we define for the end-effector; the procedure presented here could be used, for instance, for the most general IA_1 system.

The Plücker conditions on the moment of the lines, $s_i s_i^0 = 0$, yield three homogeneous quadratic equations with identical coefficients when substituting the values of Equation (24),

$$k_1 s_{ix}^2 + k_2 s_{iy}^2 + k_3 s_{iz}^2 + k_4 s_{ix} s_{iy} + k_5 s_{ix} s_{iz} + k_6 s_{iy} s_{iz} = 0, \quad i = 1,2,3. \tag{25}$$

This corresponds to the three directions lying on a real elliptic cone. In addition, the directions must be unit, so they also lie on the unit sphere,

$$s_{ix}^2 + s_{iy}^2 + s_{iz}^2 = 1, \quad i = 1,2,3. \tag{26}$$

The intersection of these two surfaces is a spheroconic, an elliptic quartic curve, and it is shown in Figure 2. Notice that the directions can be arbitrarily assigned to either one or the other intersection curves.

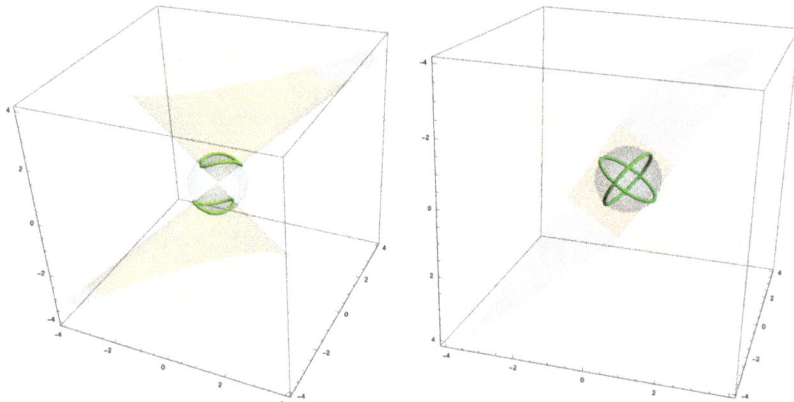

Figure 2. Locus of the direction of the axes from the twist conditions. A general case (**left**) and a case in which the elliptic cone degenerates to two planes (**right**).

The system of equations for one relative position in (23) contains eight quartic equations, six of them being independent once the Plücker conditions for the axes are imposed.

In all, we obtain a set of 12 equations (15 if we make the cosine and sine of the angles algebraic by imposing $c_{\theta_i}^2 + s_{\theta_i}^2 = 1$), which can be solved to obtain the solutions of the system. The system is summarized in Equation (27), where the forward kinematics are separated in two sets, one for the real and another one for the dual part of the equations:

$$
\begin{aligned}
& k_1 s_{ix}^2 + k_2 s_{iy}^2 + k_3 s_{iz}^2 + k_4 s_{ix} s_{iy} + k_5 s_{ix} s_{iz} + k_6 s_{iy} s_{iz} = 0, \\
& s_{ix}^2 + s_{iy}^2 + s_{iz}^2 = 1, \\
& c_{\theta_i}^2 + s_{\theta_i}^2 = 1, \quad i = 1,2,3; \\
& s_1 s_{\theta_1} c_{\theta_2} + s_2 c_{\theta_1} s_{\theta_2} + s_1 \times s_2 s_{\theta_1} s_{\theta_2} = p c_{\theta_3} - (p_w s_3 + p \times s_3) s_{\theta_3}, \\
& s_1^0 s_{\theta_1} c_{\theta_2} + s_2^0 c_{\theta_1} s_{\theta_2} + (s_1 \times s_2^0 + s_1^0 \times s_2) s_{\theta_1} s_{\theta_2} = \\
& \qquad p^0 c_{\theta_3} - (p_{w0} s_3 + p^0 \times s_3 + p \times s_3^0) s_{\theta_3},
\end{aligned}
\tag{27}
$$

where we substitute s_i^0 with their expressions as a function of s_i in Equation (24). Variables can also be further eliminated using standard techniques for a graphical representation of the solutions, see Figure 3.

The multiple solutions contain a subset, which generates the twist system at the first position but does not reach the second position. Those are due to the lack of information in the selected equations about the conjugation of the third axes, and can be eliminated by checking the forward kinematics. They account for a reduction to half the number of solutions. The remaining set contains repeated solutions due to the sign duplication in each axes, and to the cosine and sine of the half angles. The final set corresponds to $1/2^6$ of the initial solution set.

The final number of solutions obtained depends on the screw system selected; for the most general type, we obtain 4 non-repeated solutions.

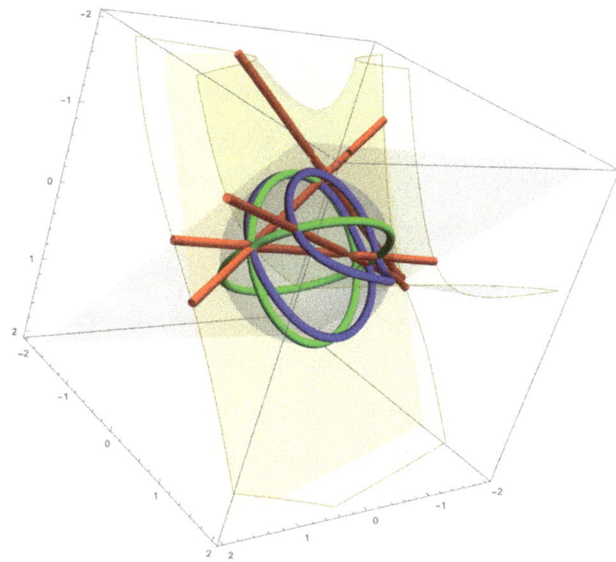

Figure 3. Solutions can be found as the intersection of the elliptic cone, the unit sphere and the relative position condition. Here, the curves are depicted for one solution value of the angle θ_2. Green curve: intersection of the elliptic cone with the unit sphere. Red curve: intersection of the elliptic cone and the relative position condition. Blue curve: intersection of the unit sphere and the relative position condition. The intersections of the curves are the solutions for the direction of the second joint axes, s_2.

5. Design for Lower-Dimensional Twist Spaces

Even though the maximum number of twists we can define per position is three, we can also specify two or one twist, for instance, for trajectory guidance purposes. We are not targeting singularity placement here, but rather incompletely defined infinitesimal motion.

In the example below (Example 1), a generic 3R robot was designed for reaching two positions; for one of them, the twist of the end-effector is generated by two perpendicular angular velocities only, and for the other one, a single twist is defined, corresponding to angular velocity about an independent axis. The positions and twists are specified in Table 2.

Table 2. Example 1: Input positions, expressed as dual quaternions, and end-effector twists.

Position	End-Effector Twists
$0.18 - 0.72i - 0.63j - 0.21k$ $+\epsilon(-0.28i + 0.33j - 0.27k - 0.28)$	$W_1^1 = (0.29, 0, 0, 0, 0, 0)$
$0.99 + 0.01i - 0.02j$ $+\epsilon(0.40i - 0.73j - 0.10k - 0.02)$	$W_1^2 = (0, 0.12, 0, 0, 0, 0)$ $W_2^2 = (0, 0, 0.27, 0, 0, 0)$

A total of 12 different solutions was obtained. The resulting joint axes at the reference configuration for the robot in Figure 4 are shown in Table 3.

In Figure 4, the 3R robot is shown at two positions. Notice how the first two of the axes intersect. It is interesting to notice that five of the solutions have the first and second axes S_1 and S_2 intersecting, while the rest have generally oriented axes for which axes S_1 and S_3 intersected at the second position.

Figure 5 shows the subspace of potential twists of the end-effector at the first and second positions for the selected solution, matching the specifications for the span of the two angular velocities at the points.

Figure 6 shows another of the solution robots for which axes S_1 and S_3 intersect at the second position only.

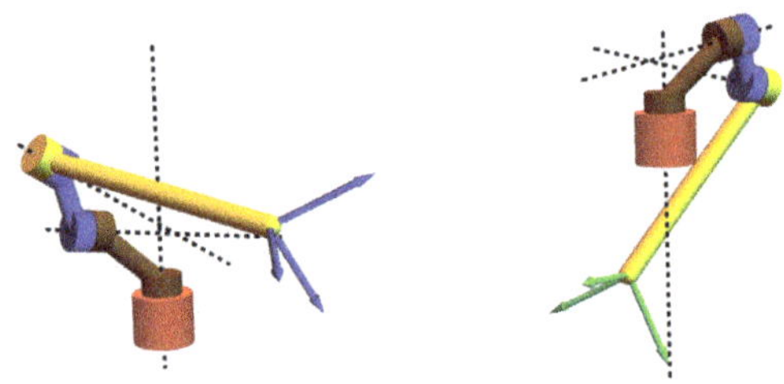

Figure 4. Example 1: The 3R robot reaching the two positions shown in Table 3. The joint axes are shown as dashed lines and link colors are those of Figure 1. Notice that two of the axes intersect at a point.

Table 3. Example 1: Plücker coordinates of the joint axes of one of the solutions of the 3R robot at the reference configuration.

Axes	Plücker Coordinates
S_1	$(0.843, -0.161, 0.514) + (-0.864, -0.750, 1.182)$
S_2	$(-0.640, 0.133, 0.757) + (-0.998, 0.634, -0.956)$
S_3	$(0.879, -0.191, -0.436) + (0.490, -0.904, 1.386)$

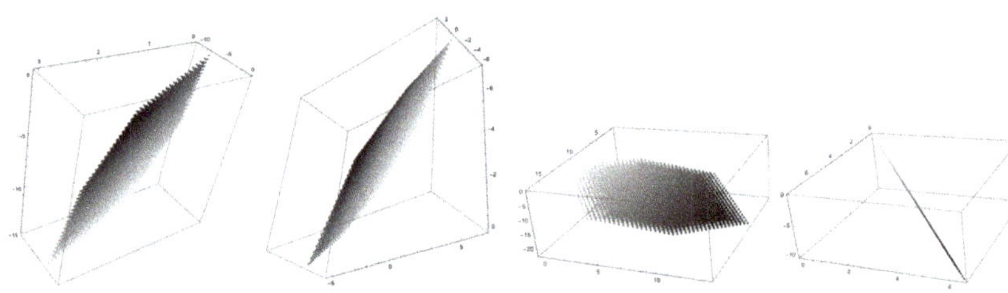

Figure 5. Example 1: Twists of the end effector. Left: subspaces of dimension 2 corresponding to the angular velocities of the end-effector at positions 1 and 2. Right: plots corresponding to the linear velocities. At position 2, the linear velocity subspace is of dimension 1.

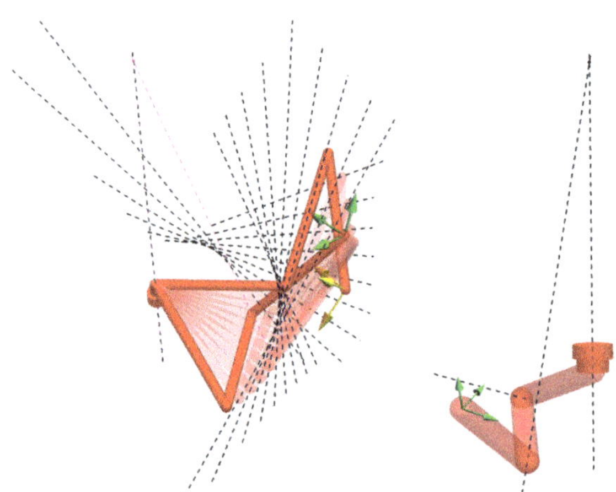

Figure 6. Example 1: A second 3R robot solution for Example 1 showing the first and third axes intersection at the second position.

6. Design for Fully Specified Twist Space

Designing for velocity subspaces allows for complete control of the motion of multi-DoF systems at particular positions. It can also be used for the design of non-general robots, for instance with parallel axes, without having possible degeneracy problems and without adding geometric constraints between axes.

The equations obtained with the elimination process can be solved with exact coefficients to obtain the complete set of solutions by using the numerical Groebner basis, and the results are obtained in a few seconds using symbolic software. Other numerical solvers, such as the Levenberg–Marquardt optimizer, can be used to obtain one solution at a time for a faster result.

6.1. Example 2: Arbitrary Finite Motion and General Twist System

For two random positions and three random twists defined at the first position, the system of equations yields four real solutions. Table 4 shows the input positions and twist system, and Figure 7 shows the motion of one of those robots from position 1 to position 2. Its resulting axes at the reference configuration are shown in Table 5.

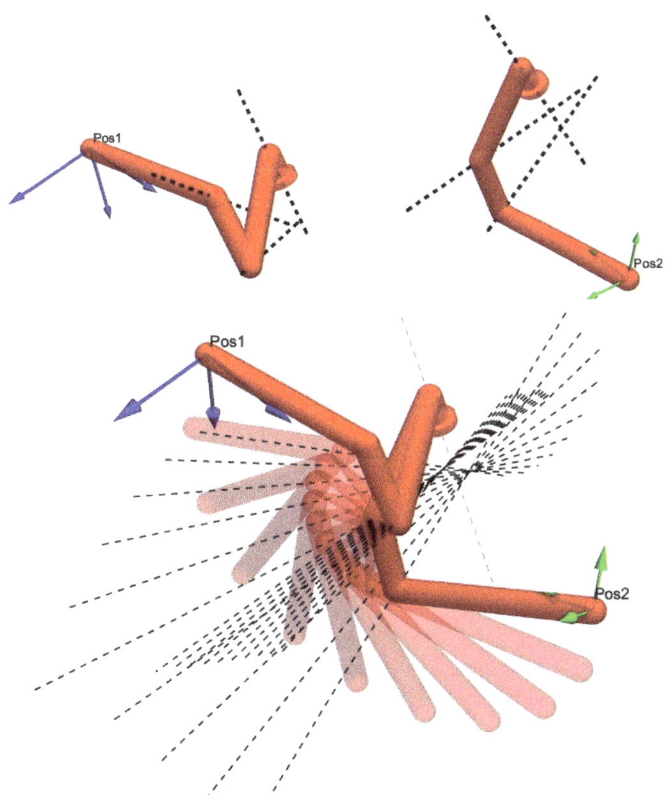

Figure 7. Example 2: One 3R robot solution at the first and second positions (**top**) and during the movement (**bottom**).

Table 4. Example 2: Input positions, expressed as dual quaternions, and end-effector twists.

Position	End-Effector Twists W_1^1, W_2^1, W_3^1
$0.18 - 0.72i - 0.63j - 0.21k$ $+\epsilon(-0.28i + 0.33j - 0.27k - 0.28)$	$(1.67, 1.84, -0.11, 0.12, 0.29, 3.42)$ $(2.39, 3.42, 0.35, 0.46, -0.72, 6.87)$ $(-4.84, 0.13, 3.86, 1.57, -9.34, 3.33)$
$1.0 + 0.01i - 0.025j - 0.004k$ $+\epsilon(0.40i - 0.73j - 0.10k - 0.02)$	

Table 5. Example 2: Plücker coordinates of the joint axes of the 3R robot at the reference configuration.

Axes	Plücker Coordinates
S_1	$(0.76, -0.09, -0.65) + \epsilon(-0.20, 1.70, -0.46)$
S_2	$(0.58, 0.81, 0.08) + \epsilon(0.08, -0.21, 1.55)$
S_3	$(0.81, 0.51, -0.30) + \epsilon(-0.19, 0.52, 0.39)$

6.2. Example 3: Invariant Twist System–Motion within a Subgroup of SE(3)

For this example, a 3R robot is to be synthesized in order to have a twist space of angular velocities only, as shown in Table 6. The compatible finite positions must belong to the subgroup of pure rotations $SO(3)$.

Table 6. Example 3: Input positions, expressed as dual quaternions, and end-effector twists.

Position	End-Effector Twists
$0.999 + 0.017i$	
$0.701 + 0.397i + 0.131j - 0.577k$	$W_1^2 = (1, 0, 0, 0, 0, 0)$ $W_2^2 = (0, 1, 0, 0, 0, 0)$ $W_3^2 = (0, 0, 1, 0, 0, 0)$

The twist system equations immediately yield $si0 = \vec{0}$ for all three axes, which only have to enforce the unit conditions. The finite position equations for the relative displacement reduce to the quaternion equations. The joint axes at the reference configuration are shown in Table 7. All three axes intersect at the origin as expected without the need to enforce this condition in the synthesis equations. Figure 8 shows the robot reaching the two positions and the intersecting joint axes that ensure the twist of the end-effector to be as defined and, in addition, invariant.

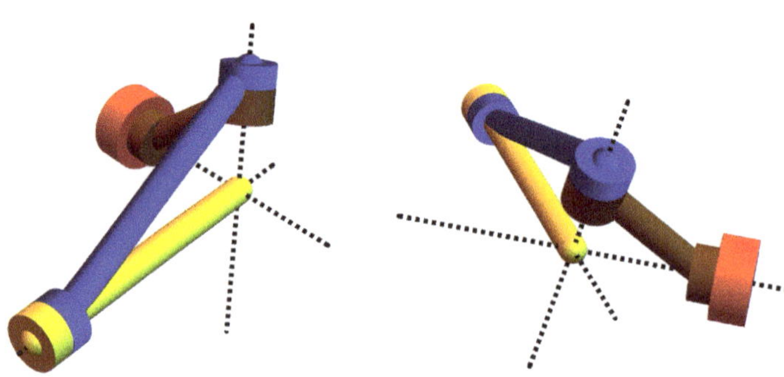

Figure 8. Example 3: Solution 3R robot for a specified angular twist space at the origin. Dotted lines correspond to the joint axes at each position.

Table 7. Example 3: Plücker coordinates of the joint axes of the 3R robot at the reference configuration.

Axes	Plücker Coordinates
S_1	$(-0.507, -0.025, 0.861) + \epsilon(0,0,0)$
S_2	$(0.704, 0.168, 0.690) + \epsilon(0,0,0)$
S_3	$(0.025, -0.973, -0.229) + \epsilon(0,0,0)$

7. Conclusions

The synthesis of spatial 3R chains with the prescribed twist space at particular configurations is studied in this work, extending the current technique of defining a single velocity for guidance along a trajectory. The mixed-position synthesis with fully specified twist spaces allows for full control of the motion in proximity of key positions. In addition, it allows for the synthesis of planar, spherical or other robots with specific geometry within the same synthesis process. Algebraic equations as well as numerical results are presented, and the compatibility between finite positions and potential twists is studied, concluding that the spatial 3R chain can have potential motions corresponding to seven different twist systems of dimension three, among them the two invariant systems corresponding to SO(3) and SE(2) finite motions; in these cases, positions for the chain cannot be defined arbitrarily, as they need to belong to the same completion group. An interesting extension could be finding the conditions to impose lower-dimensional subspaces at particular configurations in order to place singularities at desired locations.

Author Contributions: Conceptualization, N.H. and A.P.-G.; methodology, N.H. and A.P.-G.; software, N.H. and A.P.-G.; validation, N.H.; formal analysis, N.H. and A.P.-G.; investigation, N.H. and A.P.-G.; resources, N.H. and A.P.-G.; data curation, N.H.; writing—original draft preparation, A.P.-G.; writing—review and editing, N.H. and A.P.-G.; visualization, N.H.; supervision, A.P.-G.; project administration, A.P.-G.; funding acquisition, A.P.-G. All authors have read and agreed to the published version of the manuscript.

Funding: This research was partially supported by the National Science Foundation under Grant No. 1208385. The content is solely the authors' responsibility.

Informed Consent Statement: Not applicable.

Data Availability Statement: Not applicable.

Conflicts of Interest: The authors declare no conflict of interest.

References

1. Sandor, G.; Freudenstein, F. Higher-order plane motion theories in kinematic synthesis. *J. Eng. Ind.* **1967**, *89*, 223–230. [CrossRef]
2. Kramer, R.S.S. Selective precision synthesis of planar mechanisms satisfying position and velocity constraints. *Mech. Mach. Theory* **1979**, *14*, 161–170.
3. Holte, J.; Chase, T.; Erdman, A. Approximate velocities in mixed exact-approximate position synthesis of mechanisms. *AMSE J. Mech. Des.* **2001**, *123*, 388–394. [CrossRef]
4. Sathyadev, D.; Soni, A. Synthesis of planar mechanisms for coupler tangent-line envelope generation: An alternative to coupler point-path generation. *ASME J. Mech. Des.* **1988**, *110*, 122–129. [CrossRef]
5. Chen, P.; Roth, B. A unified theory for the finitely and infinitesimally separated position problems of kinematic synthesis. *ASME J. Eng. Ind.* **1969**, *91*, 203–208. [CrossRef]
6. Chen, P.; Roth, B. Design equations for finitely and infinitesimally separated position synthesis of binary link and combined link chains. *ASME J. Eng. Ind.* **1969**, *91*, 209–219. [CrossRef]
7. Tsai, L.W.; Roth, B. Design of dyads with helical, cylindrical, spherical, revolute and prismatic joints. *Mech. Mach. Theory* **1972**, *7*, 85–102. [CrossRef]
8. Robson, N.; McCarthy, J. Second order task specifications used in the geometric design of spatial mechanical linkages. *Int. J. Mod. Eng.* **2010**, *11*, 5–11.
9. Robson, N.; Chen, B. Geometric Design with Multiple Realizable Motion Directions in the Vicinity of a Planar Mechanism-Environment Contact. In Proceedings of the ASME 2018 International Design Engineering Technical Conferences and Computers and Information in Engineering Conference IDETC/CIE 2018, Quebec City, QC, Canada, 26–29 August 2018.
10. Moon, H.; Robson, N.; Langari, R. Approximating elbow constrained hand paths via kinematic synthesis with contact specifications. In *Advances in Robot Kinematics*; Lenarcic, J., Khatib, O., Eds.; Springer: Berlin/Heidelberg, Germany, 2014.

11. Robson, N.; Soh, G.S. Kinematic Synthesis of Planar Multi-limb Mechanisms for Multi-directional Interaction with Bodies in the Environment. In *Advances in Robot Kinematics 2018*; Lenarcic, J., Parenti-Castelli, V., Eds.; Springer International Publishing: Berlin/Heidelberg, Germany, 2018; pp. 356–363.
12. Hasanzadeh, N.; Gosh, S.; Robson, N.; Perez-Gracia, A. Velocity-field tasks for in-hand manipulative synthesis. In *Advances in Robot Kinematics*; Merlet, J., Lenarcic, J., Eds.; Springer: Berlin/Heidelberg, Germany, 2016.
13. Robson, N.; Ghosh, S.; Soh, G.S. Kinematic Synthesis and Design of the Robust Closed Loop Articulated Minimally actuated (CLAM) Hand. *Robotica* **2019**, *38*, 1921–1939. [CrossRef]
14. Tian, C.; Fang, Y.; Ge, Q.J. Structural synthesis of parallel manipulators with coupling sub-chains. *Mech. Mach. Theory* **2017**, *118*, 84–99. [CrossRef]
15. Lee, E.; Mavroidis, C. Solving the geometric design problem of spatial 3r robot manipulators using polynomial homotopy continuation. *ASME J. Mech. Des.* **2002**, *124*, 652–661. [CrossRef]
16. Lee, E.; Mavroidis, C. An elimination procedure for solving the geometric design of spatial 3r manipulators. *ASME J. Mech. Des.* **2006**, *128*, 142–146. [CrossRef]
17. Lee, E.; Mavroidis, C.; Merlet, J. Five precision point synthesis of spatial rrr manipulators using interval analysis. *ASME J. Mech. Des.* **2004**, *126*, 842–850. [CrossRef]
18. Hauenstein, J.D.; Wampler, C.W.; Pfurner, M. Synthesis of three-revolute spatial chains for body guidance. *Mech. Mach. Theory* **2017**, *110*, 61–72. [CrossRef]
19. Gibson, C.; Hunt, K. Geometry of screw systems–2: Classification of screw systems. *Mech. Mach. Theory* **1990**, *25*, 11–27. [CrossRef]
20. Selig, J.M. *Geometric Fundamentals of Robotics Monographs in Computer Science*, 2nd ed.; Springer: Berlin/Heidelberg, Germany, 2005.
21. Carricato, M.; Zlatanov, D. Persistent Screw Systems. *Mech. Mach. Theory* **2014**, *73*, 296–313. [CrossRef]
22. Carricato, M.; Rico-Martinez, J. Persistent screw systems. In *Advances in Robot Kinematics: Motion in Man and Machine*; Lenarcic, J., Stanisic, M., Eds.; Springer: Berlin/Heidelberg, Germany, 2010; pp. 185–194.
23. Carricato, M.; Rico-Martinez, J. Persistent screw systems of Dimension 3. In Proceedings of the 13th World Congress in Mechanism and Machine Science, Guanajuato, Mexico, 19–23 June 2011.
24. Huynh, P.; Hervé, J.M. Equivalent Kinematic Chains of Three Degree-of-Freedom Tripod Mechanisms with Planar-Spherical Bonds. *J. Mech. Des.* **2005**, *127*, 95–102. [CrossRef]
25. Tsai, M.; Lee, H. On the Special Bases of Two- and Three-Screw Systems. *ASME J. Mech. Des.* **1993**, *115*, 540–546. [CrossRef]
26. Donelan, P.; Gibson, C. On the hierarchy of Screw Systems. *Acta Appl. Math.* **1993**, *32*, 267–296. [CrossRef]

robotics

MDPI

Article

A Screw Theory Approach to Computing the Instantaneous Rotation Centers of Indeterminate Planar Linkages

Juan Ignacio Valderrama-Rodríguez [1,†], José M. Rico [2,*,†] and J. Jesús Cervantes-Sánchez [2,†] and Ricardo García-García [2,†]

1 Independent Researcher, Ciudad de Duitama 150461, Colombia; jvalderrama850@gmail.com
2 Mechanical Engineering Department, Divisi6n de Ingenierías, Campus Irapuato-Salamanca, Universidad de Guanajuato, Guanajuato 36885, Mexico; jecer@ugto.mx (J.J.C.-S.); garciar@ugto.mx (R.G.-G.)
* Correspondence: jrico@ugto.mx; Tel.: +52-461-546-5571
† These authors contributed equally to this work.

Abstract: This paper presents a screw theory approach for the computation of the instantaneous rotation centers of indeterminate planar linkages. Since the end of the 19th century, the determination of the instantaneous rotation, or velocity centers of planar mechanisms has been an important topic in kinematics that has led to the well-known Aronhold–Kennedy theorem. At the beginning of the 20th century, it was found that there were planar mechanisms for which the application of the Aronhold–Kennedy theorem was unable to find all the instantaneous rotation centers (IRCs). These mechanisms were denominated complex or indeterminate. The beginning of this century saw a renewed interest in complex or indeterminate planar mechanisms. In this contribution, a new and simpler screw theory approach for the determination of indeterminate rotation centers of planar linkages is presented. The new approach provides a simpler method for setting up the equations. Furthermore, the algebraic equations to be solved are simpler than the ones published to date. The method is based on the systematic application of screw theory, isomorphic to the Lie algebra, $se(3)$, of the Euclidean group, $SE(3)$, and the invariant symmetric bilinear forms defined on $se(3)$.

Keywords: planar linkages; indeterminate linkages; screw theory

Citation: Valderrama-Rodríguez, J.I.; Rico, J.M.; Cervantes-Sánchez, J.J.; García-García, R. A Screw Theory Approach to Computing the Instantaneous Rotation Centers of Indeterminate Planar Linkages. *Robotics* **2022**, *11*, 6. https://doi.org/10.3390/robotics11010006

Academic Editor: Marco Ceccarelli

Received: 5 November 2021
Accepted: 28 December 2021
Published: 31 December 2021

1. Introduction

The instantaneous rotation center (IRC) of two links in a one-DOF planar linkage is defined as a pair of coincident points that belong to each body and have the same velocity, with respect to another reference frame. Therefore, the relative motion between the two links is a rotation around an axis perpendicular to the plane of motion passing through the coincident points. Some IRCs can be easily determined by inspection. These IRCs are defined as primary. For example, each revolute pair constitutes the IRC of the two links that the revolute pair connects. If the IRC is not primary, then it is defined as secondary.

The Aronhold–Kennedy theorem was independently formulated by Aronhold [1], in Germany, and Kennedy [2], in England, in the second part of the 19th century. The Aronhold–Kennedy theorem is the main tool for determining secondary IRCs. The theorem indicates that the location of the IRCs associated with the relative movements of three arbitrary rigid bodies on the plane must be collinear. The great majority of planar linkages allow for the locating of all the IRCs of planar linkages using the Aronhold–Kennedy theorem. Therefore, for many decades, this theorem provided an efficient graphical method for the kinematic analysis of planar linkages; see Shigley and Uicker [3].

Nevertheless, since 1915, Klein [4] showed the existence of planar linkages for which the application of the Aronhold–Kennedy theorem was insufficient to determine the IRCs. These planar linkages were called complex planar linkages and, later, indeterminate planar linkages. Klein [4] proposed a graphical trial and error method for locating those IRCs that resisted the application of the Aronhold–Kennedy theorem.

In the middle of the 20th century, Modrey [5] employed influence coefficients to find the secondary IRCs of the so-called complex planar linkages. The posterior discussion by Goodman [6] is also quite illuminating. The method involves the solution of the velocity analysis of the indeterminate linkage and, from these results, it obtains the required IRCs. This approach reappeared in later dates, as in Yan et al. [7], and even in recent contributions, as in Kim et al. [8]. Nevertheless, in this latter case, the method was loosely applied—with many conceptual mistakes—to simple determinate planar linkages. Recently, Valderrama-Rodríguez et al. [9] generalized the approach for any spatial linkage and rebutted several statements presented in Kim et al. [8]. Moreover, from a historical point of view, the determination of the IRCs was employed precisely to carry out the velocity analysis of the linkages. In addition, the determination of the IRCs provides important insights on the performance of planar linkages, such as transmissibility, analysis of singularities (Di Gregorio [10]), and mechanical advantage (Zarkandi [11]), among others.

About 15 years ago, Pennock and Foster [12] presented a graphical technique that was able to find all the IRCs of indeterminate linkages and improved substantially the approach proposed by Klein. A few years later, Di Gregorio [13] introduced an analytical technique that generated a system of equations involving both the closure equations of the planar linkages and the location of the indeterminate IRCs, so that he was able to find the location of the indeterminate rotation centers. In 2009, Kung and Wang [14] employed graph theory to obtain a graph associated with the IRCs of a linkage. This graph allowed Kung and Wang to formulate a system of iterative equations, whose unknowns are the coordinates of the indeterminate IRCs. The solution of the problem comes down to solving a quadratic equation and then a quartic equation.

There is another line of research of the determination of the IRC of planar linkages. The origin of this line of research is due to Chang and Her [15], who in 2008 presented a virtual cam method for locating the IRCs of indeterminate linkages. This approach has been extended by Liu and Chang [16,17]. They extended the approach by introducing the "virtual cam–hexagon" method. They embarked themselves on the task of determining the IRCs of all indeterminate linkages, up to ten bar. It is important to note that the approach is almost completely graphical; Liu and Chang indicate that the [17] "...Virtual Cam–Hexagon Method is not a truly universal technique yet, but, at this moment all the other graphical and geometrical approaches are extremely limited which can only be applied on several specific constructions of kinematically indeterminate linkages, and all of those approaches either change the constructions of the original linkages dramatically, or some even alter the original system's DOF. Virtual Cam–Hexagon Method does not change the construction of the original linkage, nor alter the system's DOF in the procedure. With a nearly 75% success rate on all planar single DOF linkages up to ten-bar...".

In this contribution, which follows a similar technique for indeterminate spherical linkages, Valderrama-Rodríguez et al. [18] introduced a more efficient approach than those presented previously for the determination of the IRCs of indeterminate spherical linkages. However, in the case of spherical linkages, only the Killing form is needed, whilst in the case of planar linkages, both the Killing and Klein forms are necessary. However, the influence of the Killing form is only recognized after the general analysis presented in this paper is carried out. Therefore, the analysis and application of both the Killing and Klein forms are required.

Some of the advantages of the method presented in this contribution are:

1. The method requires a reduced set of equations. It will be shown that the development of the complete graph of the IRCs of the linkage, as proposed by Kung and Wang [14], is unnecessary. More specifically, by properly selecting a pair of sub-graphs of the complete graph, the determination of all secondary IRCs can be readily obtained;
2. The required equations are simpler. The method presented by Di Gregorio [13] requires the solution of a nonlinear set of 10 equations, in one case, and a nonlinear set of 12 equations, in another case. The method presented by Kung and Wang [14] requires first the solution of a quadratic equation and, later, the solution of a quartic

equation. The method presented in this contribution only requires the solution and comparison of 2 quadratic equations in one case, and a quadratic and a cubic equation in the other.

The rest of the contribution is revised in this paragraph. Section 2 presents the fundamental equation of the Aronhold–Kennedy theorem and the symmetric bilinear forms defined on, $se(3)$, the Lie algebra of the special Euclidean group, $SE(3)$, namely, the Killing and the Klein forms. The purpose of this section is two-fold. On the one hand, it shows that the application of the algebraic structure of the Lie algebra, $se(3)$, provides a unifying approach for the determination of the instantaneous screw axes of any spatial linkage; on the other hand, it shows that for planar linkages, the approach is reduced to the application of the Klein form to screws perpendicular to the plane motion—i.e., the classical Aronhold–Kennedy theorem. The importance of the Killing form has been obscured due to the history of kinematics, which for many decades was reduced to planar kinematics exclusively. Section 3 presents the complete approach for an eight-bar, partially indeterminate planar linkage. Section 4 solves an eight-bar, completely indeterminate planar linkage. Section 5 provides constructive proof for the validity of the method for all indeterminate linkages. Finally, some conclusions are drawn in Section 6. This paper contains a small Appendix A, which shows how to represent a secondary IRC as a linear combination of two primary IRCs.

2. The Fundamental Equation for the General Aronhold–Kennedy Theorem

This section starts with a definition of the velocity state of a body with respect to another body.

Definition 1. *Let j and m be a pair of different rigid bodies or reference frames, and let O be a point fixed in body m. The velocity state of body m with respect to body j is defined as*

$$^{j}\mathbf{V}^{m}_{O} \equiv \begin{bmatrix} ^{j}\boldsymbol{\omega}^{m} \\ ^{j}\mathbf{v}^{m}_{O} \end{bmatrix} \tag{1}$$

where $^{j}\boldsymbol{\omega}^{m}$ is the angular velocity of body m with respect to body j, and $^{j}\mathbf{v}^{m}_{O}$ is the velocity of a point O, fixed to the rigid body m as observed from the rigid body j. Ball [19] defined $^{j}\mathbf{V}^{m}_{O}$ as a twist on a screw.

Proposition 1. *Let j, k, and m, be three different bodies or reference frames, see Figure 1, and let $^{j}\mathbf{V}^{k}_{O}$, $^{j}\mathbf{V}^{m}_{O}$, and $^{k}\mathbf{V}^{m}_{O}$ be their corresponding velocity states, as given by Definition 1. Then:*

$$^{j}\mathbf{V}^{m}_{O} = {}^{j}\mathbf{V}^{k}_{O} + {}^{k}\mathbf{V}^{m}_{O}$$

or

$$\begin{bmatrix} ^{j}\boldsymbol{\omega}^{m} \\ ^{j}\mathbf{v}^{m}_{O} \end{bmatrix} = \begin{bmatrix} ^{j}\boldsymbol{\omega}^{k} \\ ^{j}\mathbf{v}^{k}_{O} \end{bmatrix} + \begin{bmatrix} ^{k}\boldsymbol{\omega}^{m} \\ ^{k}\mathbf{v}^{m}_{O} \end{bmatrix} \tag{2}$$

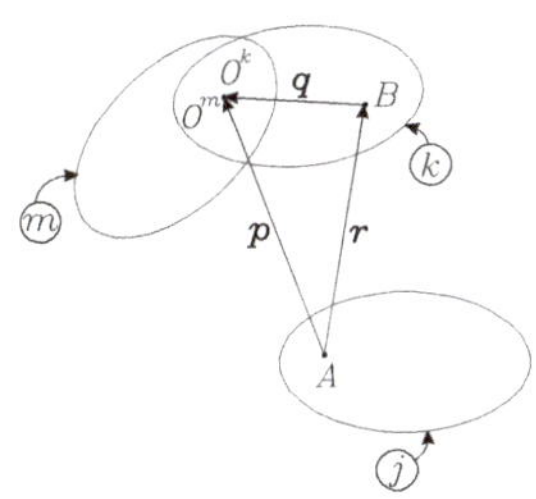

Figure 1. Three rigid bodies and the related points and vectors.

It is important to note that Equation (2) is valid whether the bodies are connected via kinematic pairs or are unconnected. This result is presented in Rico et al. [20], but it is quite likely that it was well known from the beginning of the last century, and that Ball [19] was aware of its meaning, despite not using the vector notation.

In the Lie algebra $se(3)$ of the Euclidean group $SE(3)$, two invariant symmetrical bilinear forms can be defined.

Definition 2. *Consider the Lie algebra, $se(3)$, of the Euclidean group, $SE(3)$. Then, it is possible to define a bilinear symmetric form, denoted as the Killing form for any screws $\mathbf{V}_{O1}$, $\mathbf{V}_{O2} \in se(3)$, as follows:*

$$Ki : se(3) \times se(3) \rightarrow \mathbf{R} \qquad Ki(\mathbf{V}_{O1}, \mathbf{V}_{O2}) = Ki((\boldsymbol{\omega}_1; \mathbf{v}_{O1}), (\boldsymbol{\omega}_2; \mathbf{v}_{O2})) = \boldsymbol{\omega}_1 \cdot \boldsymbol{\omega}_2$$

where "$\cdot$" is the scalar product of three-dimensional vector algebra.

Definition 3. *Consider the Lie algebra, $se(3)$, of the Euclidean group, $SE(3)$. Then, it is possible to define a bilinear symmetric form, denoted as the Klein form, for any screws $\mathbf{V}_{O1}$, $\mathbf{V}_{O2} \in se(3)$, as follows:*

$$Kl : se(3) \times se(3) \rightarrow \mathbf{R} \quad Kl(\mathbf{V}_{O1}, \mathbf{V}_{O2}) = Kl((\boldsymbol{\omega}_1; \mathbf{v}_{O1}), (\boldsymbol{\omega}_2; \mathbf{v}_{O2})) = \boldsymbol{\omega}_1 \cdot \mathbf{v}_{O2} + \boldsymbol{\omega}_2 \cdot \mathbf{v}_{O1}$$

where "$\cdot$" is the scalar product of three-dimensional vector algebra. This symmetric bilinear form is also called the reciprocal product of the screws.

It can be proved that both symmetrical bilinear forms, Killing and Klein, are well defined or invariant, i.e., the result is independent of the point O used to determine the velocity state or the selection of the coordinate system; see, for example, Brand [21], although the reader is cautioned that Brand used the "motor" terminology for screws. Moreover, it can be proved that the Klein form is a nonsingular indefinite symmetric bilinear form on $se(3)$, while the Killing form is a singular positive semi-definite symmetric bilinear form on $se(3)$ (Rico and Duffy [22]).

Finally, given a screw \$, any screw $\$_a$ that satisfies $Ki(\$, \$_a) = 0$ is called an orthogonal annihilator, regarding the Killing form of the original screw \$. It should be noted that, if $Ki(\$, \$_a) = 0$, the direction of the axes of the screws are perpendicular. Similarly, any screw $\$_a$ that satisfies $Kl(\$, \$_a) = 0$ is called an orthogonal annihilator, regarding the Klein form of the original screw \$; frequently, the screw $\$_a$, in this latter case, is also called the reciprocal screw of the original screw \$.

Symbolically, Equation (2) can be written as:

$$_j\omega_m \, {}^j\$^m = {}^j\mathbf{V}_O^m = {}^j\mathbf{V}_O^k + {}^k\mathbf{V}_O^m = {}_j\omega_k \, {}^j\$^k + {}_k\omega_m \, {}^k\$^m \tag{3}$$

where the infinitesimal screw, let us say, ${}^j\$^m$, is given by:

$${}^j\$^m = \begin{bmatrix} {}^j\mathbf{u}^m \\ {}^j\mathbf{r}^m \times {}^j\mathbf{u}^m + {}_jh_m \, {}^j\mathbf{u}^m \end{bmatrix} \tag{4}$$

with ${}^j\mathbf{u}^m$ being a corresponding unit vector in the direction of the screw axis, ${}^j\mathbf{r}^m$ is the position vector of a point along the screw axis, about the origin of the coordinate system, ${}_jh_m$ is the pitch of the screw, and "$\times$" is the vector product of three-dimensional vector algebra.

The determination of an instantaneous screw axis, for example, ${}^j\$^m$, requires the orthogonal annihilators, regarding the Klein and Killing forms, of the subspace of $se(3)$ generated by $\{{}^j\$^k, {}^k\$^m\}$, denoted also as $\left[{}^j\$^k, {}^k\$^m\right]$. (This notation should not be confused with the Lie product, which in this contribution does not play any role.) Therefore, we will look for those screws $\$_{ajm}$ that:

1. are orthogonal concerning the Klein form (also called reciprocal):

$$Kl(\$_{ajm}, {}^j\$^k) = Kl(\$_{ajm}, {}^k\$^m) = 0; \tag{5}$$

2. are orthogonal regarding the Killing form (also called perpendicular):

$$Ki(\$_{ajm}, {}^j\$^k) = Ki(\$_{ajm}, {}^k\$^m) = 0. \tag{6}$$

Therefore, for the Klein and Killing forms, one has:

$$\begin{aligned} Kl(\$_{ajm}, {}_j\omega_m\, {}^j\$^m) &= Kl(\$_{ajm}, {}^j\mathbf{V}_O^m) = Kl(\$_{ajm}, {}^j\mathbf{V}_O^k + {}^k\mathbf{V}_O^m) \\ &= Kl(\$_{ajm}, {}_j\omega_k\, {}^j\$^k + {}_k\omega_m\, {}^k\$^m) \\ &= {}_j\omega_k\, Kl(\$_{ajm}, {}^j\$^k) + {}_k\omega_m\, Kl(\$_{ajm}, {}^k\$^m) = {}_j\omega_k\,(0) + {}_k\omega_m\,(0) = 0 \end{aligned} \tag{7}$$

and

$$\begin{aligned} Ki(\$_{ajm}, {}_j\omega_m\, {}^j\$^m) &= Ki(\$_{ajm}, {}^j\mathbf{V}_O^m) = Ki(\$_{ajm}, {}^j\mathbf{V}_O^k + {}^k\mathbf{V}_O^m) \\ &= Ki(\$_{ajm}, {}_j\omega_k\, {}^j\$^k + {}_k\omega_m\, {}^k\$^m) \\ &= {}_j\omega_k\, Ki(\$_{ajm}, {}^j\$^k) + {}_k\omega_m\, Ki(\$_{ajm}, {}^k\$^m) = {}_j\omega_k\,(0) + {}_k\omega_m\,(0) = 0 \end{aligned} \tag{8}$$

Summarizing, one may conclude that the orthogonal annihilators of the subspace generated for $\{{}^j\$^k, {}^k\$^m\}$ annihilate also the screw ${}^j\$^m$ associated with the instantaneous screw axis of the relative movement between links j and m. These two Equations (7) and (8), are the fundamental equations of the Aronhold–Kennedy theorem for general spatial linkages.

2.1. The Simplification for Planar Linkages

It is important to note that the role of the Klein and Killing forms can be drastically simplified if the linkages to be analyzed are planar. Without loss of generality, it will be assumed that the common plane of motion is perpendicular to the Z-axis. In planar linkages, kinematic pairs can be revolute or prismatic joints. The axes of the revolute pairs are all parallel to the Z-axis, while the directions of prismatic pairs must be perpendicular to the Z-axis; therefore, the screws representing revolute and prismatic pairs, Equation (4), can be, respectively, reduced as:

$${}^j\$^m = \begin{bmatrix} \mathbf{k} \\ {}^j\mathbf{r}^m \times \mathbf{k} \end{bmatrix}, \text{ and } {}^n\$^r = \begin{bmatrix} \mathbf{0} \\ {}^n\mathbf{u}^r \end{bmatrix}, \text{ where } {}^n\mathbf{u}^r \cdot \mathbf{k} = 0. \tag{9}$$

Without loss of generality, it can be assumed that $\mathbf{k} \cdot {}^j\mathbf{r}^m = 0$, for any ${}^j\mathbf{r}^m$, i.e., ${}^j\mathbf{r}^m$ and $\mathbf{k}$, are perpendicular.

It follows that, for planar linkages, the corresponding Lie algebra is not $se(3)$, but the subalgebra $g_\mathbf{k}$ is also denominated $se(2)$. This is the Lie subalgebra associated with the planar displacement subgroup, $G_\mathbf{k}$; it is also called the planar subgroup. Furthermore, $g_\mathbf{k}$ continues to be an orthogonal space under the restrictions of the Klein and Killing forms to $g_\mathbf{k}$ itself. However, the restricted Klein and Killing forms, in $g_\mathbf{k}$, have properties that are different from those in $se(3)$.

The Killing form for any pair of screws, ${}^j\$^m$ and ${}^n\$^r$, associated with the revolute pairs of any planar linkage, becomes

$$Ki \;:\; g_\mathbf{k} \times g_\mathbf{k} \to \mathbf{R} \quad Ki({}^j\$^m, {}^n\$^r) = Ki\left(\begin{bmatrix} \mathbf{k} \\ {}^j\mathbf{r}^m \times \mathbf{k} \end{bmatrix}, \begin{bmatrix} \mathbf{k} \\ {}^n\mathbf{r}^r \times \mathbf{k} \end{bmatrix}\right) = \mathbf{k} \cdot \mathbf{k} = 1. \tag{10}$$

If one of the screws, namely, ${}^j\$^m$, represents a prismatic pair, it follows that:

$$Ki \;:\; g_\mathbf{k} \times g_\mathbf{k} \to \mathbf{R} \quad Ki({}^j\$^m, {}^n\$^r) = Ki\left(\begin{bmatrix} \mathbf{0} \\ {}^j\mathbf{u}^m \end{bmatrix}, \begin{bmatrix} \mathbf{k} \\ {}^n\mathbf{r}^r \times \mathbf{k} \end{bmatrix}\right) = \mathbf{0} \cdot \mathbf{k} = 0. \tag{11}$$

In the language of orthogonal spaces, the restriction of the Killing form to the subalgebra $g_{\mathbf{k}}$ becomes a singular symmetric bilinear form.

Similarly, the Klein form for any pair of screws, ${}^{j}\$^{m}$ and ${}^{n}\$^{r}$, associated with the revolute pairs of any planar linkage, becomes:

$$
\begin{aligned}
Kl \;:\; g_{\mathbf{k}} \times g_{\mathbf{k}} \to \mathbf{R} \quad Kl({}^{j}\$^{m}, {}^{n}\$^{r}) &= Kl\left(\begin{bmatrix} \mathbf{k} \\ {}^{j}\mathbf{r}^{m} \times \mathbf{k} \end{bmatrix}, \begin{bmatrix} \mathbf{k} \\ {}^{n}\mathbf{r}^{r} \times \mathbf{k} \end{bmatrix} \right) \\
&= \mathbf{k} \cdot ({}^{n}\mathbf{r}^{r} \times \mathbf{k}) + \mathbf{k} \cdot ({}^{j}\mathbf{r}^{m} \times \mathbf{k}) = 0
\end{aligned} \tag{12}
$$

for any ${}^{j}\$^{m}, {}^{n}\$^{r} \in g_{\mathbf{k}}$. This result is consistent with a well-known result in screw theory that states that any pair of parallel 0-pitch screws are reciprocal, i.e., they are orthogonal regarding the Klein form.

If one of the screws, namely, ${}^{j}\$^{m}$, represents a prismatic pair, it follows that:

$$
\begin{aligned}
Kl \;:\; g_{\mathbf{k}} \times g_{\mathbf{k}} \to \mathbf{R} \quad Kl({}^{j}\$^{m}, {}^{n}\$^{r}) = Kl\left(\begin{bmatrix} \mathbf{0} \\ {}^{j}\mathbf{u}^{m} \end{bmatrix}, \begin{bmatrix} \mathbf{k} \\ {}^{n}\mathbf{r}^{r} \times \mathbf{k} \end{bmatrix} \right) \\
= \mathbf{0} \cdot ({}^{n}\mathbf{r}^{r} \times \mathbf{k}) + \mathbf{k} \cdot {}^{j}\mathbf{u}^{m} = 0 + 0 = 0.
\end{aligned} \tag{13}
$$

Thus, the restriction of the Klein form to the subalgebra $g_{\mathbf{k}}$ becomes also a lsingular symmetric bilinear form.

2.2. Application of the Killing and Klein Forms to Planar Linkages

This section shows the detailed process to obtain the equations that allow determining the secondary centers using the Aronhold–Kennedy theorem, as it was outlined in Equations (7) and (8).

Consider a pair of screws associated with a revolute joint and a prismatic joint of a planar linkage, as shown in Equation (9), repeated here:

$$
{}^{j}\$^{m} = \begin{bmatrix} \mathbf{k} \\ {}^{j}\mathbf{r}^{m} \times \mathbf{k} \end{bmatrix} \quad {}^{n}\$^{r} = \begin{bmatrix} \mathbf{0} \\ {}^{n}\mathbf{u}^{r} \end{bmatrix}, \text{ with } {}^{n}\mathbf{u}^{r} \cdot \mathbf{k} = 0
$$

As indicated previously, without loss of generality, it will be assumed that the common plane of displacements is perpendicular to the Z-axis. Therefore, the axes of the revolute pairs are all parallel to the *Z*-axis, while the directions of the prismatic pairs must be perpendicular to the Z-axis. Additionally, without loss of generality, it can be assumed that $\mathbf{k} \cdot {}^{j}\mathbf{r}^{m} = 0$, namely, ${}^{j}\mathbf{r}^{m}$ and $\mathbf{k}$, are perpendicular. Moreover, the vector ${}^{j}\mathbf{r}^{m}$ provides the location of the instantaneous rotation, or velocity center represented by ${}^{j}\$^{m}$. The process must consider two cases.

First Case. The two kinematic pairs are revolute joints. Consider a subspace of $g_{\mathbf{k}}$, given by $[{}^{j}\$^{m}, {}^{n}\$^{r}]$, where:

$$
{}^{j}\$^{m} = \begin{bmatrix} \mathbf{k} \\ {}^{j}\mathbf{r}^{m} \times \mathbf{k} \end{bmatrix} \text{ and } {}^{n}\$^{r} = \begin{bmatrix} \mathbf{k} \\ {}^{n}\mathbf{r}^{r} \times \mathbf{k} \end{bmatrix}.
$$

Let

$$
\$_{a} = \begin{bmatrix} \mathbf{u} \\ \mathbf{r}_{a} \times \mathbf{u} \end{bmatrix}
$$

be an orthogonal annihilator of the subspace $[{}^{j}\$^{m}, {}^{n}\$^{r}]$ regarding both the Killing and Klein forms. Then:

1. applying the Killing form, it follows that:

$$\begin{aligned} Ki(\$_a, {}^j\$^m) &= Ki\left(\begin{bmatrix} \mathbf{u} \\ \mathbf{r}_a \times \mathbf{u} \end{bmatrix}, \begin{bmatrix} \mathbf{k} \\ {}^j\mathbf{r}^m \times \mathbf{k} \end{bmatrix}\right) = \mathbf{u} \cdot \mathbf{k} = 0 \\ Ki(\$_a, {}^n\$^r) &= Ki\left(\begin{bmatrix} \mathbf{u} \\ \mathbf{r}_a \times \mathbf{u} \end{bmatrix}, \begin{bmatrix} \mathbf{k} \\ {}^n\mathbf{r}^r \times \mathbf{k} \end{bmatrix}\right) = \mathbf{u} \cdot \mathbf{k} = 0 \end{aligned}$$

Therefore, $\mathbf{u}$ and $\mathbf{k}$ must be perpendicular. Hence, $\mathbf{u}$ must be of the form $\mathbf{u} = u_x\,\mathbf{i} + u_y\,\mathbf{j}$ and, without loss of generality, it will be assumed that $\mathbf{r}_a = r_{ax}\,\mathbf{i} + r_{ay}\,\mathbf{j}$. Therefore, $\$_a$, the orthogonal annihilator of the subspace $[{}^j\$^m, {}^n\$^r]$ regarding the Killing form, is given by:

$$\$_a = \begin{bmatrix} \mathbf{u} \\ \mathbf{r}_a \times \mathbf{u} \end{bmatrix} = \begin{bmatrix} u_x\,\mathbf{i} + u_y\,\mathbf{j} \\ (r_{ax}\,\mathbf{i} + r_{ay}\,\mathbf{j}) \times (u_x\,\mathbf{i} + u_y\,\mathbf{j}) \end{bmatrix};$$

2. applying the Klein form, it follows that:

$$Kl(\$_a, {}^j\$^m) = Kl\left(\begin{bmatrix} \mathbf{u} \\ \mathbf{r}_a \times \mathbf{u} \end{bmatrix}, \begin{bmatrix} \mathbf{k} \\ {}^j\mathbf{r}^m \times \mathbf{k} \end{bmatrix}\right) = \mathbf{u} \cdot \left({}^j\mathbf{r}^m \times \mathbf{k}\right) + \mathbf{k} \cdot (\mathbf{r}_a \times \mathbf{u}) = 0 \quad (14)$$

$$Kl(\$_a, {}^n\$^r) = Kl\left(\begin{bmatrix} \mathbf{u} \\ \mathbf{r}_a \times \mathbf{u} \end{bmatrix}, \begin{bmatrix} \mathbf{k} \\ {}^n\mathbf{r}^r \times \mathbf{k} \end{bmatrix}\right) = \mathbf{u} \cdot ({}^n\mathbf{r}^r \times \mathbf{k}) + \mathbf{k} \cdot (\mathbf{r}_a \times \mathbf{u}) = 0 \quad (15)$$

Using the identities of the scalar triple product on Equations (14) and (15), one obtains:

$$\begin{aligned} 0 &= \mathbf{u} \cdot \left({}^j\mathbf{r}^m \times \mathbf{k}\right) + \mathbf{k} \cdot (\mathbf{r}_a \times \mathbf{u}) = \mathbf{k} \cdot \left(\mathbf{u} \times {}^j\mathbf{r}^m\right) + \mathbf{u} \cdot (\mathbf{k} \times \mathbf{r}_a) \\ &= {}^j\mathbf{r}^m \cdot (\mathbf{k} \times \mathbf{u}) + \mathbf{r}_a \cdot (\mathbf{u} \times \mathbf{k}) = ({}^j\mathbf{r}^m - \mathbf{r}_a) \cdot (\mathbf{k} \times \mathbf{u}) \end{aligned} \quad (16)$$

and

$$\begin{aligned} 0 &= \mathbf{u} \cdot ({}^n\mathbf{r}^r \times \mathbf{k}) + \mathbf{k} \cdot (\mathbf{r}_a \times \mathbf{u}) = \mathbf{k} \cdot (\mathbf{u} \times {}^n\mathbf{r}^r) + \mathbf{u} \cdot (\mathbf{k} \times \mathbf{r}_a) \\ &= {}^n\mathbf{r}^r \cdot (\mathbf{k} \times \mathbf{u}) + \mathbf{r}_a \cdot (\mathbf{u} \times \mathbf{k}) = ({}^n\mathbf{r}^r - \mathbf{r}_a) \cdot (\mathbf{k} \times \mathbf{u}) \end{aligned} \quad (17)$$

Denoting, by $\mathbf{k} \times \mathbf{u} = \mathbf{u}^\perp$, a unit vector perpendicular to both $\mathbf{u}$ and $\mathbf{k}$, and therefore in the X–Y plane, Equations (16) and (17) can be written as:

$${}^j\mathbf{r}^m \cdot \mathbf{u}^\perp = {}^n\mathbf{r}^r \cdot \mathbf{u}^\perp = \mathbf{r}_a \cdot \mathbf{u}^\perp \qquad \text{or} \qquad \left({}^j\mathbf{r}^m - {}^n\mathbf{r}^r\right) \cdot \mathbf{u}^\perp = 0 \quad (18)$$

Equation (18) indicates that ${}^j\mathbf{r}^m - {}^n\mathbf{r}^r$ is perpendicular to $\mathbf{u}^\perp$. Since $\mathbf{u}$ and $\mathbf{u}^\perp$ must lie in the X–Y plane; then:

$$(\mathbf{u}^\perp)^\perp = \mathbf{u}$$

Therefore, ${}^j\mathbf{r}^m - {}^n\mathbf{r}^r$ must be parallel to $\mathbf{u}$; consequently:

$${}^j\mathbf{r}^m - {}^n\mathbf{r}^r = \lambda\,\mathbf{u} \quad (19)$$

Equations (15), (16) and (19) indicate that the orthogonal annihilator $\$_a$ must lay in the X–Y plane and intersect with both ${}^j\$^m$ and ${}^n\$^r$. Hence, the following proposition has been proved:

Proposition 2. *Let ${}^j\$^m$ and ${}^n\$^r$ be the screws representing a pair of revolute pairs whose common direction is the Z-axis, and let ${}^j\mathbf{r}^m$ and ${}^n\mathbf{r}^r$ be the position vectors of two points in the X–Y plane along the corresponding screw axis. Then, the orthogonal annihilator $\$_a$ must lie in the X–Y plane, and it must pass through ${}^j\mathbf{r}^m$ and ${}^n\mathbf{r}^r$.*

Second Case. One kinematic pair is a revolute joint, the other is a prismatic joint. Consider a subspace of $g_{\mathbf{k}}$, given by $\left[{}^{j}\$^{m}, {}^{n}\$^{r}\right]$:

$$ {}^{j}\$^{m} = \begin{bmatrix} \mathbf{k} \\ {}^{j}\mathbf{r}^{m} \times \mathbf{k} \end{bmatrix} \quad \text{and} \quad {}^{n}\$^{r} = \begin{bmatrix} \mathbf{0} \\ {}^{n}\mathbf{u}^{r} \end{bmatrix} $$

where ${}^{n}\mathbf{u}^{r}$ lies on the X–Y plane. Let

$$ \$_{a} = \begin{bmatrix} \mathbf{u} \\ \mathbf{r}_{a} \times \mathbf{u} \end{bmatrix} $$

be an orthogonal annihilator of the subspace $\left[{}^{j}\$^{m}, {}^{n}\$^{r}\right]$ regarding both the Killing and Klein forms. Then:
1. applying the Killing form, it follows that:

$$ Ki(\$_{a}, {}^{j}\$^{m}) = Ki\left(\begin{bmatrix} \mathbf{u} \\ \mathbf{r}_{a} \times \mathbf{u} \end{bmatrix}, \begin{bmatrix} \mathbf{k} \\ {}^{j}\mathbf{r}^{m} \times \mathbf{k} \end{bmatrix} \right) = \mathbf{u} \cdot \mathbf{k} = 0 $$
$$ Ki(\$_{a}, {}^{n}\$^{r}) = Ki\left(\begin{bmatrix} \mathbf{u} \\ \mathbf{r}_{a} \times \mathbf{u} \end{bmatrix}, \begin{bmatrix} \mathbf{0} \\ {}^{n}\mathbf{u}^{r} \end{bmatrix} \right) = \mathbf{u} \cdot \mathbf{0} = 0 $$

Therefore, $\mathbf{u}$ and $\mathbf{k}$ must be perpendicular. Hence, $\mathbf{u}$ must be of the form $\mathbf{u} = u_{x}\,\mathbf{i} + u_{y}\,\mathbf{j}$ and, without loss of generality, it will be assumed that $\mathbf{r}_{a} = r_{ax}\,\mathbf{i} + r_{ay}\,\mathbf{j}$. Therefore, $\$_{a}$, the orthogonal annihilator of the subspace $\left[{}^{j}\$^{m}, {}^{n}\$^{r}\right]$ regarding the Killing form, is given by:

$$ \$_{a} = \begin{bmatrix} \mathbf{u} \\ \mathbf{r}_{a} \times \mathbf{u} \end{bmatrix} = \begin{bmatrix} u_{x}\,\mathbf{i} + u_{y}\,\mathbf{j} \\ (r_{ax}\,\mathbf{i} + r_{ay}\,\mathbf{j}) \times (u_{x}\,\mathbf{i} + u_{y}\,\mathbf{j}) \end{bmatrix}; $$

2. applying the Klein form, it follows that:

$$ Kl(\$_{a}, {}^{j}\$^{m}) = Kl\left(\begin{bmatrix} \mathbf{u} \\ \mathbf{r}_{a} \times \mathbf{u} \end{bmatrix}, \begin{bmatrix} \mathbf{k} \\ {}^{j}\mathbf{r}^{m} \times \mathbf{k} \end{bmatrix} \right) = \mathbf{u} \cdot \left({}^{j}\mathbf{r}^{m} \times \mathbf{k} \right) + \mathbf{k} \cdot (\mathbf{r}_{a} \times \mathbf{u}) = 0 \quad (20) $$

Using the identities of the scalar triple product on Equation (20), it yields:

$$ \begin{aligned} 0 &= \mathbf{u} \cdot \left({}^{j}\mathbf{r}^{m} \times \mathbf{k} \right) + \mathbf{k} \cdot (\mathbf{r}_{a} \times \mathbf{u}) = \mathbf{k} \cdot \left(\mathbf{u} \times {}^{j}\mathbf{r}^{m} \right) + \mathbf{u} \cdot (\mathbf{k} \times \mathbf{r}_{a}) \\ &= {}^{j}\mathbf{r}^{m} \cdot (\mathbf{k} \times \mathbf{u}) + \mathbf{r}_{a} \cdot (\mathbf{u} \times \mathbf{k}) = ({}^{j}\mathbf{r}^{m} - \mathbf{r}_{a}) \cdot (\mathbf{k} \times \mathbf{u}) \end{aligned} \quad (21) $$

Denoting, by $\mathbf{k} \times \mathbf{u} = \mathbf{u}^{\perp}$, a unit vector perpendicular to both $\mathbf{u}$ and $\mathbf{k}$, and therefore in the X–Y plane, Equation (21) can be written as:

$$ {}^{j}\mathbf{r}^{m} \cdot \mathbf{u}^{\perp} = \mathbf{r}_{a} \cdot \mathbf{u}^{\perp} \quad \text{or} \quad \left({}^{j}\mathbf{r}^{m} - \mathbf{r}_{a} \right) \cdot \mathbf{u}^{\perp} = 0 \quad (22) $$

Equation (22) indicates that ${}^{j}\mathbf{r}^{m} - \mathbf{r}_{a}$ is perpendicular to $\mathbf{u}^{\perp}$. Since $\mathbf{u}$ and $\mathbf{u}^{\perp}$ must lie in the X–Y plane, then:

$$ (\mathbf{u}^{\perp})^{\perp} = \mathbf{u} $$

Therefore, ${}^{j}\mathbf{r}^{m} - \mathbf{r}_{a}$ must be parallel to $\mathbf{u}$; consequently:

$$ {}^{j}\mathbf{r}^{m} - \mathbf{r}_{a} = \lambda\,\mathbf{u} \quad \text{or} \quad {}^{j}\mathbf{r}^{m} = \mathbf{r}_{a} + \lambda\,\mathbf{u} \quad (23) $$

and

$$ Kl(\$_{a}, {}^{n}\$^{r}) = Kl\left(\begin{bmatrix} \mathbf{u} \\ \mathbf{r}_{a} \times \mathbf{u} \end{bmatrix}, \begin{bmatrix} \mathbf{0} \\ {}^{n}\mathbf{u}^{r} \end{bmatrix} \right) = \mathbf{u} \cdot {}^{n}\mathbf{u}^{r} + \mathbf{0} \cdot (\mathbf{r}_{a} \times \mathbf{u}) = \mathbf{u} \cdot {}^{n}\mathbf{u}^{r} = 0 \quad (24) $$

Therefore, **u** must be perpendicular to ${}^{n}\mathbf{u}^{r}$. These two last results provide proof of the following proposition.

Proposition 3. *Let ${}^{j}\$^{m}$ and ${}^{n}\$^{r}$ be the screws representing a revolute pair and a prismatic pair, respectively. The direction of the revolute pair is parallel to the Z-axis and ${}^{j}\mathbf{r}^{m}$ locates a point in the X–Y plane lying on the revolute axis. Let ${}^{n}\mathbf{u}^{r}$ be the direction of the prismatic pair on the X–Y plane. Then, the orthogonal annihilator $\$_a$ must lie on the X–Y plane, pass through the point given by ${}^{j}\mathbf{r}^{m}$, and be perpendicular to the unit vector ${}^{n}\mathbf{u}^{r}$.*

These two last propositions provide graphical techniques used for finding secondary centers of rotation, in planar linkages, using the Aronhold–Kennedy theorem. It should be noted that the orthogonal annihilators are precisely the lines that, in the planar case, must be intersected to find the secondary IRCs. Propositions 2 and 3 are the working tools of the Aronhold–Kennedy theorem. It is important to note that even recent papers wrongly indicate the foundations of the generalization of the Aronhold–Kennedy theorem to the spherical and spatial cases, called the Three-Axes Theorem. Zhang et al. [23] indicate "...Sugimoto and Duffy firstly obtained the IS of planar 4R and spatial RCCC mechanisms by the reciprocal screw theory. Aronhold–Kennedy Theorem is an effective geometrical method computing the IS of planar mechanism and was generalized to the spatial case using the reciprocal screw theory...". There is no reference to the important role played by the Killing form, for example, in spherical linkages, where the reciprocal product or Klein form is useless.

3. First Case Study: A "Single Flyer" Eight-Bar Linkage

Consider the planar linkage shown in Figure 2. The linkage is known as the eight-bar "single flyer" mechanism. This linkage was proposed by Klein [4], who determined the secondary IRCs using a trial-and-error method. Foster and Pennock [12] determined the secondary IRCs by applying a graphical method. Di Gregorio [13] computed the secondary IRCs using an analytical method that combined the closure equation of the linkage and the unknown locations of the centers. This linkage had eight links, three of them ternary and the remaining links binary. The linkage is a trivial one, and it can be proven that its mobility is 1.

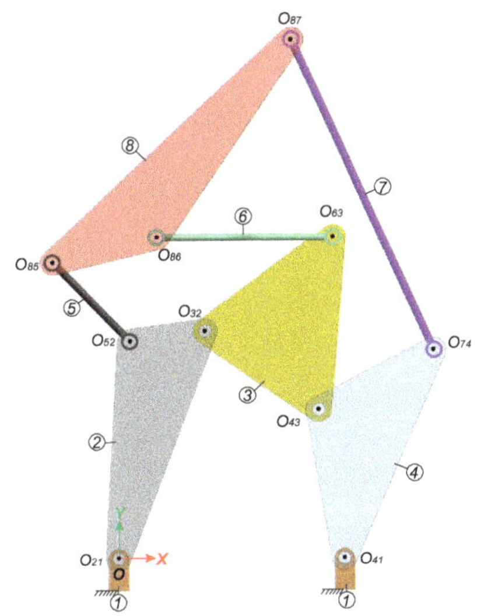

Figure 2. A, eight-bar "single flyer" undetermined planar linkage.

The first step in the process is to obtain the screws associated with the linkage's kinematic pairs with respect to the origin of the coordinate system, O. To that end, the common direction of all revolute pairs, $\mathbf{u}$, is required, given by:

$$\mathbf{u} = \begin{bmatrix} 0 & 0 & 1 \end{bmatrix}^T$$

and by the position vectors of points located along the revolute axes, with respect to the origin, O. They are given in terms of an arbitrary unit of length, by:

$$\begin{array}{lll} \mathbf{r}_{21} = \begin{bmatrix} 0 & 0 & 0 \end{bmatrix}^T, & \mathbf{r}_{32} = \begin{bmatrix} 70 & 184 & 0 \end{bmatrix}^T, & \mathbf{r}_{43} = \begin{bmatrix} 160 & 120 & 0 \end{bmatrix}^T, \\ \mathbf{r}_{41} = \begin{bmatrix} 180 & 0 & 0 \end{bmatrix}^T, & \mathbf{r}_{52} = \begin{bmatrix} 10 & 176 & 0 \end{bmatrix}^T, & \mathbf{r}_{85} = \begin{bmatrix} -52 & 240 & 0 \end{bmatrix}^T, \\ \mathbf{r}_{86} = \begin{bmatrix} 32 & 260 & 0 \end{bmatrix}^T, & \mathbf{r}_{63} = \begin{bmatrix} 172 & 260 & 0 \end{bmatrix}^T, & \mathbf{r}_{87} = \begin{bmatrix} 140 & 420 & 0 \end{bmatrix}^T, \\ \mathbf{r}_{74} = \begin{bmatrix} 252 & 168 & 0 \end{bmatrix}^T. & & \end{array}$$

From these data, the screws associated with the kinematic pairs, with respect to the origin of the coordinate system, O, are given by:

$$\begin{array}{lll} {}^{2}\$^{1}_{O} = \begin{bmatrix} 1; & 0 & 0 \end{bmatrix}^T, & {}^{3}\$^{2}_{O} = \begin{bmatrix} 1; & 184 & -70 \end{bmatrix}^T, & {}^{4}\$^{3}_{O} = \begin{bmatrix} 1; & 120 & -160 \end{bmatrix}^T, \\ {}^{4}\$^{1}_{O} = \begin{bmatrix} 1; & 0 & -180 \end{bmatrix}^T, & {}^{5}\$^{2}_{O} = \begin{bmatrix} 1; & 176 & -10 \end{bmatrix}^T, & {}^{8}\$^{5}_{O} = \begin{bmatrix} 1; & 240 & 52 \end{bmatrix}^T, \\ {}^{8}\$^{7}_{O} = \begin{bmatrix} 1; & 420 & -140 \end{bmatrix}^T, & {}^{7}\$^{4}_{O} = \begin{bmatrix} 1; & 168 & -252 \end{bmatrix}^T, & {}^{8}\$^{6}_{O} = \begin{bmatrix} 1; & 260 & -32 \end{bmatrix}^T, \\ {}^{6}\$^{3}_{O} = \begin{bmatrix} 1; & 260 & -172 \end{bmatrix}^T. & & \end{array}$$

It must be noted that, due to space considerations, only the non-zero terms are shown separated by a semicolon to indicate the different units.

3.1. *Equations for the Location of Secondary IRCs*

Regarding Figure 2, the primary IRCs are:

$$O_{21},\ O_{32},\ O_{43},\ O_{41},\ O_{52},\ O_{85},\ O_{87},\ O_{74},\ O_{86},\ O_{63}.$$

The upper triangular arrangement in Table 1 indicates the primary centers of the linkage with the number -1 and the secondary centers with the number -2.

Table 1. IRCs associated with the eight-bar planar linkage.

$O_{21}(-1)$	$O_{31}(-2)$	$O_{41}(-1)$	$O_{51}(-2)$	$O_{61}(-2)$	$O_{71}(-2)$	$O_{81}(-2)$
	$O_{32}(-1)$	$O_{42}(-2)$	$O_{52}(-1)$	$O_{62}(-2)$	$O_{72}(-2)$	$O_{82}(-2)$
		$O_{43}(-1)$	$O_{53}(-2)$	$O_{63}(-1)$	$O_{73}(-2)$	$O_{83}(-2)$
			$O_{54}(-2)$	$O_{64}(-2)$	$O_{74}(-1)$	$O_{84}(-2)$
				$O_{65}(-2)$	$O_{75}(-2)$	$O_{85}(-1)$
					$O_{76}(-2)$	$O_{86}(-1)$
						$O_{87}(-1)$

The linkage is denominated as partially undetermined because the direct application of the Aronhold–Kennedy theorem allows determining two secondary IRCs. Throughout this paper, whenever a sequence for determining an IRC is presented, the primary and secondary IRCs will be enclosed in boxes and circles, respectively. From Table 1, it follows that O_{31} can be found using links 2 and 4.

$$\bigcirc\!\!\!\!\!\!O_{31} \Longrightarrow \boxed{O_{21}}\ \boxed{O_{32}} \quad {}_{1}\omega_{3}\,{}^{3}\$^{1}_{O} = {}_{1}\omega_{2}\,{}^{2}\$^{1}_{O} + {}_{2}\omega_{3}\,{}^{3}\$^{2}_{O}$$

$$\boxed{O_{31}} \Longrightarrow \boxed{O_{43}}\ \boxed{O_{41}} \quad {}_1\omega_3{}^3\$_O^1 = {}_3\omega_4{}^4\$_O^3 + {}_1\omega_4{}^4\$_O^1$$

The equations that determine the secondary center, and therefore, the screw characteristics, are:

$$Ki(\$^2_{a,Ki,Kl}, {}^3\$^1) = 0 \qquad Kl(\$^2_{a,Ki,Kl}, {}^3\$^1) = 0 \tag{25}$$

and

$$Ki(\$^4_{a,Ki,Kl}, {}^3\$^1) = 0 \qquad Kl(\$^4_{a,Ki,Kl}, {}^3\$^1) = 0 \tag{26}$$

where $\{\$^2_{a1,Ki,Kl}\}$ is a basis for the orthogonal annihilator of the subspace $[{}^2\$^1,\ {}^3\$^2]$, and $\{\$^4_{a1,Ki,Kl}\}$ is a basis for the orthogonal annihilator of the subspace $[{}^4\$^3,\ {}^4\$^1]$, regarding the Killing and Klein forms.

From Equations (25) and (26), and applying the Aronhold–Kennedy theorem with the Klein form, the equations necessary to determine the location of the secondary center O_{31} are given by:

$$Eq31_2 = \left[0, 180 - \frac{1}{6}y_{31} - x_{31}, 0, 0\right] \tag{27}$$

$$Eq31_4 = \left[0, \frac{35}{92}y_{31} - x_{31}, 0, 0\right]. \tag{28}$$

The vector expression $Eqij_k$ indicates that it was obtained while finding the center ij and using link k as a third link. The expression is implicitly equated to the vector $\mathbf{0}$ to form a vector equation; however, only the second component is different from zero. Hence, only one scalar equation is obtained. From Equations (27) and (28), one obtains:

$$0 = 180 - \frac{1}{6}y_{31} - x_{31} \tag{29}$$

$$0 = \frac{35}{92}y_{31} - x_{31} \tag{30}$$

Equations (29) and (30) form a linear system of two equations in two unknowns, the coordinates x_{31} and y_{31} associated with the secondary center O_{31}. Solving the linear system, the position vector $\mathbf{r}_{31}$ and the infinitesimal screw associated with the secondary center O_{31} are given by:

$$\mathbf{r}_{31} = \begin{bmatrix} \frac{18900}{151} & \frac{49680}{151} & 0 \end{bmatrix}^T, \qquad {}^3\$_O^1 = \begin{bmatrix} 1; & \frac{49680}{151} & -\frac{18900}{151} \end{bmatrix}^T$$

Similarly, the secondary center O_{42} can be found by using links 1 and 3, Table 1.

$$\boxed{O_{42}} \Longrightarrow \boxed{O_{21}}\ \boxed{O_{41}} \quad {}_2\omega_4{}^4\$_O^2 = {}_2\omega_3{}^3\$_O^2 + {}_1\omega_4{}^4\$_O^1$$

$$\boxed{O_{42}} \Longrightarrow \boxed{O_{32}}\ \boxed{O_{43}} \quad {}_2\omega_4{}^4\$_O^2 = {}_2\omega_3{}^3\$_O^2 + {}_3\omega_4{}^4\$_O^3$$

Following a similar process as in the secondary center O_{31}, the position vector and the screw associated with the secondary center O_{42}, are found as:

$$\mathbf{r}_{42} = \begin{bmatrix} \frac{1315}{4} & 0 & 0 \end{bmatrix}^T, \qquad {}^4\$_O^2 = \begin{bmatrix} 0 & 0 & 1 & 0 & -\frac{1315}{4} & 0 \end{bmatrix}^T.$$

There is no other additional secondary IRC that can be determined by the direct application of the Aronhold–Kennedy theorem. The IRCs O_{31} and O_{42} will also be marked with the number -1. Moreover, they will be regarded as primary centers for all purposes. Further, the upper triangular array of the IRCs, after this step, is given in Table 2.

Table 2. IRCs associated with the eight-bar planar linkage after determining O_{31} and O_{42}.

$O_{21}(-1)$	$O_{31}(-1)$	$O_{41}(-1)$	$O_{51}(-2)$	$O_{61}(-2)$	$O_{71}(-2)$	$O_{81}(-2)$
	$O_{32}(-1)$	$O_{42}(-1)$	$O_{52}(-1)$	$O_{62}(-2)$	$O_{72}(-2)$	$O_{82}(-2)$
		$O_{43}(-1)$	$O_{53}(-2)$	$O_{63}(-1)$	$O_{73}(-2)$	$O_{83}(-2)$
			$O_{54}(-2)$	$O_{64}(-2)$	$O_{74}(-1)$	$O_{84}(-2)$
				$O_{65}(-2)$	$O_{75}(-2)$	$O_{85}(-1)$
					$O_{76}(-2)$	$O_{86}(-1)$
						$O_{87}(-1)$

The flow chart of the algorithm proposed for solving indeterminate linkages is presented in Figure 3. The computation of the undetermined IRCs starts at step 4. The first 3 steps are simple and they do not require additional explanation.

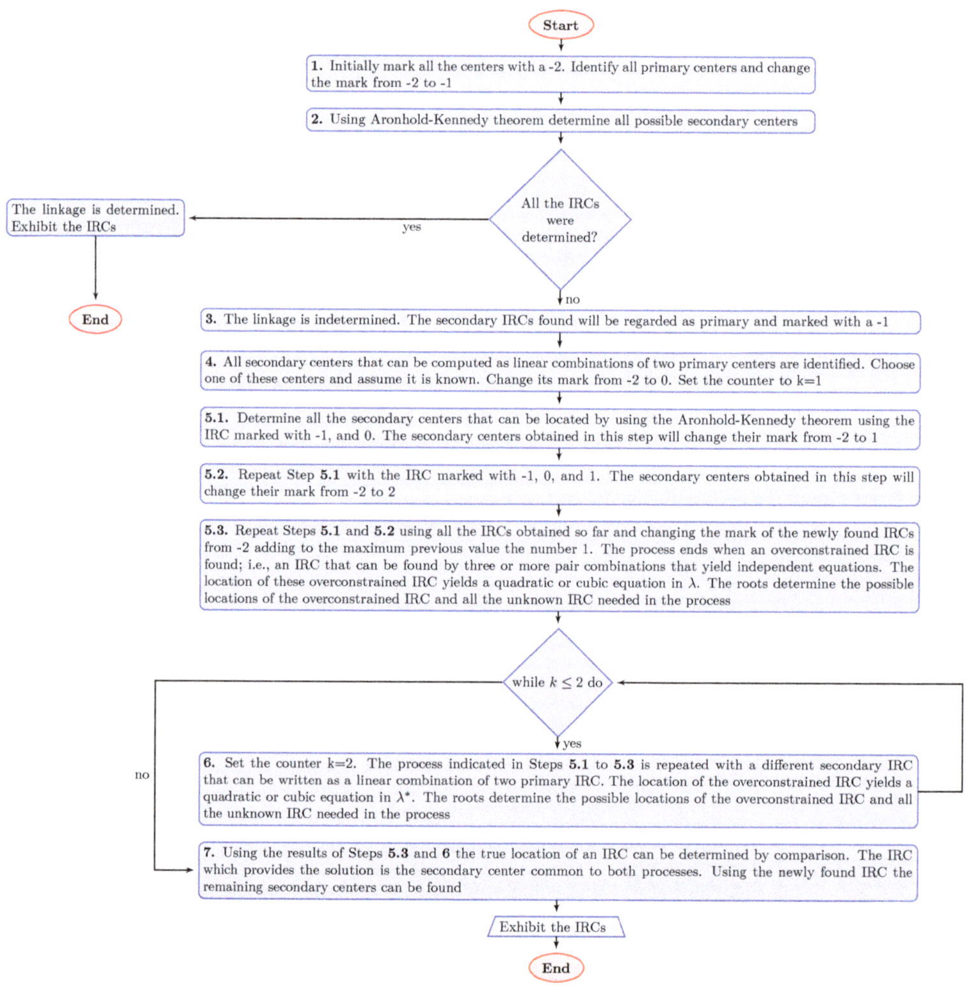

Figure 3. Flow diagram of the proposed algorithm.

Step 4. Choose a secondary center, ${}^{i}\$^{j}$, whose location can be written as a linear combination of the position of two primary centers, namely, those marked with the number -1. Appendix A shows the procedure to accomplish this task; the secondary center can be written in terms of only one variable, denoted λ or λ^*. From a theoretical point of view, any such center can be employed. However, it is convenient to choose secondary centers that yield a minimum number of equations. The marking of this secondary center will change from -2 to 0, and it will be assumed that its location is known. Set the counter $k = 1$.

Step 5.1. Determine all the secondary centers that can be located by the Aronhold–Kennedy theorem using the centers marked with -1 and 0, namely, the center chosen in step 1. The secondary centers obtained in this step will change their mark from -2 to 1.

Step 5.2. Repeat **Step 5.1**, but now, it is necessary to locate all the secondary centers that can be determined, by using the Aronhold–Kennedy theorem, on the IRCs marked with the numbers -1, 0, and 1. The IRCs obtained in this step will change their mark from -2 to 2.

Step 5.3. Repeat **Step 5.2** using all the secondary centers obtained so far and changing the mark of the newly found IRCs, adding to the maximum previous value the number 1. The process finishes when an overconstrained center is found, namely, a rotation center that can be determined by using more than two pairs of already-known IRCs that yield independent equations. Simplifying the resulting system of equations yields a quadratic or cubic equation in λ. The roots of the equation determine the possible locations of the secondary center.

Step 6. Set the counter $k = 2$. **Steps 5.1 to 5.3** must be repeated with a new secondary center that can be expressed as a linear combination of two primary centers. The process eventually yields an overconstrained secondary center, whose location can be obtained through a system of equations that yields a quadratic or cubic equation in λ^*. The roots of the equation determine the possible locations of the secondary center.

Step 7. The correct location of the secondary center, obtained through the values of λ and λ^*, is determined by choosing the location of a secondary center common to the two processes outlined in the two previous paragraphs. In the process, the location of many other secondary IRCs, usually all, are also found.

The proposed algorithm is used in the rest of the paper.

3.2. Determination and Solution of the System of Equations

The objective of this analysis is to determine the position vectors, in the plane $z = 0$, of a point along the rotation axes of the relative movements associated with the secondary centers shown in Table 2. The secondary centers that can be written as a linear combination of two primary centers are O_{51}, O_{61}, O_{71}, O_{62}, O_{72}, O_{82}, O_{53}, O_{73}, O_{83}, O_{54}, O_{64}, O_{84}, O_{65}, O_{75}, and O_{76}. Table 3 shows only the secondary centers used in the process.

Table 3. Secondary centers whose location can be written as a linear combination of the location of two primary centers.

O_{62} is found on the line determined by O_{32} and O_{63}
O_{72} is found on the line determined by O_{42} and O_{74}

3.2.1. Selection of a First Arbitrary Center and Determination of an Overconstrained Center

From Table 3, the secondary center O_{62} is chosen. This center can be written as:

$$\mathbf{r}_{62} \quad = \quad \lambda\, \mathbf{r}_{32} + (1-\lambda)\mathbf{r}_{63} \tag{31}$$

The upper triangular arrangement of centers, after this point, is given by:

$$\begin{array}{ccccccc}
O_{21}(-1) & O_{31}(-1) & O_{41}(-1) & O_{51}(-2) & O_{61}(-2) & O_{71}(-2) & O_{81}(-2) \\
 & O_{32}(-1) & O_{42}(-1) & O_{52}(-1) & O_{62}(0) & O_{72}(-2) & O_{82}(-2) \\
 & & O_{43}(-1) & O_{53}(-2) & O_{63}(-1) & O_{73}(-2) & O_{83}(-2) \\
 & & & O_{54}(-2) & O_{64}(-2) & O_{74}(-1) & O_{84}(-2) \\
 & & & & O_{65}(-2) & O_{75}(-2) & O_{85}(-1) \\
 & & & & & O_{76}(-2) & O_{86}(-1) \\
 & & & & & & O_{87}(-1)
\end{array}$$

From this selection, it is possible to find the **secondary centers**. (The text shows only the required centers. However, the subsequent upper triangular arrangements show all centers that can be found. This convention will be followed in the rest of the contribution.) O_{82} and O_{64} are indicated, from now on, in blue color and dashed lines.

$$O_{82} \Longrightarrow \boxed{O_{52}}\ \boxed{O_{85}} \qquad\qquad O_{64} \Longrightarrow \boxed{O_{42}}\ O_{62}$$

$$O_{82} \Longrightarrow O_{62}\ \boxed{O_{86}} \qquad\qquad O_{64} \Longrightarrow \boxed{O_{43}}\ \boxed{O_{63}}$$

After identifying the secondary centers O_{82} and O_{64} with the number 1, the upper triangular arrangement of the IRCs is given by:

$$\begin{array}{ccccccc}
O_{21}(-1) & O_{31}(-1) & O_{41}(-1) & O_{51}(-2) & O_{61}(1) & O_{71}(-2) & O_{81}(-2) \\
 & O_{32}(-1) & O_{42}(-1) & O_{52}(-1) & O_{62}(0) & O_{72}(-2) & O_{82}(1) \\
 & & O_{43}(-1) & O_{53}(-2) & O_{63}(-1) & O_{73}(-2) & O_{83}(-2) \\
 & & & O_{54}(-2) & O_{64}(1) & O_{74}(-1) & O_{84}(-2) \\
 & & & & O_{65}(1) & O_{75}(-2) & O_{85}(-1) \\
 & & & & & O_{76}(-2) & O_{86}(-1) \\
 & & & & & & O_{87}(-1)
\end{array}$$

The next step is to employ all the centers found so far to find all additional secondary centers. Therefore,

$$O_{84} \Longrightarrow O_{42}\ \boxed{O_{82}} \qquad O_{84} \Longrightarrow O_{64}\ \boxed{O_{86}} \qquad O_{84} \Longrightarrow \boxed{O_{74}}\ \boxed{O_{87}}$$

After identifying the **overconstrained secondary center**, indicated from now on in red color and dotted lines, O_{84}, with the number 2, the final upper triangular arrangement of IRCs is given by:

$$\begin{array}{ccccccc}
O_{21}(-1) & O_{31}(-1) & O_{41}(-1) & O_{51}(2) & O_{61}(1) & O_{71}(-2) & O_{81}(2) \\
 & O_{32}(-1) & O_{42}(-1) & O_{52}(-1) & O_{62}(0) & O_{72}(2) & O_{82}(1) \\
 & & O_{43}(-1) & O_{53}(2) & O_{63}(-1) & O_{73}(-2) & O_{83}(2) \\
 & & & O_{54}(2) & O_{64}(1) & O_{74}(-1) & O_{84}(2) \\
 & & & & O_{65}(1) & O_{75}(-2) & O_{85}(-1) \\
 & & & & & O_{76}(2) & O_{86}(-1) \\
 & & & & & & O_{87}(-1)
\end{array}$$

From Equation (31), the position vector associated with the chosen secondary center O_{62} is given by:

$$\mathbf{r}_{62} = \left[\begin{array}{ccc} 172 - 102\,\lambda & 260 - 76\,\lambda & 0 \end{array}\right]^T \quad (32)$$

The position vectors associated with the secondary centers involved in this step are:

$$\mathbf{r}_{82} = \left[\begin{array}{ccc} x_{82} & y_{82} & 0 \end{array}\right], \quad \mathbf{r}_{64} = \left[\begin{array}{ccc} x_{64} & y_{64} & 0 \end{array}\right], \quad \mathbf{r}_{84} = \left[\begin{array}{ccc} x_{84} & y_{84} & 0 \end{array}\right]. \quad (33)$$

The required equations are obtained applying systematically the Aronhold–Kennedy theorem together with the Klein form, Equations (34)–(40). These are vector equations; however, only one component is different from zero. Due to space considerations, only the non-zero component is indicated. (The two numbers of the equation indicate the IRC to be calculated, and the subscript indicates the third link required to apply the Aronhold–Kennedy theorem.).

$$Eq82_5 = \frac{361}{2} - \frac{31}{32}\,y_{82} - x_{82} \quad (34)$$

$$Eq82_6 = \frac{38\lambda\,x_{82} - 51\lambda\,y_{82} + 12044\lambda + 70\,y_{82} - 18200}{\lambda} \quad (35)$$

$$Eq64_3 = \frac{1048}{7} + \frac{3}{35}\,y_{64} - x_{64} \quad (36)$$

$$Eq64_2 = \frac{304\lambda\,x_{64} - 408\lambda\,y_{64} - 99940\lambda - 1040\,x_{64} - 627\,y_{64} + 341900}{19\lambda - 65} \quad (37)$$

$$Eq84_7 = \frac{980}{3} - \frac{4}{9}\,y_{84} - x_{84} \quad (38)$$

$$Eq84_2 = \frac{4\,x_{82}\,y_{84} - 4\,x_{84}\,y_{82} + 1315\,y_{82} - 1315\,y_{84}}{y_{82}} \quad (39)$$

$$Eq84_6 = \frac{x_{64}\,y_{84} - x_{84}\,y_{64} + 260\,(x_{84} - x_{64}) + 32\,(y_{64} - y_{84})}{y_{64} - 260} \quad (40)$$

From Equations (34)–(40), the final set of equations is:

$$0 = \frac{361}{2} - \frac{31}{32}\,y_{82} - x_{82} \quad (41)$$

$$0 = \frac{38\lambda\,x_{82} - 51\lambda\,y_{82} + 12044\lambda + 70\,y_{82} - 18200}{\lambda} \quad (42)$$

$$0 = \frac{1048}{7} + \frac{3}{35}\,y_{64} - x_{64} \quad (43)$$

$$0 = \frac{304\lambda\,x_{64} - 408\lambda\,y_{64} - 99940\lambda - 1040\,x_{64} - 627\,y_{64} + 341900}{19\lambda - 65} \quad (44)$$

$$0 = \frac{980}{3} - \frac{4}{9}\,y_{84} - x_{84} \quad (45)$$

$$0 = \frac{4\,x_{82}\,y_{84} - 4\,x_{84}\,y_{82} + 1315\,y_{82} - 1315\,y_{84}}{4\,y_{82}} \quad (46)$$

$$0 = \frac{x_{64}\,y_{84} - x_{84}\,y_{64} + 260\,(x_{84} - x_{64}) + 32\,(y_{64} - y_{84})}{y_{64} - 260} \quad (47)$$

The coordinates y_{ij} of each linear equation are solved in terms of the corresponding coordinates x_{ij}. Thus, Equations (41), (43) and (45) yield the following coordinates: y_{82}, y_{64}, and y_{82}:

$$y_{82} = \frac{5776 - 32\,x_{82}}{31}, \quad y_{64} = \frac{35\,x_{64} - 5240}{3}, \quad y_{84} = 735 - \frac{9}{4}\,x_{84}. \quad (48)$$

Substituting the results of Equations (48) into Equations (42), (44), (46) and (47), it follows that:

$$0 = \frac{-1405\,x_{82}\,\lambda - 39394\,\lambda + 1120\,x_{82} + 79940}{\lambda} \quad (49)$$

$$0 = \frac{8\,x_{64}\,\lambda - 1100\,\lambda + 15\,x_{64} - 2580}{19\,\lambda - 65} \quad (50)$$

$$0 = \frac{604\,x_{82}\,x_{84} - 196240\,x_{82} - 274469\,x_{84} + 89467340}{-361 + 2\,x_{82}} \quad (51)$$

$$0 = \frac{167\,x_{64}\,x_{84} - 10180\,x_{64} - 24944\,x_{84} + 952960}{172 - x_{64}} \quad (52)$$

From Equation (52), x_{84} is solved in terms of x_{64} to obtain:

$$x_{84} = \frac{20(509\,x_{64} - 47648)}{167\,x_{64} - 24944} \quad (53)$$

The solution for x_{84}, from Equation (53), is substituted into Equation (51) to solve x_{82} also in terms of x_{64}, so that:

$$x_{82} = \frac{1}{2}\,\frac{37959223\,x_{64} - 6156610471}{41599\,x_{64} - 6749098} \quad (54)$$

This result for x_{82}, from Equation (54), is substituted into Equation (49) to obtain a solution for x_{64} in terms of λ. Therefore,

$$x_{64} = \frac{296186633709\,\lambda - 257240306960}{-1585973480 + 1826135817\,\lambda} \quad (55)$$

Substituting the solution for x_{64} into Equation (50), a quadratic equation in λ is obtained:

$$0 = \frac{40082630108\,\lambda^2 - 64664725545\,\lambda + 25911886000}{(-1585973480 + 1826135817\,\lambda)(19\,\lambda - 65)} \quad (56)$$

After clearing the denominator, the two possible solutions of Equation (56) are:

$$\lambda_a = \frac{711865}{816796}, \qquad \lambda_b = \frac{36400}{49073}. \quad (57)$$

The two roots of Equation (56) are substituted into (32), thus obtaining the possible position vectors of the secondary center O_{62}:

$$\mathbf{r}_{62a} = \begin{bmatrix} \frac{33939341}{408398} & \frac{39566305}{204199} & 0 \end{bmatrix}^T, \qquad \mathbf{r}_{62b} = \begin{bmatrix} \frac{4727756}{49073} & \frac{9992580}{49073} & 0 \end{bmatrix}^T. \quad (58)$$

Finally, position vectors in Equation (58) can be used to compute the two possible position vectors associated with the secondary center O_{82}:

$$\mathbf{r}_{82a} = \begin{bmatrix} \frac{1286974}{2949} & -\frac{779024}{2949} & 0 \end{bmatrix}^T \qquad \mathbf{r}_{82b} = \begin{bmatrix} -\frac{935697}{1436} & \frac{308360}{359} & 0 \end{bmatrix}^T \quad (59)$$

3.2.2. Selection of a Second Arbitrary Center and Determination of an Overconstrained Center

The procedure must be repeated for another secondary center. From Table 3, the secondary center O_{72} is chosen. The location of this center can be written as:

$$\mathbf{r}_{O72} = \lambda^*\,\mathbf{r}_{O42} + (1 - \lambda^*)\mathbf{r}_{O74} \quad (60)$$

The upper triangular arrangement of centers becomes:

$$\begin{array}{ccccccc}
O_{21}(-1) & O_{31}(-1) & O_{41}(-1) & O_{51}(-2) & O_{61}(-2) & O_{71}(-2) & O_{81}(-2) \\
 & O_{32}(-1) & O_{42}(-1) & O_{52}(-1) & O_{62}(-2) & O_{72}(0) & O_{82}(-2) \\
 & & O_{43}(-1) & O_{53}(-2) & O_{63}(-1) & O_{73}(-2) & O_{83}(-2) \\
 & & & O_{54}(-2) & O_{64}(-2) & O_{74}(-1) & O_{84}(-2) \\
 & & & & O_{65}(-2) & O_{75}(-2) & O_{85}(-1) \\
 & & & & & O_{76}(-2) & O_{86}(-1) \\
 & & & & & & O_{87}(-1)
\end{array}$$

Using the centers found up to this point, it is possible to find the following secondary centers: O_{82}, O_{73}.

After identifying the secondary centers O_{82} and O_{73} with the number 1, the upper diagonal arrangement becomes:

$$\begin{array}{ccccccc}
O_{21}(-1) & O_{31}(-1) & O_{41}(-1) & O_{51}(-2) & O_{61}(-2) & O_{71}(1) & O_{81}(-2) \\
 & O_{32}(-1) & O_{42}(-1) & O_{52}(-1) & O_{62}(-2) & O_{72}(0) & O_{82}(1) \\
 & & O_{43}(-1) & O_{53}(-2) & O_{63}(-1) & O_{73}(1) & O_{83}(-2) \\
 & & & O_{54}(-2) & O_{64}(-2) & O_{74}(-1) & O_{84}(-2) \\
 & & & & O_{65}(-2) & O_{75}(1) & O_{85}(-1) \\
 & & & & & O_{76}(-2) & O_{86}(-1) \\
 & & & & & & O_{87}(-1)
\end{array}$$

The final step requires employing all the secondary centers found so far to find additional secondary centers, in particular, an overconstrained secondary center. Therefore,

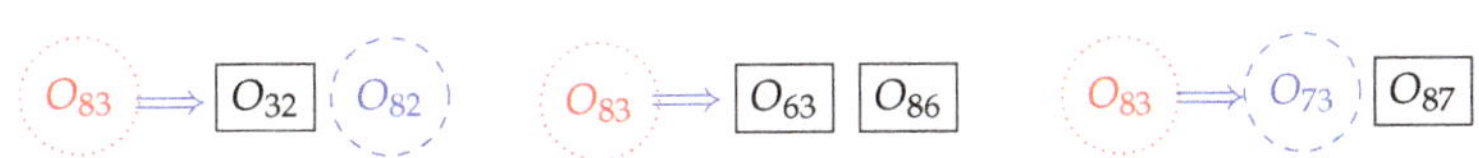

After identifying the overconstrained secondary center O_{83} with the number 2, the final upper triangular arrangement of centers becomes:

$$\begin{array}{ccccccc}
O_{21}(-1) & O_{31}(-1) & O_{41}(-1) & O_{51}(2) & O_{61}(-2) & O_{71}(1) & O_{81}(2) \\
 & O_{32}(-1) & O_{42}(-1) & O_{52}(-1) & O_{62}(2) & O_{72}(0) & O_{82}(1) \\
 & & O_{43}(-1) & O_{53}(2) & O_{63}(-1) & O_{73}(1) & O_{83}(2) \\
 & & & O_{54}(2) & O_{64}(-2) & O_{74}(-1) & O_{84}(2) \\
 & & & & O_{65}(-2) & O_{75}(1) & O_{85}(-1) \\
 & & & & & O_{76}(2) & O_{86}(-1) \\
 & & & & & & O_{87}(-1)
\end{array}$$

Now, it is possible to generate the equations required for the location of the secondary centers involved in this process. From Equation (60), the position vector of the center O_{72}, the secondary center chosen, is given by:

$$\mathbf{r}_{72} = \begin{bmatrix} 252 + \frac{307\,\lambda^*}{4} & 168 - 168\,\lambda^* & 0 \end{bmatrix}^T \tag{61}$$

The position vectors of the centers O_{82}, O_{73}, and O_{83} involved in this process are:

$$\mathbf{r}_{82} = \begin{bmatrix} x_{82} & y_{82} & 0 \end{bmatrix}^T, \quad \mathbf{r}_{73} = \begin{bmatrix} x_{73} & y_{73} & 0 \end{bmatrix}^T, \quad \mathbf{r}_{83} = \begin{bmatrix} x_{83} & y_{83} & 0 \end{bmatrix}^T.$$

The equations are obtained by applying systematically the Aronhold–Kennedy theorem together with the Klein form. It should be noted that these expressions are equated to the vector $\mathbf{0}$, generating four scalar equations, only one of which is non-trivial. Due to space considerations, only the non-zero component is shown.

$$Eq82_5 = \frac{361}{2} - \frac{31}{32}\,y_{82} - x_{82}, \tag{62}$$

$$Eq82_7 = \frac{672\lambda^*\,x_{82} + 307\lambda^*\,y_{82} - 223020\lambda^* + 1008\,x_{82} + 448\,y_{82} - 329280}{3 + 2\lambda^*}, \tag{63}$$

$$Eq73_4 = -70 + \frac{23}{12}\,y_{73} - x_{73}, \tag{64}$$

$$Eq73_2 = \frac{672\lambda^*\,x_{73} + 307\lambda^*\,y_{73} - 103528\lambda^* + 64\,x_{73} + 728\,y_{73} - 138432}{2 + 21\lambda^*}, \tag{65}$$

$$Eq83_6 = -260 + y_{83}, \tag{66}$$

$$Eq83_2 = \frac{x_{82}\,y_{83} - x_{83}\,y_{82} - 184\,x_{82} + 184\,x_{83} + 70\,y_{82} - 70\,y_{83}}{y_{82} - 184}, \tag{67}$$

$$Eq83_7 = \frac{x_{73}\,y_{83} - x_{83}\,y_{73} - 420\,x_{73} + 420\,x_{83} + 140\,y_{73} - 140\,y_{83}}{y_{73} - 420}, \tag{68}$$

From Equations (62)–(68), a final set of equations is obtained:

$$0 = \frac{361}{2} - \frac{31}{32}\,y_{82} - x_{82} \tag{69}$$

$$0 = \frac{672\lambda^*\,x_{82} + 307\lambda^*\,y_{82} - 223020\lambda^* + 1008\,x_{82} + 448\,y_{82} - 329280}{3 + 2\lambda^*}, \tag{70}$$

$$0 = -70 + \frac{23}{12}\,y_{73} - x_{73} \tag{71}$$

$$0 = \frac{672\lambda^*\,x_{73} + 307\lambda^*\,y_{73} - 103528\lambda^* + 64\,x_{73} + 728\,y_{73} - 138432}{2 + 21\lambda^*} \tag{72}$$

$$0 = -260 + y83 \tag{73}$$

$$0 = \frac{x_{82}\,y_{83} - x_{83}\,y_{82} - 184\,x_{82} + 184\,x_{83} + 70\,y_{82} - 70\,y_{83}}{y_{82} - 184} \tag{74}$$

$$0 = \frac{x_{73}\,y_{83} - x_{83}\,y_{73} - 420\,x_{73} + 420\,x_{83} + 140\,y_{73} - 140\,y_{83}}{y_{73} - 420} \tag{75}$$

Systematically, the coordinates y_{ij} are solved, from the linear equations, in terms of the coordinates x_{ij}. Hence, the coordinates y_{82}, y_{73}, and y_{83}, solved from Equations (69), (71) and (73), are given, respectively, by:

$$y_{82} = \frac{5776 - 32\,x_{82}}{31}, \quad y_{73} = \frac{840 + 12\,x_{73}}{23}, \quad y_{83} = 260. \tag{76}$$

Substituting the results of Equation (76) into Equations (70), (72), (74) and (75), it follows that:

$$0 = \frac{1285097\,\lambda^* - 2752\,\lambda^*\,x_{82} - 4228\,x_{82} + 1905008}{7812 + 5208\,\lambda^*} \tag{77}$$

$$0 = \frac{1664\,\lambda^* - 15\,\lambda^*\,x_{73} - 8\,x_{73} + 2016}{2 + 21\,\lambda^*} \tag{78}$$

$$0 = \frac{8\,x_{82}\,x_{83} + 29\,x_{82} - 18\,x_{83} - 39970}{-9 + 4\,x_{82}} \tag{79}$$

$$0 = \frac{3\,x_{73}\,x_{83} + 500\,x_{73} - 2205\,x_{83} + 179900}{-735 + x_{73}} \tag{80}$$

From Equation (80), x_{83} is solved in terms of x_{73}, to obtain:

$$x_{83} = -\frac{500\,x_{73} + 179900}{-2205 + 3\,x_{73}} \tag{81}$$

The solution for x_{83}, obtained from Equation (81), is substituted into Equation (79) to solve x_{73} in terms of x_{82}, thus obtaining:

$$x_{73} = (-35)\,\frac{42947\,x_{82} - 2610630}{3913\,x_{82} + 110910} \tag{82}$$

The result for x_{73}, obtained from Equation (82), is substituted into Equation (78) to solve x_{82} in terms of λ^*:

$$x_{82} = \frac{51566370\,\lambda^* + 22060080}{1263409\,\lambda^* + 865816} \tag{83}$$

Finally, the solution for x_{82}, obtained from Equation (83), is substituted into Equation (77) to obtain a quadratic equation in λ^*:

$$0 = \frac{47796531143\,\lambda^{*2} + 104539672384\,\lambda^* + 50197302848}{(106126356\,\lambda^* + 72728544)\,(3 + 2\,\lambda^*)} \tag{84}$$

After clearing the denominator, the possible solutions of the quadratic Equation (84) are given by:

$$\lambda_a^* = -\frac{1138984}{1599991}, \qquad \lambda_b^* = -\frac{44072}{29873}. \tag{85}$$

The two roots of Equation (85) are substituted into Equation (61) to obtain the two possible locations of the secondary center O_{72}:

$$\mathbf{r}_{72a} = \begin{bmatrix} \frac{315780710}{1599991} & \frac{460147800}{1599991} & 0 \end{bmatrix}^T, \qquad \mathbf{r}_{72b} = \begin{bmatrix} \frac{4145470}{29873} & \frac{12422760}{29873} & 0 \end{bmatrix}^T. \tag{86}$$

Finally, from the position vectors $\mathbf{r}_{72a}, \mathbf{r}_{72b}$, the two possible position vectors of the secondary center O_{82} are given by:

$$\mathbf{r}_{82a} = \begin{bmatrix} \frac{1286974}{2949} & -\frac{779024}{2949} & 0 \end{bmatrix}^T, \qquad \mathbf{r}_{82b} = \begin{bmatrix} \frac{129290}{2389} & \frac{311664}{2389} & 0 \end{bmatrix}^T. \tag{87}$$

Summarizing, in the first process, the secondary IRC O_{62} is assumed to be known, and the overconstrained rotation center is O_{84}. In the second process, the secondary IRC O_{72} is assumed to be known, and the overconstrained rotation center is O_{83}. The subgraphs of the secondary rotation centers needed to accomplish this task is shown in Figure 4.

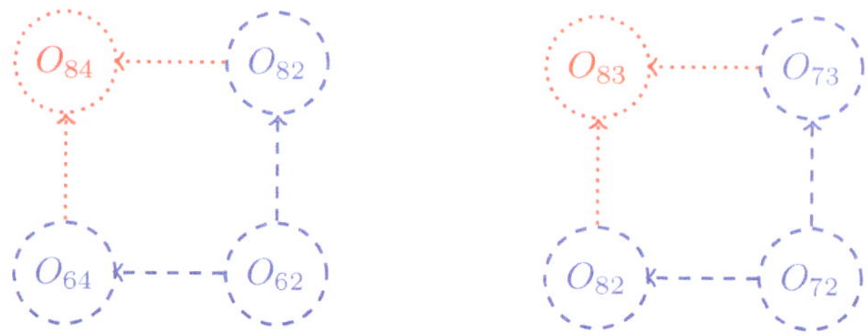

Figure 4. The subgraph of secondary rotation centers required to solve the "single-flyer" indeterminate planar linkage.

Comparing the results obtained in both processes for the position vector associated with IRC O_{82}, the only coincident result is given by:

$$\mathbf{r}_{82a} = \left[\begin{array}{ccc} \frac{1286974}{2949} & -\frac{779024}{2949} & 0 \end{array} \right]^T$$

Therefore, the infinitesimal screw associated with the secondary center O_{82} is given by:

$$^{8}\$_{O}^{2} = \left[\begin{array}{ccc} 1; & -\frac{779024}{2949} & -\frac{1286974}{2949} \end{array} \right]^T \tag{88}$$

Hence, the position vector associated with the secondary center O_{62} is given by:

$$\mathbf{r}_{62} = \left[\begin{array}{ccc} \frac{33939341}{408398} & \frac{39566305}{204199} & 0 \end{array} \right]^T$$

Furthermore, the infinitesimal screw associated with the secondary center O_{62} is given by:

$$^{6}\$_{O}^{2} = \left[\begin{array}{ccc} 1; & \frac{39566305}{204199} & -\frac{33939341}{408398} \end{array} \right]^T \tag{89}$$

Finally from the result of Equation (89), together with the primary centers, applying the Aronhold–Kennedy theorem and using the Klein form, it is possible to find the remaining secondary centers following the procedure illustrated for the determination of the centers O_{31} and O_{42}, Equations (25) and (26). Therefore, the position vectors corresponding to the remaining secondary centers are given by:

$$\mathbf{r}_{72} = \left[\begin{array}{ccc} \frac{315780710}{1599991} & \frac{460147800}{1599991} & 0 \end{array} \right]^T, \quad \mathbf{r}_{64} = \left[\begin{array}{ccc} \frac{144519259}{897343} & \frac{118698915}{897343} & 0 \end{array} \right]^T,$$

$$\mathbf{r}_{84} = \left[\begin{array}{ccc} \frac{72796180}{206947} & -\frac{11685360}{206947} & 0 \end{array} \right]^T, \quad \mathbf{r}_{73} = \left[\begin{array}{ccc} -\frac{41572265}{133901} & -\frac{16799580}{133901} & 0 \end{array} \right]^T,$$

$$\mathbf{r}_{83} = \left[\begin{array}{ccc} \frac{68378}{8695} & 260 & 0 \end{array} \right]^T, \quad \mathbf{r}_{76} = \left[\begin{array}{ccc} -\frac{112144664}{850397} & \frac{14647860}{850397} & 0 \end{array} \right]^T,$$

$$\mathbf{r}_{71} = \left[\begin{array}{ccc} \frac{5684052780}{11857451} & \frac{8282660400}{11857451} & 0 \end{array} \right]^T, \quad \mathbf{r}_{81} = \left[\begin{array}{ccc} -\frac{347482980}{1624111} & \frac{210336480}{1624111} & 0 \end{array} \right]^T,$$

$$\mathbf{r}_{75} = \left[\begin{array}{ccc} -\frac{9105880}{26227} & -\frac{3854865}{104908} & 0 \end{array} \right]^T, \quad \mathbf{r}_{51} = \left[\begin{array}{ccc} \frac{62723700}{3852029} & \frac{1103937120}{3852029} & 0 \end{array} \right]^T$$

$$\mathbf{r}_{53} = \left[\begin{array}{ccc} -\frac{99285}{241} & \frac{86570}{723} & 0 \end{array} \right]^T, \quad \mathbf{r}_{54} = \left[\begin{array}{ccc} \frac{56976220}{511177} & \frac{61329840}{511177} & 0 \end{array} \right]^T,$$

$$\mathbf{r}_{61} = \left[\begin{array}{ccc} \frac{3665448828}{27164597} & \frac{8546321880}{27164597} & 0 \end{array} \right]^T, \quad \mathbf{r}_{64} = \left[\begin{array}{ccc} \frac{144519259}{897343} & \frac{118698915}{897343} & 0 \end{array} \right]^T.$$

The determination of the remaining infinitesimal screw representing the secondary IRCs is straightforward and, due to space considerations, is not presented here.

The location of the secondary IRCs, obtained from the position vectors, is shown in Figure 5. Therein, one can observe that the three centers associated with three relative movements, namely, $^{i}\$^{k}$, $^{i}\$^{j}$, and $^{j}\$^{k}$, between three arbitrary links lie on a straight line, according to the Aronhold–Kennedy theorem.

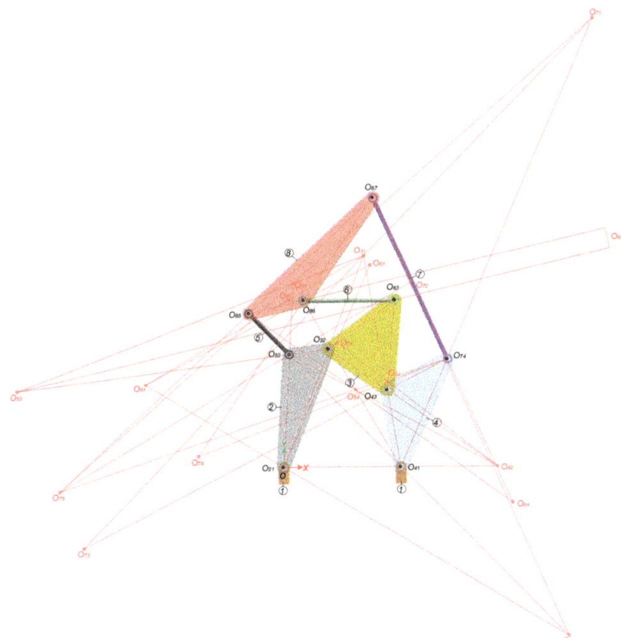

Figure 5. All rotation centers of a "single flyer" planar eight-bar linkage.

4. Second Case Study: Eight-Bar "Double-Butterfly" Planar Linkage

Consider the linkage shown in Figure 6, which is known as a "double-butterfly" eight-bar undetermined planar linkage. The linkage was proposed by Klein [4], who determined the secondary centers by using a trial-and-error method. Foster and Pennock [12] determined the secondary centers using a graphical method. Di Gregorio [13] determined the secondary centers by developing an algebraic method that combines the closure equations and the location of the secondary centers. This linkage has eight links, four of them ternary, and the remaining four binary, and it has ten revolutes. The linkage is trivial, and its mobility is 1.

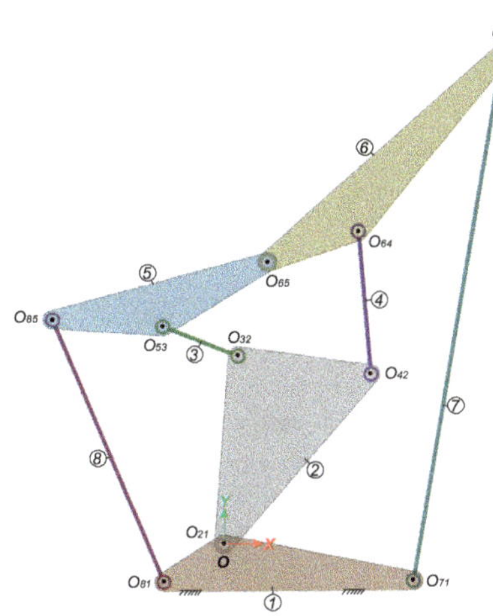

Figure 6. Eight-bar "double-butterfly" undetermined planar linkage.

The first step in the process is to obtain the screws associated with the linkage's kinematic pairs about the origin of the coordinate system, O. To that end, the common direction of all revolute pairs, $\mathbf{u}$, is required, given by:

$$\mathbf{u} = \begin{bmatrix} 0 & 0 & 1 \end{bmatrix}^T$$

and the position vectors of points located along the revolute axes, about the origin, O. They are given in terms of an arbitrary unit of length, by:

$$\mathbf{r}_{21} = \begin{bmatrix} 0 & 0 & 0 \end{bmatrix}^T, \quad \mathbf{r}_{81} = \begin{bmatrix} -80 & -50 & 0 \end{bmatrix}^T, \quad \mathbf{r}_{71} = \begin{bmatrix} 250 & -50 & 0 \end{bmatrix}^T,$$
$$\mathbf{r}_{32} = \begin{bmatrix} 20 & 250 & 0 \end{bmatrix}^T, \quad \mathbf{r}_{53} = \begin{bmatrix} -80 & 290 & 0 \end{bmatrix}^T, \quad \mathbf{r}_{85} = \begin{bmatrix} -225 & 300 & 0 \end{bmatrix}^T,$$
$$\mathbf{r}_{65} = \begin{bmatrix} 60 & 375 & 0 \end{bmatrix}^T, \quad \mathbf{r}_{42} = \begin{bmatrix} 195 & 225 & 0 \end{bmatrix}^T, \quad \mathbf{r}_{64} = \begin{bmatrix} 180 & 415 & 0 \end{bmatrix}^T,$$
$$\mathbf{r}_{76} = \begin{bmatrix} 370 & 650 & 0 \end{bmatrix}^T.$$

From this data, the screws associated with the kinematic pairs, about the origin of the coordinate system, O, are given by:

$${}^2\$^1_O = \begin{bmatrix} 1; & 0 & 0 \end{bmatrix}^T, \quad {}^3\$^2_O = \begin{bmatrix} 1; & 250 & -20 \end{bmatrix}^T, \quad {}^5\$^3_O = \begin{bmatrix} 1; & 290 & 80 \end{bmatrix}^T,$$
$${}^8\$^5_O = \begin{bmatrix} 1; & 300 & 225 \end{bmatrix}^T, \quad {}^8\$^1_O = \begin{bmatrix} 1; & -50 & 80 \end{bmatrix}^T, \quad {}^4\$^2_O = \begin{bmatrix} 1; & 225 & -195 \end{bmatrix}^T,$$
$${}^6\$^4_O = \begin{bmatrix} 1; & 415 & -180 \end{bmatrix}^T, \quad {}^6\$^5_O = \begin{bmatrix} 1; & 375 & -60 \end{bmatrix}^T, \quad {}^7\$^6_O = \begin{bmatrix} 1; & 650 & -370 \end{bmatrix}^T,$$
$${}^7\$^1_O = \begin{bmatrix} 1; & -50 & -250 \end{bmatrix}^T.$$

4.1. Determination of the Secondary Rotation Axes

From the enumeration of the links shown in Figure 6, the primary IRCs, namely, those that can be determined by inspection of the linkage, are:

$$O_{21},\ O_{32},\ O_{53},\ O_{85},\ O_{81},\ O_{42},\ O_{64},\ O_{65},\ O_{76},\ O_{71}$$

The upper diagonal arrangement of IRCs, where the primary centers are marked with the number −1, and the secondary centers are marked with the number −2, are shown in Table 4.

Table 4. Instantaneous rotation axes associated with the planar eight-bar linkage.

$O_{21}(-1)$	$O_{31}(-2)$	$O_{41}(-2)$	$O_{51}(-2)$	$O_{61}(-2)$	$O_{71}(-1)$	$O_{81}(-1)$
	$O_{32}(-1)$	$O_{42}(-1)$	$O_{52}(-2)$	$O_{62}(-2)$	$O_{72}(-2)$	$O_{82}(-2)$
		$O_{43}(-2)$	$O_{53}(-1)$	$O_{63}(-2)$	$O_{73}(-2)$	$O_{83}(-2)$
			$O_{54}(-2)$	$O_{64}(-1)$	$O_{74}(-2)$	$O_{84}(-2)$
				$O_{65}(-1)$	$O_{75}(-2)$	$O_{85}(-1)$
					$O_{76}(-1)$	$O_{86}(-2)$
						$O_{87}(-2)$

This linkage is completely undetermined since no secondary center can be found by the direct application of the Aronhold–Kennedy theorem. The search for these secondary IRCs will require two steps. In the first step, an arbitrary secondary center is chosen and the location of an overconstrained secondary center is determined up to the two possible roots of a quadratic equation. In the second step, another arbitrary secondary center is chosen and the location of the same overconstrained secondary center is determined up to the two possible roots of a quadratic equation. The comparison between the solutions of the two quadratic roots yields the correct solution and, from the intermediate results, the location of many other secondary centers will be accomplished. Therefore, Table 5 shows the secondary centers, which can be written as a linear combination of two primary centers employing a single variable λ, and which are used in this example.

Table 5. Secondary centers that can be written as a linear combination of two primary centers.

O_{51} is found along the line determined by O_{81} O_{85}
O_{74} is found along the line determined by O_{76} O_{64}

4.1.1. Selection of a First Arbitrary Center and Determination of an Overconstrained Center

From Table 5, the secondary center O_{51} is chosen. This center can be written as:

$$\mathbf{r}_{51} = \lambda\,\mathbf{r}_{81} + (1-\lambda)\,\mathbf{r}_{85} \tag{90}$$

Thus, the upper triangular arrangement of IRC becomes:

$$\begin{array}{ccccccc}
O_{21}(-1) & O_{31}(-2) & O_{41}(-2) & O_{51}(0) & O_{61}(-2) & O_{71}(-1) & O_{81}(-1) \\
 & O_{32}(-1) & O_{42}(-1) & O_{52}(-2) & O_{62}(-2) & O_{72}(-2) & O_{82}(-2) \\
 & & O_{43}(-2) & O_{53}(-1) & O_{63}(-2) & O_{73}(-2) & O_{83}(-2) \\
 & & & O_{54}(-2) & O_{64}(-1) & O_{74}(-2) & O_{84}(-2) \\
 & & & & O_{65}(-1) & O_{75}(-2) & O_{85}(-1) \\
 & & & & & O_{76}(-1) & O_{86}(-2) \\
 & & & & & & O_{87}(-2)
\end{array}$$

Therefore, following the approach indicated in Section 3.1, the application of the Aronhold–Kennedy theorem renders the location of the IRCs O_{61}, O_{52}

$O_{61} \Longrightarrow O_{51}\ \boxed{O_{65}}$ $\qquad$ $O_{52} \Longrightarrow \boxed{O_{21}}\ O_{51}$

$O_{61} \Longrightarrow \boxed{O_{71}}\ \boxed{O_{76}}$ $\qquad$ $O_{52} \Longrightarrow \boxed{O_{32}}\ \boxed{O_{53}}$

After marking the secondary IRCs O_{61} and O_{52} with the number 1, the upper triangular arrangement of the IRCs becomes:

$$\begin{array}{ccccccc}
O_{21}(-1) & O_{31}(1) & O_{41}(-2) & O_{51}(0) & O_{61}(1) & O_{71}(-1) & O_{81}(-1) \\
 & O_{32}(-1) & O_{42}(-1) & O_{52}(1) & O_{62}(-2) & O_{72}(-2) & O_{82}(-2) \\
 & & O_{43}(-2) & O_{53}(-1) & O_{63}(-2) & O_{73}(-2) & O_{83}(-2) \\
 & & & O_{54}(-2) & O_{64}(-1) & O_{74}(-2) & O_{84}(-2) \\
 & & & & O_{65}(-1) & O_{75}(1) & O_{85}(-1) \\
 & & & & & O_{76}(-1) & O_{86}(-2) \\
 & & & & & & O_{87}(-2)
\end{array}$$

In the next step, all the IRCs found so far will be employed to find additional centers. In this case, the overconstrained center O_{62} is found.

$O_{62} \Longrightarrow \boxed{O_{21}}\ O_{61}$ $\qquad$ $O_{62} \Longrightarrow \boxed{O_{42}}\ \boxed{O_{64}}$ $\qquad$ $O_{62} \Longrightarrow O_{52}\ \boxed{O_{65}}$

After marking the overconstrained IRC O_{62} with the number 2, the upper triangular arrangement of the IRC is given by:

$$\begin{matrix} O_{21}(-1) & O_{31}(1) & O_{41}(2) & O_{51}(0) & O_{61}(1) & O_{71}(-1) & O_{81}(-1) \\ & O_{32}(-1) & O_{42}(-1) & O_{52}(1) & O_{62}(2) & O_{72}(2) & O_{82}(2) \\ & & O_{43}(-2) & O_{53}(-1) & O_{63}(2) & O_{73}(2) & O_{83}(2) \\ & & & O_{54}(2) & O_{64}(-1) & O_{74}(-2) & O_{84}(-2) \\ & & & & O_{65}(-1) & O_{75}(1) & O_{85}(-1) \\ & & & & & O_{76}(-1) & O_{86}(2) \\ & & & & & & O_{87}(2) \end{matrix}$$

From Equation (90), the position vector associated with the secondary IRC O_{51} is:

$$\mathbf{r}_{51} = \begin{bmatrix} 145\,\lambda - 225 & -350\,\lambda + 300 & 0 \end{bmatrix}^T \tag{91}$$

Furthermore, the position vectors associated with the involved IRC are:

$$\mathbf{r}_{61} = \begin{bmatrix} x_{61} & y_{61} & 0 \end{bmatrix}^T, \quad \mathbf{r}_{52} = \begin{bmatrix} x_{52} & y_{52} & 0 \end{bmatrix}^T, \quad \mathbf{r}_{62} = \begin{bmatrix} x_{62} & y_{62} & 0 \end{bmatrix}^T.$$

The necessary equations to solve the problem are obtained by systematically applying the Aronhold–Kennedy theorem employing the Klein form, Equations (25) and (26). It should be noted that these expressions are equated to the vector $\mathbf{0}$, generating four scalar equations, of which only one is non-trivial. Due to space considerations, only the non-zero component is shown.

$$Eq61_7 = \frac{1810}{7} + \frac{6}{35}\,y_{61} - x_{61}, \tag{92}$$

$$Eq61_5 = \frac{70\lambda\,x_{61} + 29\lambda\,y_{61} - 15075\lambda + 15\,x_{61} - 57\,y_{61} + 20475}{3 + 14\lambda} \tag{93}$$

$$Eq52_3 = 645 - \frac{5}{2}\,y_{52} - x_{52}, \tag{94}$$

$$Eq52_1 = \frac{70\lambda\,x_{52} + 29\lambda\,y_{52} - 60\,x_{52} - 45\,y_{52}}{7\lambda - 6}, \tag{95}$$

$$Eq62_4 = \frac{8085}{38} - \frac{3}{38}\,y_{62} - x_{62}, \tag{96}$$

$$Eq62_1 = \frac{x_{61}\,y_{62} - x_{62}\,y_{61}}{y_{61}}, \tag{97}$$

$$Eq62_5 = \frac{x_{52}\,y_{62} - x_{62}\,y_{52} - 375\,x_{52} + 375\,x_{62} + 60\,y_{52} - 60\,y_{62}}{y_{52} - 375} \tag{98}$$

From Equations (92)–(98), the final set of equations is formed:

$$0 = \frac{1810}{7} + \frac{6}{35}\,y_{61} - x_{61} \tag{99}$$

$$0 = \frac{70\lambda\,x_{61} + 29\lambda\,y_{61} - 15075\lambda + 15\,x_{61} - 57\,y_{61} + 20475}{3 + 14\lambda} \tag{100}$$

$$0 = 645 - \frac{5}{2}\,y_{52} - x_{52} \tag{101}$$

$$0 = \frac{70\lambda\,x_{52} + 29\lambda\,y_{52} - 60\,x_{52} - 45\,y_{52}}{7\lambda - 6} \tag{102}$$

$$0 = \frac{8085}{38} - \frac{3}{38}\,y_{62} - x_{62} \tag{103}$$

$$0 = \frac{x_{61}\,y_{62} - x_{62}\,y_{61}}{y_{61}} \tag{104}$$

$$0 = \frac{x_{52}\,y_{62} - x_{62}\,y_{52} - 375\,x_{52} + 375\,x_{62} + 60\,y_{52} - 60\,y_{62}}{y_{52} - 375} \tag{105}$$

Systematically, the coordinates y_{ij} from each linear equation are solved in terms of the coordinates x_{ij}. Therefore, the coordinates y_{61}, y_{52}, and y_{62} from Equations (99), (101) and (103) are written as:

$$y_{61} = -\frac{4525}{3} + \frac{35\,x_{61}}{6}, \quad y_{52} = 258 - \frac{2\,x_{52}}{5}, \quad y_{62} = 2695 - \frac{38\,x_{62}}{3}. \tag{106}$$

Substituting the results of Equation (106) into Equations (100), (102), (104) and (105), it follows that:

$$0 = \frac{287\,\lambda\,x_{61} - 70580\,\lambda - 381\,x_{61} + 127740}{3 + 14\,\lambda} \tag{107}$$

$$0 = \frac{146\,\lambda\,x_{52} + 18705\,\lambda - 105\,x_{52} - 29025}{7\,\lambda - 6} \tag{108}$$

$$0 = \frac{111\,x_{61}\,x_{62} - 16170\,x_{61} - 9050\,x_{62}}{-1810 + 7\,x_{61}} \tag{109}$$

$$0 = \frac{184\,x_{52}\,x_{62} - 34440\,x_{52} - 13155\,x_{62} + 2193300}{585 + 2\,x_{52}} \tag{110}$$

From Equation (110), x_{62} is solved in terms of x_{52} to obtain:

$$x_{62} = \frac{60\,(574\,x_{52} - 36555)}{184\,x_{52} - 13155} \tag{111}$$

The result for x_{62}, obtained from Equation (111), is substituted into Equation (109) to solve x_{52} in terms of x_{61}, so that:

$$x_{52} = \frac{1024665\,x_{61} - 661645500}{28252\,x_{61} - 10389400} \tag{112}$$

Similarly, the result for x_{52} of Equation (112) is substituted into Equation (108) to solve x_{61} in terms of λ, therefore:

$$x_{61} = \frac{3620\,(214316\,\lambda - 273315)}{-2473611 + 1808146\,\lambda} \tag{113}$$

Finally, the result for x_{61}, from Equation (113), is substituted into Equation (107) to obtain a quadratic equation in λ, given by:

$$0 = \frac{-60\,(264007001\,\lambda^2 - 483296570\,\lambda + 169395681)}{(-2473611 + 1808146\,\lambda)(3 + 14\,\lambda)}$$

After clearing the denominator, the two solutions of λ are given by:

$$\lambda_a = \frac{1861491}{3940403}, \qquad \lambda_b = \frac{91}{67}. \tag{114}$$

These solutions are substituted into Equation (91) to obtain the two possible locations of the center O_{51}.

$$\mathbf{r}_{51a} = \begin{bmatrix} -\frac{616674480}{3940403} & \frac{530599050}{3940403} & 0 \end{bmatrix}^T, \qquad \mathbf{r}_{51b} = \begin{bmatrix} -\frac{1880}{67} & -\frac{11750}{67} & 0 \end{bmatrix}^T \tag{115}$$

4.1.2. Selection of a Second Arbitrary Center and Determination of an Overconstrained Center

The process must be repeated with another secondary center. From Table 5, the center O_{74} is chosen, and its position vector can be written as:

$$\mathbf{r}_{74} = \lambda^*\,\mathbf{r}_{76} + (1 - \lambda^*)\,\mathbf{r}_{64} = \begin{bmatrix} 180 + 190\,\lambda^* & 415 + 235\,\lambda^* & 0 \end{bmatrix}^T \tag{116}$$

The upper triangular array of IRCs, with the center O_{74} marked with the number 0, is given by:

$$
\begin{array}{ccccccc}
O_{21}(-1) & O_{31}(-2) & O_{41}(-2) & O_{51}(-2) & O_{61}(-2) & O_{71}(-1) & O_{81}(-1) \\
 & O_{32}(-1) & O_{42}(-1) & O_{52}(-2) & O_{62}(-2) & O_{72}(-2) & O_{82}(-2) \\
 & & O_{43}(-2) & O_{53}(-1) & O_{63}(-2) & O_{73}(-2) & O_{83}(-2) \\
 & & & O_{54}(-2) & O_{64}(-1) & O_{74}(0) & O_{84}(-2) \\
 & & & & O_{65}(-1) & O_{75}(-2) & O_{85}(-1) \\
 & & & & & O_{76}(-1) & O_{86}(-2) \\
 & & & & & & O_{87}(-2)
\end{array}
$$

Using this secondary center O_{74}, the secondary centers O_{41} and O_{72} can be found:

$$
\begin{array}{ll}
O_{41} \Longmapsto \boxed{O_{21}}\ \boxed{O_{42}} & \qquad O_{72} \Longmapsto \boxed{O_{21}}\ \boxed{O_{71}} \\
O_{41} \Longmapsto \boxed{O_{71}}\ O_{74} & \qquad O_{72} \Longmapsto \boxed{O_{42}}\ O_{74}
\end{array}
$$

After identifying the secondary centers O_{41} and O_{72} with the number 1, the upper triangular arrangement of centers becomes:

$$
\begin{array}{ccccccc}
O_{21}(-1) & O_{31}(-2) & O_{41}(1) & O_{51}(-2) & O_{61}(1) & O_{71}(-1) & O_{81}(-1) \\
 & O_{32}(-1) & O_{42}(-1) & O_{52}(-2) & O_{62}(-2) & O_{72}(1) & O_{82}(-2) \\
 & & O_{43}(-2) & O_{53}(-1) & O_{63}(-2) & O_{73}(-2) & O_{83}(-2) \\
 & & & O_{54}(-2) & O_{64}(-1) & O_{74}(0) & O_{84}(-2) \\
 & & & & O_{65}(-1) & O_{75}(-2) & O_{85}(-1) \\
 & & & & & O_{76}(-1) & O_{86}(-2) \\
 & & & & & & O_{87}(1)
\end{array}
$$

The next step employs all the centers found so far to find additional centers. The relevant ones are:

$$
\begin{array}{ll}
O_{61} \Longmapsto O_{41}\ \boxed{O_{64}} & \qquad O_{62} \Longmapsto \boxed{O_{42}}\ \boxed{O_{64}} \\
O_{61} \Longmapsto \boxed{O_{71}}\ \boxed{O_{76}} & \qquad O_{62} \Longmapsto O_{72}\ \boxed{O_{76}}
\end{array}
$$

After this step, the upper triangular arrangement of centers becomes:

$$
\begin{array}{ccccccc}
O_{21}(-1) & O_{31}(-2) & O_{41}(1) & O_{51}(-2) & O_{61}(2) & O_{71}(-1) & O_{81}(-1) \\
 & O_{32}(-1) & O_{42}(-1) & O_{52}(-2) & O_{62}(2) & O_{72}(1) & O_{82}(-2) \\
 & & O_{43}(-2) & O_{53}(-1) & O_{63}(-2) & O_{73}(-2) & O_{83}(-2) \\
 & & & O_{54}(-2) & O_{64}(-1) & O_{74}(0) & O_{84}(-2) \\
 & & & & O_{65}(-1) & O_{75}(-2) & O_{85}(-1) \\
 & & & & & O_{76}(-1) & O_{86}(-2) \\
 & & & & & & O_{87}(-2)
\end{array}
$$

The following step employs all the IRCs marked with the numbers $-1, 0, 1$, and 2 to find additional IRCs; the relevant ones are:

Marking the newly found IRCs with the number 3, the upper triangular arrangement of centers becomes:

$$\begin{array}{ccccccc}
O_{21}(-1) & O_{31}(-2) & O_{41}(1) & O_{51}(3) & O_{61}(2) & O_{71}(-1) & O_{81}(-1) \\
 & O_{32}(-1) & O_{42}(-1) & O_{52}(3) & O_{62}(2) & O_{72}(1) & O_{82}(-2) \\
 & & O_{43}(-2) & O_{53}(-1) & O_{63}(3) & O_{73}(-2) & O_{83}(-2) \\
 & & & O_{54}(-2) & O_{64}(-1) & O_{74}(0) & O_{84}(-2) \\
 & & & & O_{65}(-1) & O_{75}(-2) & O_{85}(-1) \\
 & & & & & O_{76}(-1) & O_{86}(3) \\
 & & & & & & O_{87}(-2)
\end{array}$$

The final step is to find an overconstrained IRC. In this case, the IRC O_{31} can be found using the following three IRC combinations:

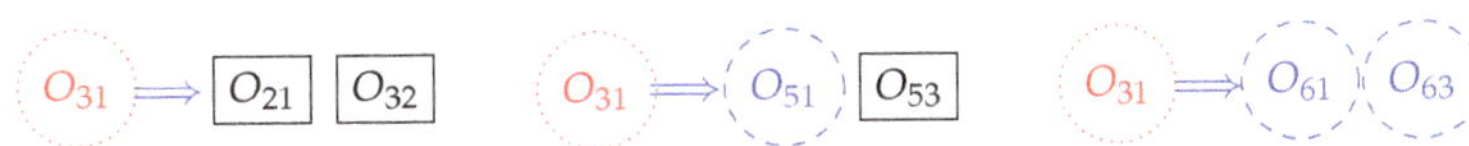

The position vectors of the secondary centers found in this process are:

$$\mathbf{r}_{31} = \begin{bmatrix} x_{31} & y_{31} & 0 \end{bmatrix}^T, \quad \mathbf{r}_{41} = \begin{bmatrix} x_{41} & y_{41} & 0 \end{bmatrix}^T, \mathbf{r}_{51} = \begin{bmatrix} x_{51} & y_{51} & 0 \end{bmatrix}^T,$$
$$\mathbf{r}_{61} = \begin{bmatrix} x_{61} & y_{61} & 0 \end{bmatrix}^T, \quad \mathbf{r}_{62} = \begin{bmatrix} x_{62} & y_{62} & 0 \end{bmatrix}^T, \mathbf{r}_{63} = \begin{bmatrix} x_{63} & y_{63} & 0 \end{bmatrix}^T.$$
$$\mathbf{r}_{72} = \begin{bmatrix} x_{72} & y_{72} & 0 \end{bmatrix}^T.$$

The equations required to solve the problem are obtained applying the Aronhold–Kennedy theorem and using systematically the Klein form, Equations (25) and (26). It should be noted that these expressions are equated to the vector **0**, generating four scalar equations, of which only one is non-trivial. Due to space considerations, only the non-zero component is shown.

$$Eq41_2 = 225\,x_{41} - 195\,y_{41} = 0, \tag{117}$$
$$Eq41_7 = 235\,\lambda^*\,x_{41} - 190\,\lambda^*\,y_{41} - 68250\,\lambda^* + 465\,x_{41} + 70\,y_{41} - 112750 = 0, \tag{118}$$
$$Eq72_1 = -50\,x_{72} - 250\,y_{72} \tag{119}$$
$$Eq72_4 = 235\,\lambda^*\,x_{72} - 190\,\lambda^*\,y_{72} - 3075\,\lambda^* + 190\,x_{72} + 15\,y_{72} - 40425 = 0 \tag{120}$$
$$Eq61_7 = 700\,x_{61} - 120\,y_{61} - 181000 = 0, \tag{121}$$
$$Eq61_4 = x_{41}\,y_{61} - x_{61}\,y_{41} - 415\,x_{41} + 415\,x_{61} + 180\,y_{41} - 180\,y_{61} = 0, \tag{122}$$
$$Eq62_4 = 190\,x_{62} + 15\,y_{62} - 40425 = 0, \tag{123}$$
$$Eq62_7 = -x_{62}\,y_{72} + x_{72}\,y_{62} + 650\,x_{62} - 650\,x_{72} - 370\,y_{62} + 370\,y_{72} = 0, \tag{124}$$
$$Eq63_2 = -x_{62}\,y_{63} + x_{63}\,y_{62} + 250\,x_{62} - 250\,x_{63} - 20\,y_{62} + 20\,y_{63} = 0, \tag{125}$$
$$Eq63_5 = 85\,x_{63} - 140\,y_{63} + 47400 = 0, \tag{126}$$
$$Eq51_6 = -x_{51}\,y_{61} + x_{61}\,y_{51} + 375\,x_{51} - 375\,x_{61} - 60\,y_{51} + 60\,y_{61} = 0, \tag{127}$$
$$Eq51_8 = 350\,x_{51} + 145\,y_{51} + 35250 = 0, \tag{128}$$
$$Eq31_2 = 250\,x_{31} - 20\,y_{31} = 0, \tag{129}$$
$$Eq31_5 = -x_{31}\,y_{51} + x_{51}\,y_{31} + 290\,x_{31} - 290\,x_{51} + 80\,y_{31} - 80\,y_{51} = 0, \tag{130}$$
$$Eq31_6 = -x_{31}\,y_{61} + x_{31}\,y_{63} + x_{61}\,y_{31} - x_{61}\,y_{63} - x_{63}\,y_{31} + x_{63}\,y_{61} = 0. \tag{131}$$

Equations (117)–(131) represent a system of 15 equations in 15 unknowns. The solution requires that the coordinates y_{ij} are solved in terms of the coordinates x_{ij} from the linear equations. Hence, the coordinates y_{41}, y_{72}, y_{61}, y_{62}, y_{63}, y_{51}, and y_{31} are solved from Equations (117), (119), (121), (123), (126), (128) and (129), and one obtains:

$$y_{41} = \frac{15}{13}\,x_{41}, \qquad y_{72} = -\tfrac{1}{5}\,x_{72}, \qquad y_{61} = \frac{35}{6}\,x_{61} - \frac{4525}{3} \tag{132}$$

$$y_{62} = -\frac{38}{3}\,x_{62} + 2695, \qquad y_{51} = -\tfrac{70}{29}\,x_{51} - \tfrac{7050}{29} \qquad y_{63} = \frac{17}{28}\,x_{63} + \frac{2370}{7} \tag{133}$$

$$y_{31} = \frac{25}{2}\,x_31 \tag{134}$$

Substituting the results of Equations (132)–(134) into the remaining non-linear equations, it follows that:

$$0 = \frac{205}{13}\,x_{41}\,\lambda^* - 68250\,\lambda^* + \frac{7095}{13}\,x_{41} - 112750 \tag{135}$$

$$0 = -\frac{1435}{174}\,x_{51}\,x_{61} + \frac{176450}{87}\,x_{51} - \frac{7775}{29}\,x_{61} - \frac{2201500}{29} \tag{136}$$

$$0 = \frac{365}{78}\,x_{61}\,x_{41} - \frac{66910}{39}\,x_{41} - 635\,x_{61} + 271500 \tag{137}$$

$$0 = -\frac{187}{15}\,x_{62}\,x_{72} + 1971\,x_{72} + \frac{16010}{3}\,x_{62} - 997150 \tag{138}$$

$$0 = -\frac{1115}{84}\,x_{62}\,x_{63} + \frac{3460}{21}\,x_{62} + \frac{17200}{7}\,x_{63} - \frac{329900}{7} \tag{139}$$

$$0 = 273\,\lambda^*\,x_{72} - 3075\,\lambda^* + 187\,x_{72} - 40425 \tag{140}$$

$$0 = \frac{865}{58}\,x_{31}\,x_{51} + \frac{44460}{29}\,x_{31} - \frac{2810}{29}\,x_{51} + \frac{564000}{29} \tag{141}$$

$$0 = \frac{20}{3}x_{31}x_{61} + \frac{38785}{21}x_{31} - \frac{333}{28}x_{31}x_{63} + \frac{439}{84}x_{61}x_{63} - \frac{2370}{7}x_{61} - \frac{4525}{3}x_{63} \tag{142}$$

From Equation (135), x_{41} is solved in terms of λ^*. The result given by

$$x_{41} = \frac{650\,(273\,\lambda^* + 451)}{41\,\lambda^* + 1419}, \tag{143}$$

is substituted into Equation (137). Furthermore, solving the resulting equation for x_{61}, one obtains:

$$0 = \frac{921\,\lambda^*\,x_{61} - 335850\,\lambda^* + 539\,x_{61} - 134750}{41\,\lambda^* + 1419} \qquad x_{61} = \frac{50\,(6717\,\lambda^* + 2695)}{921\,\lambda^* + 539}. \tag{144}$$

The result for x_{61} is substituted into Equation (136), to obtain:

$$0 = \frac{1046159\,\lambda^*\,x_{51} + 185552610\,\lambda^* + 21021\,x_{51} + 89371590}{921\,\lambda^* + 539}. \tag{145}$$

Solving Equation (145) for x_{51}, it follows that:

$$x_{51} = -\frac{30\,(6185087\,\lambda^* + 2979053)}{1046159\,\lambda^* + 21021} \tag{146}$$

The value of x_{61}, from Equation (143), is also substituted into Equation (137). The result is given by:

$$0 = \frac{35382\,\lambda^*\,x_{62} - 7437465\,\lambda^* - 24973\,x_{62} + 4201260}{381\,\lambda^* - 94}. \tag{147}$$

The resulting equation, (147), is solved for x_{62} in terms of λ^*; hence:

$$x_{62} = \frac{735\,(10119\,\lambda^* - 5716)}{35382\,\lambda^* - 24973}. \tag{148}$$

In the next step, x_{72} is solved from Equation (140), and the result is given by:

$$x_{72} = \frac{75\,(41\,\lambda^* + 539)}{273\,\lambda^* + 187}. \tag{149}$$

The value of x_{72} is substituted into Equation (138). Furthermore, solving the resulting equation for x_{62}, one obtains:

$$0 = \frac{537\,\lambda^*\,x_{62} - 100755\,\lambda^* + 187\,x_{62} - 40425}{273\,\lambda^* + 187} \qquad x_{62} = \frac{15\,(6717\,\lambda^* + 2695)}{537\,\lambda^* + 187}. \tag{150}$$

Repeating the process by substituting the value of x_{62} into (139), and then solving for x_{63}, it follows that:

$$0 = \frac{6689\,\lambda^*\,x_{63} + 3250784\,\lambda^* + 28787\,x_{63} + 803616}{537\,\lambda^* + 187} \qquad x_{63} = -\frac{32\,(101587\,\lambda^* + 25113)}{6689\,\lambda^* + 28787}. \tag{151}$$

Before the final substitutions, the value of x_{51} is substituted into (141) to obtain:

$$0 = -\frac{77561551\,\lambda^*\,x_{31} - 2555026460\,\lambda^* + 86709469\,x_{31} - 604574740}{1046159\,\lambda^* + 21021}.$$

Then, solving for x_{31} a solution in terms of λ^* is given by:

$$x_{31} = -\frac{140\,(18250189\,\lambda^* + 4318391)}{77561551\,\lambda^* + 86709469}. \tag{152}$$

Similarly, the value of x_{61} is substituted into Equation (131) to obtain:

$$0 = \frac{c_1\,\lambda^{*2}\,x_{31} - c_2\,\lambda^{*2} + c_3\,\lambda^*\,x_{31} - c_4\,\lambda^* + c_5\,x_{31} - c_6}{(921\,\lambda^* + 539)\,(6689\,\lambda^* + 28787)}, \tag{153}$$

where

$$c_1 = 61961537269 \quad c_2 = 1950521392700 \quad c_3 = 155729090462$$
$$c_4 = 3519109121000 \quad c_5 = 59668631869 \quad c_6 = 1225931622300$$

Substituting the value of x_{31} obtained in Equation (152) into Equation (153), a final cubic equation in λ^* is obtained:

$$0 = \frac{d_1\,\lambda^{*3} - d_2\,\lambda^{*2} - D_3\,\lambda^* - d_4}{(77561551\,\lambda^* + 86709469)\,(921\,\lambda^* + 539)\,(6689\,\lambda^* + 28787)}. \tag{154}$$

where

$$d_1 = 175697568701976501 \qquad d_2 = 168097719765363443$$
$$d_3 = 384051084116763817 \qquad d_4 = 175564331003962924 1$$

After clearing the denominator of (153), its solutions are given by:

$$\lambda^*_a = \frac{815507}{151807}, \qquad \lambda^*_b = -\frac{31449671}{7974909}, \qquad \lambda^*_c = -\frac{68453}{145127} \tag{155}$$

The results for λ^* are substituted into Equation (116) to obtain the three possible position vectors of the secondary center O_{74}:

$$\mathbf{r}_{74a} = \begin{bmatrix} \frac{182271590}{151807} \\ \frac{254644050}{151807} \end{bmatrix} \quad \mathbf{r}_{74b} = \begin{bmatrix} -\frac{4539953870}{7974909} \\ -\frac{4081085450}{7974909} \\ 0 \end{bmatrix} \quad \mathbf{r}_{74c} = \begin{bmatrix} \frac{13116790}{145127} \\ \frac{44141250}{145127} \\ 0 \end{bmatrix}. \tag{156}$$

From the results of Equation (156), it is possible to find the three possible locations of the secondary centers O_{41} and O_{61}:

$$\mathbf{r}_{41a} = \begin{bmatrix} \frac{6719245}{8837} \\ \frac{7752975}{8837} \\ 0 \end{bmatrix} \quad \mathbf{r}_{41b} = \begin{bmatrix} -\frac{115159785}{356071} \\ -\frac{132876675}{356071} \\ 0 \end{bmatrix} \quad \mathbf{r}_{41c} = \begin{bmatrix} \frac{664270}{4439} \\ \frac{9964050}{57707} \\ 0 \end{bmatrix}, \tag{157}$$

and

$$\mathbf{r}_{61a} = \begin{bmatrix} \frac{155140}{439} \\ \frac{242825}{439} \\ 0 \end{bmatrix} \quad \mathbf{r}_{61b} = \begin{bmatrix} \frac{898461460}{2335859} \\ \frac{5153313575}{7007577} \\ 0 \end{bmatrix} \quad \mathbf{r}_{61c} = \begin{bmatrix} -\frac{905}{4} \\ -\frac{22625}{8} \\ 0 \end{bmatrix}. \tag{158}$$

Finally, from the results in Equation (158), the three possible locations of the secondary centers O_{51} are given by:

$$\mathbf{r}_{51a} = \begin{bmatrix} -\frac{472320}{2453} \\ \frac{543750}{2453} \\ 0 \end{bmatrix} \quad \mathbf{r}_{51b} = \begin{bmatrix} -\frac{616674480}{3940403} \\ \frac{530599050}{3940403} \\ 0 \end{bmatrix} \quad \mathbf{r}_{51c} = \begin{bmatrix} \frac{141570}{36137} \\ \frac{9126750}{36137} \\ 0 \end{bmatrix}. \tag{159}$$

Comparing the results for the position vector of IRC O_{51}, obtained in Equations (115) and (159), it follows that the unique coincident result is:

$$\mathbf{r}_{51a} = \begin{bmatrix} -\frac{616674480}{3940403} & \frac{530599050}{3940403} & 0 \end{bmatrix}^T \tag{160}$$

Therefore, the screw associated with the secondary center O_{51} is given by:

$$^5\$^1_O = \begin{bmatrix} 1; & \frac{530599050}{3940403} & \frac{616674480}{3940403} \end{bmatrix}^T$$

Finally, from the results of Equation (160), together with the primary centers, applying the Aronhold–Kennedy theorem and using the Klein form, the remaining secondary centers can be computed following the procedure followed to find the secondary centers O_{31} and O_{42}, Equations (27) and (28). Therefore, the position vectors for the secondary centers are:

$$\mathbf{r}_{31} = \begin{bmatrix} \frac{52863440}{1223221} & \frac{660793000}{1223221} & 0 \end{bmatrix}^T, \qquad \mathbf{r}_{41} = \begin{bmatrix} -\frac{115159785}{356071} & -\frac{132876675}{356071} & 0 \end{bmatrix}^T,$$
$$\mathbf{r}_{61} = \begin{bmatrix} \frac{898461460}{2335859} & \frac{5153313575}{7007577} & 0 \end{bmatrix}^T, \qquad \mathbf{r}_{52} = \begin{bmatrix} -\frac{616674480}{1100501} & \frac{530599050}{1100501} & 0 \end{bmatrix}^T,$$
$$\mathbf{r}_{62} = \begin{bmatrix} \frac{2695384380}{14580649} & \frac{5153313575}{14580649} & 0 \end{bmatrix}^T, \qquad \mathbf{r}_{72} = \begin{bmatrix} -\frac{34193630}{1074917} & \frac{6838726}{1074917} & 0 \end{bmatrix}^T,$$
$$\mathbf{r}_{82} = \begin{bmatrix} \frac{49639760}{326137} & \frac{31024850}{326137} & 0 \end{bmatrix}^T, \qquad \mathbf{r}_{43} = \begin{bmatrix} -\frac{47950495}{702931} & \frac{184591195}{702931} & 0 \end{bmatrix}^T,$$
$$\mathbf{r}_{63} = \begin{bmatrix} \frac{1448067620}{290239} & \frac{977450545}{290239} & 0 \end{bmatrix}^T, \qquad \mathbf{r}_{73} = \begin{bmatrix} \frac{5947782410}{88544233} & \frac{41777847550}{88544233} & 0 \end{bmatrix}^T,$$
$$\mathbf{r}_{83} = \begin{bmatrix} -\frac{43192400}{4307933} & \frac{1228511450}{4307933} & 0 \end{bmatrix}^T, \qquad \mathbf{r}_{54} = \begin{bmatrix} -\frac{54239025}{574438} & \frac{185845815}{574438} & 0 \end{bmatrix}^T,$$
$$\mathbf{r}_{75} = \begin{bmatrix} -\frac{2027100510}{10530437} & \frac{1590188550}{10530437} & 0 \end{bmatrix}^T, \qquad \mathbf{r}_{74} = \begin{bmatrix} -\frac{4539953870}{7974909} & -\frac{4081085450}{7974909} & 0 \end{bmatrix}^T,$$
$$\mathbf{r}_{84} = \begin{bmatrix} \frac{65520025}{264426} & \frac{101851825}{264426} & 0 \end{bmatrix}^T, \qquad \mathbf{r}_{86} = \begin{bmatrix} \frac{208933300}{1088323} & \frac{445919525}{1088323} & 0 \end{bmatrix}^T,$$
$$\mathbf{r}_{87} = \begin{bmatrix} -\frac{74039790}{498077} & -50 & 0 \end{bmatrix}^T.$$

The location of all primary and secondary centers is shown in Figure 7. Therein, it can be illustrated that, ${}^i\$^k$, ${}^i\$^j$, and ${}^j\$^k$, the three centers associated with three relative movements between three arbitrary links, lie on a straight line, as indicated by the Aronhold–Kennedy theorem. Figure 8 shows the subgraphs of the secondary instantaneous centers involved in the process.

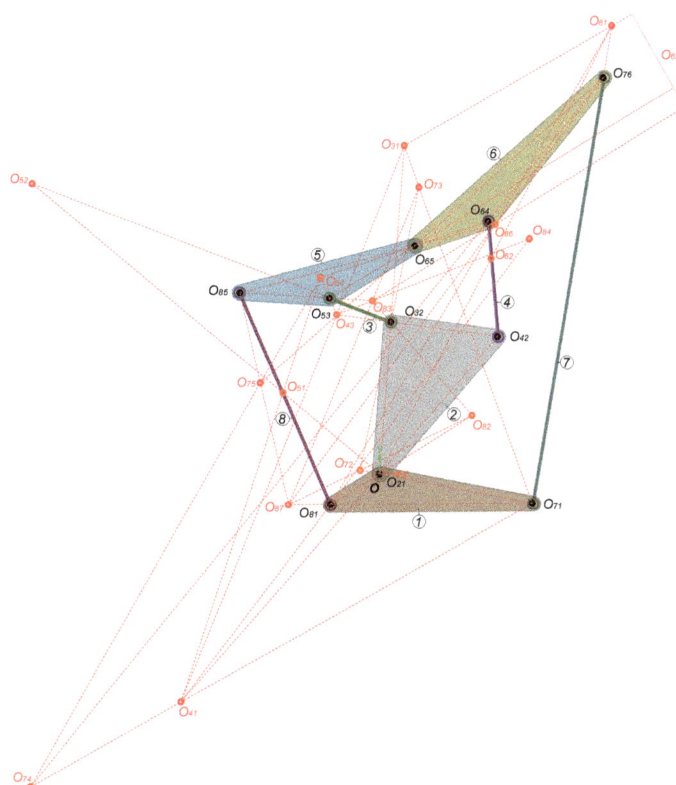

Figure 7. All rotation centers of an eight-bar "double-butterfly" planar linkage.

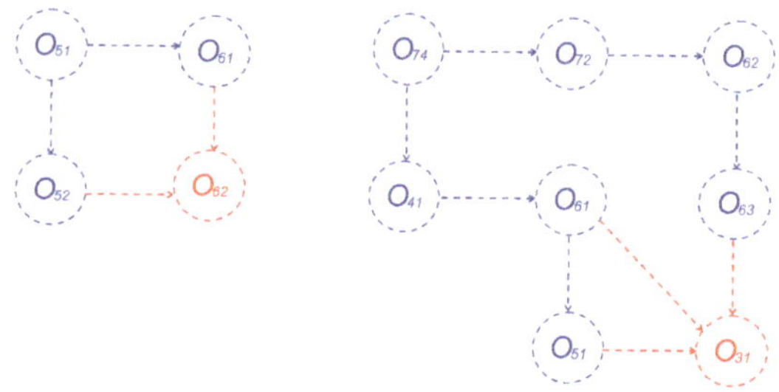

Figure 8. Subgraphs of secondary rotation centers required to solve the "double-butterfly" indeterminate planar linkage.

5. Proof of the Method

This short section presents constructive proof of the method. It should be noted that the number of IRCs is finite, and it is given by:

$$N = \frac{n\,(n-1)}{2},$$

where N is the number of IRCs, and n is the number of links in the linkage. Since both linkages have eight links, the number of IRCs is also, in both cases, twenty-eight.

- The single-butterfly linkage has nine primary centers. Two additional centers can be obtained by a straightforward application of the Aronhold–Kennedy theorem, O_{31}, and O_{42}. During the two-step process, the location of the following IRCs, O_{62}, O_{64}, O_{72}, O_{73}, O_{82}, O_{83}, and O_{84}, are found. Hence, after the process, the users will know the location of 18 out of 28 centers.
- The double-butterfly has 10 primary centers. During the two-step process, the location of the following IRCs, O_{31}, O_{41}, O_{51}, O_{52}, O_{61}, O_{62}, O_{63}, O_{72}, and O_{74}, are found. Hence, after the process, the users will know the location of 19 out of 28 centers.

Any person with some familiarity with the graphical techniques of the Aronhold–Kennedy theorem would recognize that, most likely, the location of the remaining IRC can be determined by the simple application of the Aronhold–Kennedy theorem. Even if this is not the case, the process can be repeated by choosing another secondary center. The location of all IRCs will be finished after a finite, usually small—no more than two- or three-steps—process. (In the examples shown in this contribution, it was possible to determine all the IRCs. There was no need to repeat the process.)

6. Conclusions

This paper introduced a novel and simpler method to obtain the secondary instantaneous centers of rotation of indeterminate planar linkages. The method only requires the solution and comparison of two quadratic equations or a quadratic equation and a cubic equation. They are simpler than the previous works reported in the literature. In addition, the method does not require the construction of the complete graph of the secondary IRCs, as indicated by Kung and Wang [14]. Motivated by the reviewers' comments, the authors will embark in the search of a method applicable to indeterminate linkages containing gear pairs and up to 10 links. Finally, the results were verified by carrying out the velocity analyses of the linkages involved and by simulation using Adams©. The verification is not included here due to space considerations. However all the details are included in the M.Sc. thesis of the first author [24].

Author Contributions: Conceptualization, J.I.V.-R. and J.M.R.; methodology, J.M.R.; software, J.I.V.-R.; validation, J.I.V.-R. and J.M.R.; formal analysis, J.M.R.; resources, J.I.V.-R.; curation, J.J.C.-S.; writing—original draft preparation, J.M.R., J.J.C.-S. and R.G.-G.; writing—review and editing, J.J.C.-S. and

R.G.-G.; visualization, J.I.V.-R. and R.G.-G.; supervision, J.M.R. All authors have read and agreed to the published version of the manuscript.

Funding: Valderrama-Rodríguez was funded by CONACYT, the Mexican National Council of Science and Technology, grant number 458523, to pursue a M.Sc. degree.

Data Availability Statement: The complete thesis written for this study, Valderrama-Rodríguez [24], is publicly available in http://repositorio.ugto.mx/handle/20.500.12059/2643 (accessed on 29 October 2021).

Acknowledgments: The first author thanks the Mexican National Council of Science and Technology (CONACyT) for support through a scholarship, grant number 458523, to pursue an M.Sc. degree at the Universidad of Guanajuato. The authors thank CONACyT, and the Universidad of Guanajuato, for their continuous support.

Conflicts of Interest: The authors declare no conflict of interest. The funders had no role in the design of the study; in the collection, analyses, or interpretation of data; in the writing of the manuscript, or in the decision to publish the results.

Appendix A. Secondary Rotation Axis as a Linear Combination of Two Primary Rotation Centers

This appendix shows the process of expressing a secondary rotation center as a linear combination of two primary centers.

Proposition A1. *Let O_{jk} be a secondary center that, according to the Aronhold–Kennedy theorem, is collinear with the primary centers O_{ij} and O_{ik}. Then, their position vectors are related by:*

$$\mathbf{r}_{jk/O} = \lambda\, \mathbf{r}_{ik/O} + (1-\lambda)\, \mathbf{r}_{ij/O}. \tag{A1}$$

where $\lambda \in \mathbf{R}$.

Proof. Assume that O_{jk} is a secondary center that, according to the Aronhold–Kennedy theorem, is collinear with the primary centers O_{ij} and O_{ik}. Then, the two possible configurations of the three centers are:

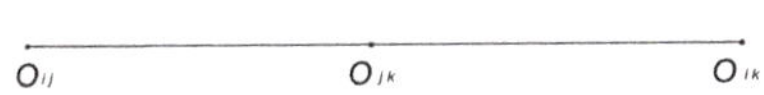

Figure A1. A first configuration of the three centers.

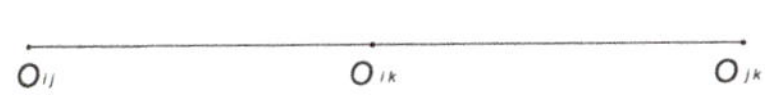

Figure A2. A second configuration of the three centers.

In both cases, the following relationship holds true:

$$\mathbf{r}_{jk/O} - \mathbf{r}_{ij/O} = \lambda(\mathbf{r}_{ik/O} - \mathbf{r}_{ij/O}) \tag{A2}$$

In the first configuration, Figure A1, $0 < \lambda < 1$, whereas in the second configuration, Figure A2, $\lambda > 1$. In any case, it follows that:

$$\mathbf{r}_{jk/O} = \lambda\, \mathbf{r}_{ik/O} + (1-\lambda)\, \mathbf{r}_{ij/O}$$

It can be observed that the position vector of the secondary rotation center can be written as a linear combination of the primary centers. Therefore, the location of the secondary rotation center can be expressed in terms of a unique variable λ, since $\mathbf{r}_{ik/O}$ and $\mathbf{r}_{ij/O}$ are known. This result finishes the proof. A procedure implemented in Maple© computes all the instantaneous secondary centers that fulfill this condition. □

References

1. Aronhold, S. Gründzuge der Kinematischen Geometrie. *Verh. Ver. Z. Beförderung des Gewerbefleiss Preuss.* **1872**, *51*, 129–155.
2. Kennedy, A.B.W. *The Mechanics of Machinery*; Macmillan: London, UK, 1886; pp. 60–83.
3. Shigley, J.E.; Uicker, J.J., Jr. *Theory of Machines and Mechanisms*; McGraw-Hill Inc.: Singapore, 1980.
4. Klein, A.W. *Kinematics of Machinery*; Mc-Graw Hill Book Co.: New York, NY, USA, 1917.
5. Modrey, J. Analysis of Complex Kinematic Chains with Influence Coefficients. *J. Appl. Mech.* **1959**, *26*, 184–188. [CrossRef]
6. Goodman, T.P. Discussion to the paper Analysis of Complex Kinematic Chains with Influence Coefficients. *J. Appl. Mech.* **1960**, *27*, 215–216. [CrossRef]
7. Yan, H.S.; Hsu, M.H. An Analytical Method for Locating Instantaneous Velocity Centers. In *DE-Vol.47 Flexible Mechanisms, Dynamics and Analysis, Proceedings of the ASME 1992 Design Technical Conferences, 22nd Biennial Mechanisms Conference, Scottsdale, AZ, USA, 13–16 September 1992*; Kinzel, G., Reinholtz, C., Eds.; ASME: New York, NY, USA, 1992; pp. 353–359.
8. Kim, M.; Han, M.S.; Seo, T.; Lee, J.W. A new instantaneous centers analysis methodology for planar closed chains via graphical representation. *Int. J. Control Autom. Syst.* **2016**, *14*, 1528–1534. [CrossRef]
9. Valderrama-Rodríguez, J.I.; Rico, J.M.; Cervantes-Sánchez, J.J. A general method for the determination of the instantaneous screw axes of one-degree-of-freedom spatial mechanisms. *Mech. Sci.* **2020**, *11*, 91–99. [CrossRef]
10. Di Gregorio, R. A novel method for the singularity analysis of planar mechanisms with more than one degree of freedom. *Mech. Mach. Theory* **2009**, *44*, 83–102. [CrossRef]
11. Zarkandi, S. Application of mechanical advantage and instant centers on singularity analysis of single-DOF planar mechanisms. *Trans. Mech. Eng. Div. Inst. Eng. Bangladesh* **2010**, *41*, 50–57. [CrossRef]
12. Foster, D.E.; Pennock, G.R. Graphical methods to locate the secondary instant centers of single-degree-of-freedom indeterminate linkages. *J. Mech. Des.* **2003**, *125*, 268–274. [CrossRef]
13. Di Gregorio, R. An algorithm for analytically calculating the positions of the secondary instant centers of indeterminate linkages. *J. Mech. Des.* **2008**, *130*, 1–9. [CrossRef]
14. Kung, C.M.; Wang, L.C.T. Analytical method for locating the secondary instant centers of indeterminate planar linkages. *Proc. Inst. Mech. Eng. Part C J. Mech. Eng. Sci.* **2009**, *223*, 491–502. [CrossRef]
15. Chang, Y.P.; Her, C. Locating instant centers of kinematically indeterminate linkages. *J. Mech. Des.* **2008**, *130*, 062304. [CrossRef]
16. Liu, Z.; Chang, Y.P. The virtual cam method application and verification on locating key instant centers of ten-bars 1-DOF indeterminate linkages. In Proceedings of the ASME 2017 International Mechanical Engineering Congress and Exposition, Tampa, FL, USA, 3–9 November 2017; ASME: New York, NY, USA, 2017.
17. Liu, Z.; Chang, Y.P. The virtual cam-hexagon method authentication on locating key instant centers of all planar single degree of freedom kinematically indeterminate linkages up to ten-bar. In Proceedings of the ASME 2019 International Mechanical Engineering Congress and Exposition, Pittsburgh, PA, USA, 9–15 November 2017; ASME: New York, NY, USA, 2017.
18. Valderrama-Rodríguez, J.I.; Rico, J.M.; Cervantes-Sánchez, J.J. A screw theory approach to compute instantaneous rotation axes of indeterminate spherical linkages. *Mech. Based Des. Struct. Mach.* **2020**, 1–41. [CrossRef]
19. Ball, R.S. *A Treatise on the Theory of Screws*; Cambridge University Press: Cambridge, UK, 1900.
20. Rico, J.M.; Gallardo-Alvarado, J.; Duffy, J. Screw theory and higher order kinematic analysis of open serial and closed chains. *Mech. Mach. Theory* **1999**, *34*, 559–586. [CrossRef]
21. Brand, L. *Vector and Tensor Analysis*; John Wiley and Sons: New York, NY, USA, 1947.
22. Rico, J.M.; Duffy, J. Orthogonal spaces and screw systems. *Mech. Mach. Theory* **1992**, *27*, 451–458. [CrossRef]
23. Zhang, L.; Zhao, Y.; Ren, J. Screw rolling between moving and fixed axodes traced by lower-mobility parallel mechanism. *Mech. Mach. Theory* **2021**, *163*, 104354. [CrossRef]
24. Valderrama-Rodríguez, J.I. Localización de los Ejes de Tornillos Instantáneos de Mecanismos Espaciales y sus Aplicaciones. Master's Thesis, Universidad de Guanajuato, Guanajuato, Mexico, December 2018.

robotics

MDPI

Article

Dimensional Synthesis of a Novel 3-URU Translational Manipulator Implemented through a Novel Method

Raffaele Di Gregorio

LaMaViP, Department of Engineering, University of Ferrara, 44121 Ferrara, Italy; raffaele.digregorio@unife.it; Tel.: +39-0532-974828

Abstract: A dimensional synthesis of parallel manipulators (PMs) consists of determining the values of the geometric parameters that affect the platform motion so that a useful workspace with assigned sizes can be suitably located in a free-from-singularity region of its operational space. The main goal of this preliminary dimensioning is to keep the PM far enough from singularities to avoid high internal loads in the links and guarantee a good positioning precision (i.e., for getting good kinematic performances). This paper presents a novel method for the dimensional synthesis of translational PMs (TPMs) and applies it to a TPM previously proposed by the author. The proposed method, which is based on Jacobians' properties, exploits the fact that TPM parallel Jacobians are block diagonal matrices to overcome typical drawbacks of indices based on Jacobian properties. The proposed method can be also applied to all the lower-mobility PMs with block diagonal Jacobians that separate platform rotations from platform translations (e.g., parallel wrists).

Keywords: machine design; dimensional synthesis; parallel manipulator; useful workspace; performance index; kinetostatics

Citation: Di Gregorio, R. Dimensional Synthesis of a Novel 3-URU Translational Manipulator Implemented through a Novel Method. *Robotics* **2022**, *11*, 10. https://doi.org/10.3390/robotics11010010

Academic Editor: Giuseppe Carbone

Received: 19 November 2021
Accepted: 29 December 2021
Published: 5 January 2022

1. Introduction

Designing the mechanical structure of a general-purpose machine, like a manipulator, has three mandatory steps that transform the original intuition of the designer into a real device: type synthesis [1–4], dimensional synthesis [1,2,4], and machine-element design [5–7]. The type synthesis identifies the most suitable topology (machine type) that can meet the motion requirements. The dimensional synthesis determines the values of the geometric parameters that affect the motion of the previously selected machine type. Eventually, the machine-element design determines the materials and actual sizes of the machine components that make them carry the nominal loads applied to the machine.

In the parallel-manipulator (PM) design, the dimensional synthesis [8–10] consists of determining the values of the above-mentioned geometric parameters (i.e., in this case, the ones that affect the platform motion) so that a useful workspace with assigned sizes can be located in a region of the operational space that is far from singularities. Indeed, satisfying this condition makes it possible to avoid high internal loads in the links and to guarantee a good positioning precision (i.e., to get good kinematic performances). This type of preliminary dimensioning is usually addressed [10–13] by using condition number [14–16] or manipulability [17] or transmission indices [18–21] as objective functions to optimize.

Condition number and manipulability, which are defined through the Jacobians appearing in the input–output instantaneous relationship (i.e., the linear mapping that analytically states a relationship between the actuated-joint rates (input) and the end-effector (platform) twist (output)), can identify the best performances at a given PM configuration (i.e., they are local indices) that are referable to this relationship. The duality between statics and kinematics makes these indices also carry pieces of information usually related to transmission indices [11,22,23]. Unfortunately, if the input and/or the output variables are not dimensionally homogeneous the proposed indices are not usable in their original form, since they refer to vector norms of the inputs and the outputs. Thus, in these cases, either some difficult-to-determine constants that homogenize the variables by taking into account the platform sizes must be included or some other analytic tricks must be conceived [11,16,24–27]. In addition, since, in general, these indices are not related to geometric or physics interpretations, even though they are able to identify configuration that are far from singularities, they are not able to quantify how far they are, which makes them difficult to use in a design context.

Transmission indices are extensions of the transmission/pressure angles of linkages [28,29]. Such extensions are based on the concept of virtual coefficient [30], which is defined as the virtual power delivered by a unit transmission wrench (transmission wrench screw (TWS)) on the corresponding unit output twist (output twist screw (OTS)) of the output link. They are defined as the ratio of the configuration-dependent (local) virtual coefficient and its maximum value obtained when the TWS is (virtually) rotated about a suitable characteristic point placed on the TWS axis. This definition has the merit of providing dimensionless indices (i.e., there is no homogeneity issue) that are frame independent, and the drawback of depending on the choices both of the characteristic point and of the adopted procedure for computing the maximum virtual coefficient. Moreover, despite the many definitions and extensions [18–21], their actual meanings when used in a design optimization still need further investigations.

With reference to the input–output instantaneous relationship, singularities are manipulator configurations where the one-to-one correspondence between actuated-joint rates and platform twist fails. PMs' singularities are mainly collected into three groups [31,32]: type-I (serial) singularities, where the actuated-joint rates are indeterminate even if the platform is locked, type-II (parallel) singularities, where the platform twist is indeterminate even if the actuated joints are locked, and type-III singularities, where both the actuated-joint rates and the platform twist are indeterminate. The farther is the manipulator configuration from singularities, the higher are its positioning precision and the reduction of the internal loads in its components.

In lower-mobility PMs (LPMs), constraint singularities [33] may be present among their type-II singularities. Such singularities occur in LPMs where the connectivity of the kinematic chains (limbs)[1] that in parallel join the platform to the frame (base) is higher than the degrees of freedom (DOF) of the LPM. The platform can change its motion type when the LPM is at a constraint singularity. Translational PMs (TPMs) are LPMs where, out of constraint singularities, the platform can only translate with respect to the base; accordingly, at a constraint singularity, the platform of a TPM may rotate. Evaluating the kinetostatic performances of TPMs must take into account the possible presence of constraint singularities (also named rotation singularities [35] in TPMs). TPM performances have been evaluated by using about all the above-mentioned types of indices (see [36,37] for Refs.).

Here, a novel method for the dimensional synthesis of TPMs is presented and applied to a TPM previously proposed by this author [38]. The proposed method, which is based on Jacobians' properties, exploits the fact that TPM parallel Jacobians are block diagonal matrices [35] to overcome typical drawbacks of indices based on Jacobian properties. In particular, the proposed method introduces three novel indices: two of them are dimensionless and have clear geometric and static meanings and the remaining third rates the quality of the load transmission from the actuators to the platform. Consequently, the proposed methodology does not have homogeneity issues and indices' minimum values to adopt during design come easily out from simple static and geometric considerations. The proposed method can be also applied to all the lower-mobility PMs with block diagonal Jacobians that separate platform rotations from platform translations (e.g., parallel wrists).

1 According to [34], here, the term "limb connectivity" denotes the DOF number the platform would have if it were connected to the base only through that limb.

The paper is organized as follows. Section 2 presents the adopted methodology. Section 3 presents the previously proposed TPM and applies the novel method to its dimensional synthesis. Then, Section 4 discusses the obtained results, and Section 5 draws the conclusions.

2. Materials and Methods

The canonic form of the input–output instantaneous relationship of a non-redundant PM is [31]

$$A_{6\times 6}\hat{\$} = B_{6\times n}\dot{\mathbf{q}} \tag{1}$$

in which $A_{6\times 6}$ is the 6 × 6 Jacobian that multiplies the platform twist, $\hat{\$} = (\dot{\mathbf{p}}^{T}, \boldsymbol{\omega}^{T})^{T}$, where $\dot{\mathbf{p}}$ is the velocity of a reference platform point and $\boldsymbol{\omega}$ is platform's angular velocity; whereas, $B_{6\times n}$ is the 6 × n Jacobian that multiplies the n-tuple, $\dot{\mathbf{q}} = (\dot{q}_1, \ldots, \dot{q}_n)^{T}$, collecting all the actuated-joint rates, where $n \le 6$ is the number of PM's DOF. Both the Jacobians depend on the PM configurations.

With reference to Equation (1), the PM configurations that make the determinant of $A_{6\times 6}$ equal to zero (in formula: det($A_{6\times 6}$) = 0) are type-II (parallel) singularities. The PM configurations that make the rank of $B_{6\times n}$ lower than n (in formula: rank($B_{6\times n}$) $\le n$) are type-I (serial) singularities; whereas, those that simultaneously make det($A_{6\times 6}$) = 0 and rank($B_{6\times n}$) $\le n$ are type-III singularities. In LPMs (i.e., if $n < 6$), by using the Gauss–Jordan elimination [39], Equation (1) can always be transformed in the following canonic form

$$\begin{bmatrix} V_{n\times n} & T_{n\times(6-n)} \\ 0_{(6-n)\times n} & H_{(6-n)\times(6-n)} \end{bmatrix} \begin{pmatrix} \xi_{n\times 1} \\ \xi_{(6-n)\times 1} \end{pmatrix} = \begin{bmatrix} G_{n\times n} \\ F_{(6-n)\times n} \end{bmatrix} \dot{\mathbf{q}} \tag{2}$$

which, if the submatrix $H_{(6-n)\times(6-n)}$ is not singular (i.e., if det($H_{(6-n)\times(6-n)}$) ≠ 0), can be further reduced to the form

$$\begin{bmatrix} V_{n\times n} & 0_{n\times(6-n)} \\ 0_{(6-n)\times n} & H_{(6-n)\times(6-n)} \end{bmatrix} \begin{pmatrix} \xi_{n\times 1} \\ \xi_{(6-n)\times 1} \end{pmatrix} = \begin{bmatrix} G_{n\times n} - T_{n\times(6-n)} H^{-1}_{(6-n)\times(6-n)} F_{(6-n)\times n} \\ F_{(6-n)\times n} \end{bmatrix} \dot{\mathbf{q}} \tag{3}$$

where $(\cdot)_{i\times j}$ and $0_{i\times j}$ denote, respectively, an $i \times j$ matrix and an $i \times j$ null matrix, $\xi_{n\times 1}$ is a n-tuple collecting a suitable selection of $\hat{\$}$'s entries, and $\xi_{(6-n)\times 1}$ is a (6–n)-tuple collecting the remaining entries of $\hat{\$}$.

The condition det($H_{(6-n)\times(6-n)}$) = 0, which analytically forbids the transformation of system (2) into system (3), identifies configurations where $\xi_{(6-n)\times 1}$ becomes indeterminate; whereas, the condition det($V_{n\times n}$) = 0 identifies configurations where $\xi_{n\times 1}$ becomes indeterminate. Both these two conditions identify LPM configurations where the platform locally either gains additional DOFs or, without gaining further DOFs, is not controllable by the limbs any longer (i.e., they are type-II singularities). In general, all the non-null submatrices appearing in Equations (2) and (3) depend on the manipulator configuration. Nevertheless, when the LPM constrains the platform motion to belong to one displacement subgroup (e.g., translational or spherical or planar, etc., motion type) of the displacement group [40,41], the submatrix $F_{(6-n)\times n}$ is always a null matrix (in formula: $F_{(6-n)\times n} = 0_{(6-n)\times n}$). In this case, $\xi_{(6-n)\times 1}$ collects $\hat{\$}$'s entries that are forbidden in the displacement subgroup the LPM refers to. Consequently, the condition det($H_{(6-n)\times(6-n)}$) = 0 identifies constraint singularities, that is, LPM configurations where the platform instantaneous motion may change its type.

In TPMs (n = 3), Equation (3) becomes

$$\begin{bmatrix} V_{3\times 3} & 0_{3\times 3} \\ 0_{3\times 3} & H_{3\times 3} \end{bmatrix} \begin{pmatrix} \dot{\mathbf{p}} \\ \boldsymbol{\omega} \end{pmatrix} = \begin{bmatrix} G_{3\times 3} \\ 0_{3\times 3} \end{bmatrix} \dot{\mathbf{q}} \tag{4}$$

which is a particular case of Equation (1), where the Jacobian $A_{6\times6}$ is a block diagonal matrix, and can be split into the two smaller subsystems

$$V_{3\times3}\dot{\mathbf{p}} = G_{3\times3}\dot{\mathbf{q}} \quad (5a)$$

$$H_{3\times3}\boldsymbol{\omega} = \boldsymbol{0}_{3\times1} \quad (5b)$$

In addition, since the platform translates, TPM configurations can be identified through the position vector $\mathbf{p} = (x, y, z)^T$ that collects the coordinates, measured in a Cartesian system fixed to the base, of the reference platform point the velocity $\dot{\mathbf{p}}$ refers to. Consequently, the entries of the matrices $V_{3\times3}$, $G_{3\times3}$, and $H_{3\times3}$ can be written as functions of $\dot{\mathbf{p}}$ and of the TPM's geometric constants.

Let $\mathbf{v}_i$ ($\mathbf{h}_i$), for i = 1, 2, 3, denote the tridimensional vector collecting the entries of the *i-th* row of $V_{3\times3}$ ($H_{3\times3}$) so that $V_{3\times3} = [\mathbf{v}_1, \mathbf{v}_2, \mathbf{v}_3]^T$ ($H_{3\times3} = [\mathbf{h}_1, \mathbf{h}_2, \mathbf{h}_3]^T$), system (5a), and system (5b) can always be multiplied, respectively, by the diagonal matrices D_v and D_h defined as follows:

$$D_v = \mathrm{diag}\left(\|\mathbf{v}_1\|^{-1}, \|\mathbf{v}_2\|^{-1}, \|\mathbf{v}_3\|^{-1}\right),\ D_h = \mathrm{diag}\left(\|\mathbf{h}_1\|^{-1}, \|\mathbf{h}_2\|^{-1}, \|\mathbf{h}_3\|^{-1}\right) \quad (6)$$

Such matrix products transform systems (5a) and (5b) into the equivalent form

$$V^*_{3\times3}\dot{\mathbf{p}} = D_v G_{3\times3}\dot{\mathbf{q}} \quad (7a)$$

$$H^*_{3\times3}\boldsymbol{\omega} = 0_{3\times1} \quad (7b)$$

where $V^*_{3\times3} = D_v V_{3\times3}$ and $H^*_{3\times3} = D_h H_{3\times3}$ are matrices whose entries are dimensionless and whose rows are unit vectors. With reference to systems (7a) and (7b), the TPM's analytic conditions that identify its singularities are

$$\det(H^*_{3\times3}) = \frac{\mathbf{h}_1 \cdot (\mathbf{h}_2 \times \mathbf{h}_3)}{\|\mathbf{h}_1\|\|\mathbf{h}_2\|\|\mathbf{h}_3\|} = 0 \quad (8a)$$

$$\det(V^*_{3\times3}) = \frac{\mathbf{v}_1 \cdot (\mathbf{v}_2 \times \mathbf{v}_3)}{\|\mathbf{v}_1\|\|\mathbf{v}_2\|\|\mathbf{v}_3\|} = 0 \quad (8b)$$

$$\det(D_v G_{3\times3}) = \frac{\det(G_{3\times3})}{\|\mathbf{v}_1\|\|\mathbf{v}_2\|\|\mathbf{v}_3\|} = 0 \quad (8c)$$

Condition (8a) (condition (8b)) refers to type-II singularities that are (are not) constraint singularities and imposes the coplanarity of three unit vectors, that is, it is also a geometric condition; whereas, condition (8c) refers to type-I singularities.

From a static point of view, the fact that, out of constraint singularities, the unit vectors, $\mathbf{h}_i/\|\mathbf{h}_i\|$, for i = 1, 2, 3, individuate directions around which the platform cannot rotate (see Equation (7b)) means that the platform's constraints due to the limbs generate torques parallel to those directions. Analogously, when the actuated joints are locked, the right-hand side of Equation (7a) becomes a null vector (i.e., Equation (7a) becomes similar to Equation (7b)) and the unit vectors, $\mathbf{v}_i/\|\mathbf{v}_i\|$, for i = 1, 2, 3, individuate directions along which, out of type-II singularities, the platform reference point cannot translate. Consequently, the platform's constraints due to the limbs also generate forces parallel to those directions and with lines of action passing through the platform reference point. This conclusion implies that the generic entry, t_{ij}, of matrix $D_v G_{3\times3}$ somehow has the meaning of transmission coefficient between the axis of the j-th actuated joint and the line with the direction of $\mathbf{v}_i/\|\mathbf{v}_i\|$ that passes through the reference platform point.

The above reported considerations allow the use of the left-hand sides of Equation (8) as measures of the TPM's kinetostatic performances at a given configuration by defining the following local performance indices:

$$\mathrm{k_h} = \frac{|\mathbf{h}_1 \cdot (\mathbf{h}_2 \times \mathbf{h}_3)|}{\|\mathbf{h}_1\|\|\mathbf{h}_2\|\|\mathbf{h}_3\|},\ \mathrm{k_v} = \frac{|\mathbf{v}_1 \cdot (\mathbf{v}_2 \times \mathbf{v}_3)|}{\|\mathbf{v}_1\|\|\mathbf{v}_2\|\|\mathbf{v}_3\|},\ \mathrm{k_g} = \frac{|\det(G_{3\times 3})|}{\|\mathbf{v}_1\|\|\mathbf{v}_2\|\|\mathbf{v}_3\|}. \tag{9}$$

Indeed, $\mathrm{k_h}$ and $\mathrm{k_v}$ are absolute values of mixed products of three unit vectors; consequently, they are dimensionless and range from 0 to 1 with 0 that identifies singular configurations and 1 that identifies the farthest-from-singular configurations. Differently, $\mathrm{k_g}$ is not negative and its minimum value, 0, identifies type-I singularities; in general, it has not a maximum value and is not dimensionless. Anyway, it can be stated that the higher $\mathrm{k_g}$ is the better the motion/force transmission is with the chosen limb types.

Accordingly, the dimensional synthesis of a TPM can be implemented by sizing it so that a given useful workspace can be located in a region of the operational space where $\mathrm{k_h}$, $\mathrm{k_v}$, and $\mathrm{k_g}$ are higher than assigned minimum values, that is, where the following inequalities are satisfied:

$$\mathrm{k_h}(\mathbf{p}) \geq \mathrm{k_{h,min}},\ \mathrm{k_v}(\mathbf{p}) \geq \mathrm{k_{v,min}},\ \mathrm{k_g}(\mathbf{p}) \geq \mathrm{k_{g,min}}. \tag{10}$$

The next section illustrates the application of this procedure to the dimensional synthesis of a TPM recently presented by the author [38], hereafter named LaMaViP 3-URU (Figure 1).

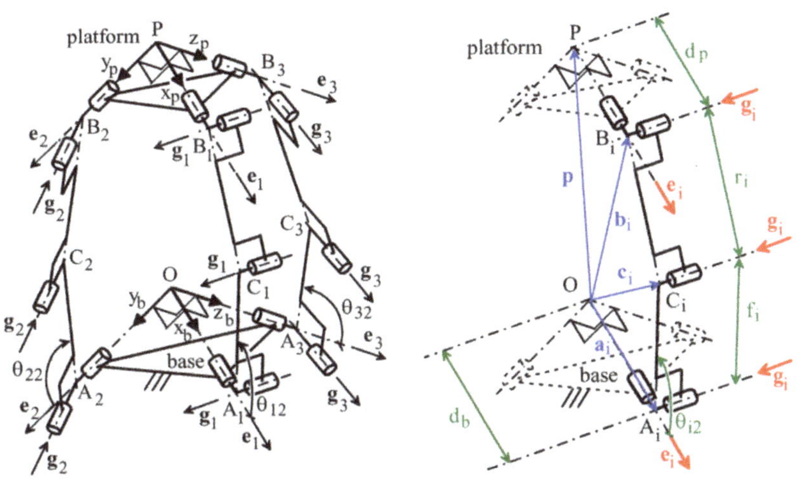

Figure 1. LaMaViP 3-URU: (**a**) overall scheme and notations, (**b**) detailed scheme of the *i-th* limb (figure reproduced from [38]).

3. Results

TPMs of the 3-URU type [35] feature three equal limbs constituted of two links, one adjacent to the base and the other to the platform, joined to one another through a revolute (R)-pair and to the platform or to the base through a universal (U)-joint, that is, they are of the URU type. Since each U-joint is constituted of two R-pairs with axes mutually perpendicular and with a common intersection point (points $\mathrm{A_i}$ and $\mathrm{B_i}$ in Figure 1b), an URU limb contains five R-pairs in series: two at the endings (one adjacent to the base and the other adjacent to the platform) and three intermediate. If, in each limb (see Figure 1), the axes of the two R-pairs at the endings are parallel to one another and the axes of the three intermediate R-pairs are all parallel and, of course, perpendicular to the axes of the two R-pairs at the endings, the platform is constrained to translate when it is out of the constraint singularities [35].

LaMaViP 3-URU (Figure 1) is a particular geometry of a TPM family recently presented in an international patent by the author (see Section 6 for Refs.). The peculiarities of this geometry are (see Figure 1):

(i). the axes of the three R-pairs (one for each limb) adjacent to the base are mutually perpendicular and share a common intersection (point O in Figure 1);

(ii). the axes of the three R-pairs (one for each limb) adjacent to the platform are mutually perpendicular and share a common intersection (point P in Figure 1);

(iii). in each URU limb, the actuated R-pair is the one not adjacent to the base in the U-joint adjacent to the base, but the actuator is located on the base.

The singularity analysis of the LaMaViP 3-URU has been presented in [38] and its position analysis has been solved in [42]. With reference to Figure 1, the following notations are introduced: $Ox_by_bz_b$ ($Px_py_pz_p$) is a Cartesian reference fixed to the base (to the platform), and $\mathbf{e}_1$, $\mathbf{e}_2$, and $\mathbf{e}_3$ are unit vectors of the coordinate axes x_b, y_b, and z_b (x_p, y_p, and z_p), respectively, and, at the same time, unit vectors of the three R-pair axes fixed to the base (to the platform). Furthermore, A_i (B_i) for $i = 1, 2, 3$ are the centers of the U-joints adjacent to the base (to the platform). In the *i-th* limb, $i = 1, 2, 3$, without losing generality [43], the points A_i and B_i are assumed to lie on the same plane perpendicular to the axes of the three intermediate R-pairs; such a plane intersects at C_i the axis of the R-pair between the two U-joints. In addition, $\mathbf{g}_i$, i = 1, 2, 3, is the unit vector parallel to the axes of the three intermediate R-pairs of the *i-th* limb.

Moreover, the following definition/choices are introduced: $d_p = B_1P = B_2P = B_3P$; $d_b = A_1O = A_2O = A_3O$; $f_i = A_iC_i$; and $r_i = B_iC_i$, for $i = 1, 2, 3$. In each URU limb, the five R-pairs are numbered with an index, j, that increases by moving from the base toward the platform; the actuated joint is the second R-pair. The angle θ_{ij}, for $i = 1, 2, 3$, and $j = 1, \ldots, 5$, is the joint variable of the j-th R-pair of the i-th limb; the actuated-joint variables are the angles θ_{i2}, $i = 1, 2, 3$ (see Figure 1). In addition, the phase reference of the angles θ_{i1}, $i = 1, 2, 3$, are given by the relationships (see Figure 1):

$$\mathbf{g}_1 = \cos\theta_{11}\,\mathbf{e}_2 + \sin\theta_{11}\,\mathbf{e}_3,\ \mathbf{g}_2 = -\cos\theta_{21}\,\mathbf{e}_1 + \sin\theta_{21}\,\mathbf{e}_3,\ \mathbf{g}_3 = \cos\theta_{31}\,\mathbf{e}_1 + \sin\theta_{31}\,\mathbf{e}_2$$

The introduced notations yield the following relationships (Figure 1):

$$\mathbf{p} = (P - O) = x\mathbf{e}_1 + y\mathbf{e}_2 + z\mathbf{e}_3$$

$$\mathbf{a}_i = (A_i - O) = d_b\mathbf{e}_i,\ \mathbf{b}_i = (B_i - O) = \mathbf{p} + d_p\mathbf{e}_i,\ \mathbf{c}_i = (C_i - O) = \mathbf{a}_i + f_i\mathbf{u}_i, \quad i = 1, 2, 3$$

where x, y, and z are the coordinates of the platform reference point (point P in Figure 1) measured in $Ox_by_bz_b$, and

$$\mathbf{u}_i = \frac{(C_i - A_i)}{f_i} = \frac{(\mathbf{c}_i - \mathbf{a}_i)}{f_i}.\ \ i = 1,\ 2,\ 3$$

With reference to Equations (5a) and (5b), this author demonstrated [38] that, for the LaMaViP 3-URU, the following relationships hold

$$\mathbf{h}_i = \mathbf{g}i \times \mathbf{e}_i,\ \mathbf{v}_i = \frac{(B_i - C_i)}{r_i} = \frac{(\mathbf{b}_i - \mathbf{c}_i)}{r_i},\ \ i = 1,\ 2,\ 3 \tag{11a}$$

$$G_{3\times3} = \mathrm{diag}([\mathbf{g}_1 \times (\mathbf{b}_1 - \mathbf{a}_1)] \cdot \mathbf{v}_1, [\mathbf{g}_2 \times (\mathbf{b}_2 - \mathbf{a}_2)] \cdot \mathbf{v}_2, [\mathbf{g}_3 \times (\mathbf{b}_3 - \mathbf{a}_3)] \cdot \mathbf{v}_3) \tag{11b}$$

which, since the vectors defined by Equation (11a) are unit vectors, bring one to conclude that $\boldsymbol{D}_v$ and $\boldsymbol{D}_h$ are both 3 × 3 identity matrices. As a consequence, for the LaMaViP 3-URU, $\mathbf{H}^*_{3\times3} = \mathbf{H}_{3\times3}$, $\mathbf{V}^*_{3\times3} = \mathbf{V}_{3\times3}$ and the above-defined indices become (see Figure 1b)

$$k_h = |\mathbf{h}_1 \cdot (\mathbf{h}_2 \times \mathbf{h}_3)| = |(\mathbf{g}_1 \times \mathbf{e}_1) \cdot [(\mathbf{g}_2 \times \mathbf{e}_2) \times (\mathbf{g}_3 \times \mathbf{e}_3)]| \tag{12a}$$

$$k_v = |\mathbf{v}_1 \cdot (\mathbf{v}_2 \times \mathbf{v}_3)| = \frac{|(\mathbf{b}_1 - \mathbf{c}_1) \cdot [(\mathbf{b}_2 - \mathbf{c}_2) \times (\mathbf{b}_3 - \mathbf{c}_3)]|}{r_1 r_2 r_3} \tag{12b}$$

$$\begin{aligned} k_g &= |[\mathbf{g}_1 \times (\mathbf{b}_1 - \mathbf{a}_1)]\cdot\mathbf{v}_1||[\mathbf{g}_2 \times (\mathbf{b}_2 - \mathbf{a}_2)]\cdot\mathbf{v}_2||[\mathbf{g}_3 \times (\mathbf{b}_3 - \mathbf{a}_3)]\cdot\mathbf{v}_3| = \\ &= \frac{|[(\mathbf{b}_1 - \mathbf{a}_1) \times (\mathbf{b}_1 - \mathbf{c}_1)]\cdot\mathbf{g}_1||[(\mathbf{b}_2 - \mathbf{a}_2) \times (\mathbf{b}_2 - \mathbf{c}_2)]\cdot\mathbf{g}_2||[(\mathbf{b}_1 - \mathbf{a}_1) \times (\mathbf{b}_1 - \mathbf{c}_1)]\cdot\mathbf{g}_1|}{r_1 r_2 r_3} = \\ &= \frac{|[(\mathbf{c}_1 - \mathbf{a}_1) \times (\mathbf{b}_1 - \mathbf{c}_1)]\cdot\mathbf{g}_1||[(\mathbf{c}_2 - \mathbf{a}_2) \times (\mathbf{b}_2 - \mathbf{c}_2)]\cdot\mathbf{g}_2||[(\mathbf{c}_1 - \mathbf{a}_1) \times (\mathbf{b}_1 - \mathbf{c}_1)]\cdot\mathbf{g}_1|}{r_1 r_2 r_3} = \\ &= f_1 f_2 f_3 |\sin\theta_{13}||\sin\theta_{23}||\sin\theta_{33}| \end{aligned} \tag{12c}$$

3.1. *Analytic Expression of the Indices*

The geometric expressions of the indices given by Equation (12) can be transformed into functions of $\mathbf{p}$ as follows.

3.1.1. Index k_h

The adopted notations bring one to write (see Figure 1b)

$$\mathbf{g}_i = \frac{\mathbf{e}_i \times (\mathbf{b}_i - \mathbf{a}_i)}{|\mathbf{e}_i \times (\mathbf{b}_i - \mathbf{a}_i)|} = \frac{\mathbf{e}_i \times [\mathbf{p} + (d_p - d_b)\mathbf{e}_i]}{|\mathbf{e}_i \times [\mathbf{p} + (d_p - d_b)\mathbf{e}_i]|} = \frac{\mathbf{e}_i \times \mathbf{p}}{|\mathbf{e}_i \times \mathbf{p}|} \quad i = 1, 2, 3 \tag{13a}$$

which, when introduced into the geometric definition (Equation (11a)) of $\mathbf{h}_i$, gives

$$\mathbf{h}_i = \mathbf{g}_i \times \mathbf{e}_i = \frac{(\mathbf{e}_i \times \mathbf{p}) \times \mathbf{e}_i}{|\mathbf{e}_i \times \mathbf{p}|} = \frac{\mathbf{p} - (\mathbf{e}_i \cdot \mathbf{p})\mathbf{e}_i}{|\mathbf{e}_i \times \mathbf{p}|} \quad i = 1, 2, 3 \tag{13b}$$

Equation (13b), after the introduction of the analytic expression of $\mathbf{p}$ (i.e., $\mathbf{p} = x\mathbf{e}_1 + y\mathbf{e}_2 + z\mathbf{e}_3$), becomes

$$\mathbf{h}_1 = \frac{y\mathbf{e}_2 + z\mathbf{e}_3}{\sqrt{y^2 + z^2}};\ \mathbf{h}_2 = \frac{x\mathbf{e}_1 + z\mathbf{e}_3}{\sqrt{x^2 + z^2}};\ \mathbf{h}_3 = \frac{x\mathbf{e}_1 + y\mathbf{e}_2}{\sqrt{x^2 + y^2}} \tag{14}$$

whose introduction into Equation (12a) yields the sought-after expression, that is:

$$k_h = |\mathbf{h}_1 \cdot (\mathbf{h}_2 \times \mathbf{h}_3)| = \frac{2|xyz|}{\sqrt{(x^2 + z^2)(x^2 + y^2)(y^2 + z^2)}} \tag{15}$$

3.1.2. Index k_v

The adopted notations (see Figure 1) bring the following relationships

$$(\mathbf{b}_i - \mathbf{c}_i) = r_i\,\mathbf{v}_i = \mathbf{p} + (d_p - d_b)\,\mathbf{e}_i - f_i\,(\cos\theta_{i2}\,\mathbf{e}_i + \sin\theta_{i2}\,\mathbf{h}_i) \quad i = 1, 2, 3 \tag{16}$$

which, after the introduction of the analytic expressions of $\mathbf{p}$ (i.e., $\mathbf{p} = x\mathbf{e}_1 + y\mathbf{e}_2 + z\mathbf{e}_3$) and of $\mathbf{h}_i$ (i.e., Equation (14)), become

$$\mathbf{b}_1 - \mathbf{c}_1 = [x + (d_p - d_b) - f_1 \cos\theta_{12}]\mathbf{e}_1 + [1 - f_1\, m_1 \sin\theta_{12}]\, y\mathbf{e}_2 + [1 - f_1\, m_1 \sin\theta_{12}]\, z\mathbf{e}_3 \tag{17a}$$

$$\mathbf{b}_2 - \mathbf{c}_2 = [1 - f_2\, m_2 \sin\theta_{22}]\, x\mathbf{e}_1 + [y + (d_p - d_b) - f_2 \cos\theta_{22}]\mathbf{e}_2 + [1 - f_2\, m_2 \sin\theta_{22}]\, z\mathbf{e}_3 \tag{17b}$$

$$\mathbf{b}_3 - \mathbf{c}_3 = [1 - f_3\, m_3 \sin\theta_{32}]\, x\mathbf{e}_1 + [1 - f_3\, m_3 \sin\theta_{32}]\, y\mathbf{e}_2 + [z + (d_p - d_b) - f_3 \cos\theta_{32}]\mathbf{e}_3 \tag{17c}$$

with

$$m_1 = \frac{1}{\sqrt{y^2 + z^2}},\ m_2 = \frac{1}{\sqrt{x^2 + z^2}},\ m_3 = \frac{1}{\sqrt{x^2 + y^2}} \tag{18}$$

which, when introduced into definition (12b), yield

$$k_v = \frac{|(x+q_1)[(y+q_2)(z+q_3) - n_2n_3yz] - n_1y[n_2x(z+q_3) - n_2n_3xz] + n_1z[n_2n_3xy - n_3x(y+q_2)]|}{r_1r_2r_3} \tag{19}$$

with

$$n_1 = [1 - f_1\, m_1 \sin\theta_{12}];\ n_2 = [1 - f_2\, m_2 \sin\theta_{22}];\ n_3 = [1 - f_3\, m_3 \sin\theta_{32}] \tag{20a}$$

$$q_1 = (d_p - d_b) - f_1 \cos\theta_{12};\ q_2 = (d_p - d_b) - f_2 \cos\theta_{22};\ q_3 = (d_p - d_b) - f_3 \cos\theta_{32} \tag{20b}$$

The actuated-joint variables, θ_{12}, θ_{22}, and θ_{32}, can be eliminated from Equation (20) by using the solution formulas of the inverse position analysis presented in [44], that is:

$$\begin{array}{l} \tan\left(\frac{\theta_{i2}}{2}\right)_j = \frac{2f_i\beta_i + (-1)^j\sqrt{4f_i^2(\alpha_i^2+\beta_i^2) - (\alpha_i^2+\beta_i^2+f_i^2-r_i^2)^2}}{(\alpha_i+f_i)^2+\beta_i^2-r_i^2} \\ \sin(\theta_{i2})_j = \frac{2\tan\left(\frac{\theta_{i2}}{2}\right)_j}{1+\tan^2\left(\frac{\theta_{i2}}{2}\right)_j};\ \cos(\theta_{i2})_j = \frac{1-\tan^2\left(\frac{\theta_{i2}}{2}\right)_j}{1+\tan^2\left(\frac{\theta_{i2}}{2}\right)_j}; \end{array} \quad i = 1,2,3;\ j = 0,\ 1 \tag{21}$$

where

$$\alpha_1 = x + d_p - d_b,\ \alpha_2 = y + d_p - d_b,\ \alpha_3 = z + d_p - d_b,\ \beta_1 = \sqrt{y^2+z^2},\ \beta_2 = \sqrt{x^2+z^2},\ \beta_3 = \sqrt{x^2+y^2} \tag{22}$$

and j might be limited to only one value according to the limb configuration selected when assembling the TPM.

3.1.3. Index k_g

The analysis of Figure 1b reveals that the closed polyline $OPB_iC_iA_iO$ always lies on a plane (i.e., it is a polygon) that is perpendicular to the unit vector $\mathbf{g}_i$. As a consequence, the following geometric relationships can be written

$$[\mathbf{p} + (d_p - d_b)\mathbf{e}_i]^2 = f_i^2 + r_i^2 + 2f_ir_i\cos\theta_{i3} \quad i = 1,\ 2,\ 3 \tag{23}$$

which gives

$$|\sin\theta_{i3}| = \sqrt{1 - \left\{\frac{[\mathbf{p} + (d_p - d_b)\mathbf{e}_i]^2 - f_i^2 - r_i^2}{2f_ir_i}\right\}^2} \quad i = 1,\ 2,\ 3 \tag{24}$$

whose introduction into Equation (12c) yields

$$k_g = f_1f_2f_3 \prod_{i=1,2,3} \sqrt{1 - \left\{\frac{[\mathbf{p} + (d_p - d_b)\mathbf{e}_i]^2 - f_i^2 - r_i^2}{2f_ir_i}\right\}^2} \tag{25}$$

3.2. Simulation Results

The values assumed by the indices k_h, k_v, and k_g in the free-from-singularity regions of the operational space have been computed as a preliminary computation to identify in the operational space where conditions (10) may be satisfied. Such computation is illustrated below.

3.2.1. Index k_h

Expression (15) of k_h shows that (see [38] for details) the three coordinate planes of reference $Ox_by_bz_b$ constitute the geometric locus of LaMaViP 3-URU's constraint singularities. In addition, its analysis reveals that it does not contain the geometric constants of the studied TPM and that the change of sign of any coordinate does not affect the value of k_h. The second observation brings the conclusion that the values assumed by k_h have the same pattern in every octant of reference $Ox_by_bz_b$; consequently, this analysis can be conducted in only one octant, hereafter the first octant is chosen.

Regarding the values of k_h, when the condition x = y = z, which identifies the line whose points are equally distant from the coordinate planes and the coordinate axes (i.e., from the surfaces of the constraint-singularity locus), is introduced into Equation (15), the constant value $k_h = 1/\sqrt{2} = 0.7071$ is obtained for all the points of that line provided the point O is excluded. As it is confirmed below, this value is also the maximum value that k_h can assume.

The circumferences centered at a generic point $D = (d, d, d)^T$ of the line x = y = z that lie on a plane perpendicular to that line (see Figure 2) have the following parametric equations

$$\begin{cases} x = d + r\sqrt{\frac{2}{3}}\sin\psi = d(1 + \lambda\sqrt{\frac{2}{3}}\sin\psi) \\ y = d - \frac{r}{\sqrt{2}}(\cos\psi + \frac{\sin\psi}{\sqrt{3}}) = d[1 - \frac{\lambda}{\sqrt{2}}(\cos\psi + \frac{\sin\psi}{\sqrt{3}})] \\ z = d + \frac{r}{\sqrt{2}}(\cos\psi - \frac{\sin\psi}{\sqrt{3}}) = d[1 + \frac{\lambda}{\sqrt{2}}(\cos\psi - \frac{\sin\psi}{\sqrt{3}})] \end{cases} \tag{26}$$

where r is the radius of the circumference, ψ is the parameter, and $\lambda = r/d$. The introduction of Equation (26) into Equation (15) yields

$$k_h = \frac{2\left|(1+\lambda\sqrt{\frac{2}{3}}\sin\psi)[1-\frac{\lambda}{\sqrt{2}}(\cos\psi+\frac{\sin\psi}{\sqrt{3}})][1+\frac{\lambda}{\sqrt{2}}(\cos\psi-\frac{\sin\psi}{\sqrt{3}})]\right|}{\sqrt{\left((1+\lambda\sqrt{\frac{2}{3}}\sin\psi)^2+[1+\frac{\lambda}{\sqrt{2}}(\cos\psi-\frac{\sin\psi}{\sqrt{3}})]^2\right)\left((1+\lambda\sqrt{\frac{2}{3}}\sin\psi)^2+[1-\frac{\lambda}{\sqrt{2}}(\cos\psi+\frac{\sin\psi}{\sqrt{3}})]^2\right)\left([1-\frac{\lambda}{\sqrt{2}}(\cos\psi+\frac{\sin\psi}{\sqrt{3}})]^2+[1+\frac{\lambda}{\sqrt{2}}(\cos\psi-\frac{\sin\psi}{\sqrt{3}})]^2\right)}} = f(\lambda,\psi) \tag{27}$$

which is a function that depends only on λ and ψ. Figure 3 shows the diagram of this function. Since the two parameters λ and ψ uniquely identify all the lines of the considered octant that pass through O, this result proves that each line passing through O collects points that have a constant value of k_h given by the diagram of Figure 3 provided that point O is excluded.

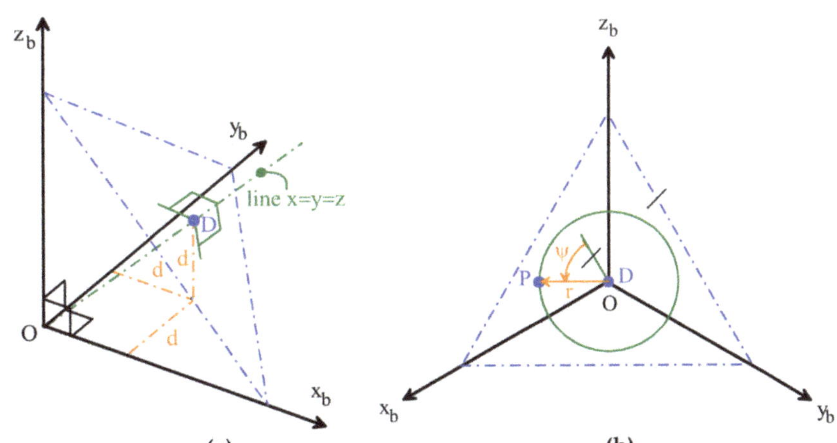

Figure 2. An octant of $Ox_by_bz_b$: (**a**) line x = y = z and plane passing through point $D = (d,d,d)^T$ perpendicular to that line, (**b**) top view along the line x = y = z containing the circumference, centered at D, with radius r, lying on the plane perpendicular to the line x = y = z.

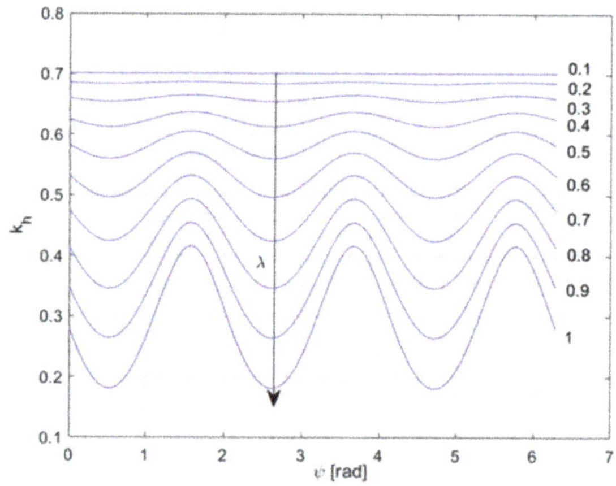

Figure 3. Diagram of the function defined by Equation (27).

3.2.2. Index k_v

Expression (19) of k_v depends on the geometric constants of the studied TPM. Such constants are the following eight: d_b, $d_p = \mu d_b$ with $\mu = d_p/d_b$, r_i and $f_i = \nu_i r_i$ with $\nu_i = f_i/r_i$ and $i = 1, 2, 3$. Since a well-sized TPM has equal limbs, hereafter, the analysis will be restricted to this case by introducing the following conditions on the geometric constants: $r_1 = r_2 = r_3 = \rho$ and $\nu_1 = \nu_2 = \nu_3 = \nu$. This restriction reduces the geometric constants to four, that is, d_b, μ, ρ, and ν.

Figure 4 shows the values assumed by k_v along the line $x = y = z = -d$, where d is the parameter of the line, for different values of ρ, μ, and ν when the limbs are all assembled so that the index j appearing in Equation (21) is equal to 0. The analysis of Figure 4 reveals that, for all the analyzed geometries, k_v reaches its maximum value (i.e., $k_v = 1$) in different positions along the line, that is, for different values, d_{max}, of the line parameter, d. In particular, d_{max} increases when ρ or μ or ν increase. In addition, Figure 4 highlights that the neighborhood, Δd, centered at d_{max}, in which k_v keeps values adequately high (e.g., greater than 0.7), increases when ρ increases and, if $\rho \geq 3d_b$, it is always wide enough for locating a useful workspace with sizes of industrial interest.

For the case $\rho = 4d_b$, Figure 5 shows the values of k_v on circumferences (see Figure 2) with radius r, centered at point $D' = (-d, -d, -d)^T$ of the line $x = y = z$, that lie on planes perpendicular to the same line. In particular, it refers to the four geometries of Figure 4 in the following three positions along the above-mentioned line: $(d/d_b) = (d_{max}/d_b) - 0.5$, $(d/d_b) = (d_{max}/d_b)$, and $(d/d_b) = (d_{max}/d_b) + 0.5$. The analysis of Figure 5 reveals that, in all the geometries, if $(r/d_{max}) \leq 0.5$, k_v is always greater than 0.75 with values that are greater than 0.95 for $(d/d_b) = (d_{max}/d_b)$ and greater than 0.86 for $(d/d_b) = (d_{max}/d_b) - 0.5$.

These results bring the conclusion that, in all the analyzed geometries, a useful workspace with the shape of a right circular cylinder having height equal to d_b and radius $r \leq 0.5d_{max}$, if it is centered at $D'_{max} = (-d_{max}, -d_{max}, -d_{max})^T$ by keeping its axis along the line $x = y = z$, will guarantee a high value of k_v everywhere inside it.

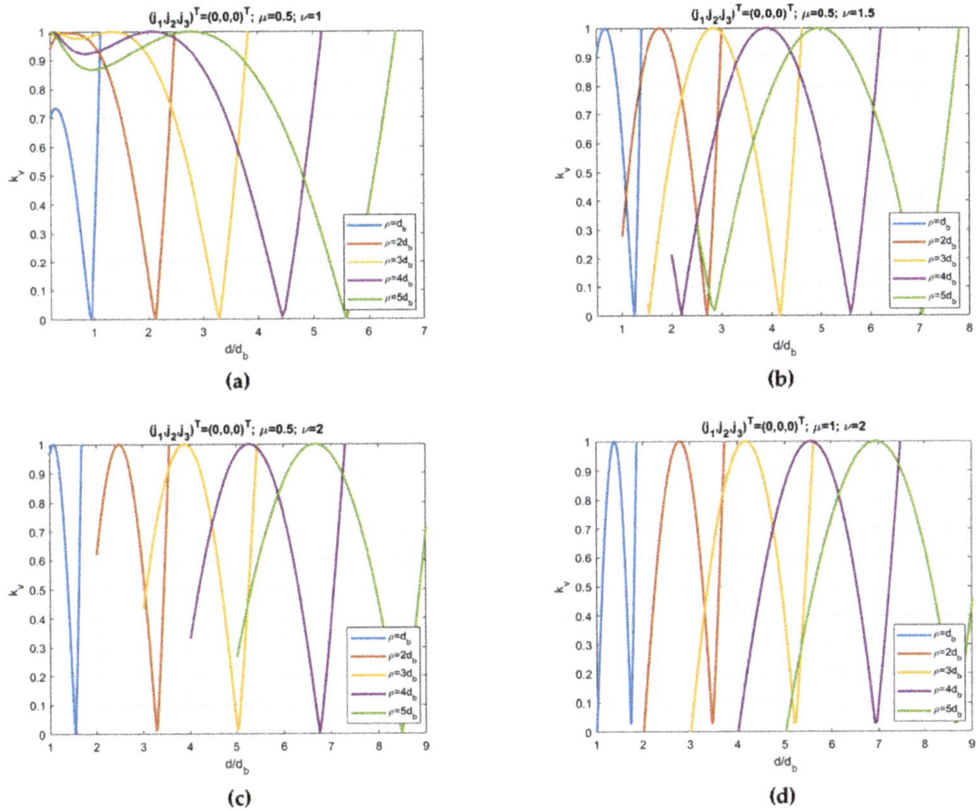

Figure 4. Values of k_V along the line x = y = z= −d for different values of ρ and limbs assembled so that the index j appearing in Equation (21) is equal to 0 in the cases: (**a**) μ = 0.5, ν = 1; (**b**) μ = 0.5, ν = 1.5; (**c**) μ = 0.5, ν = 2; (**d**) μ = 1, ν = 2.

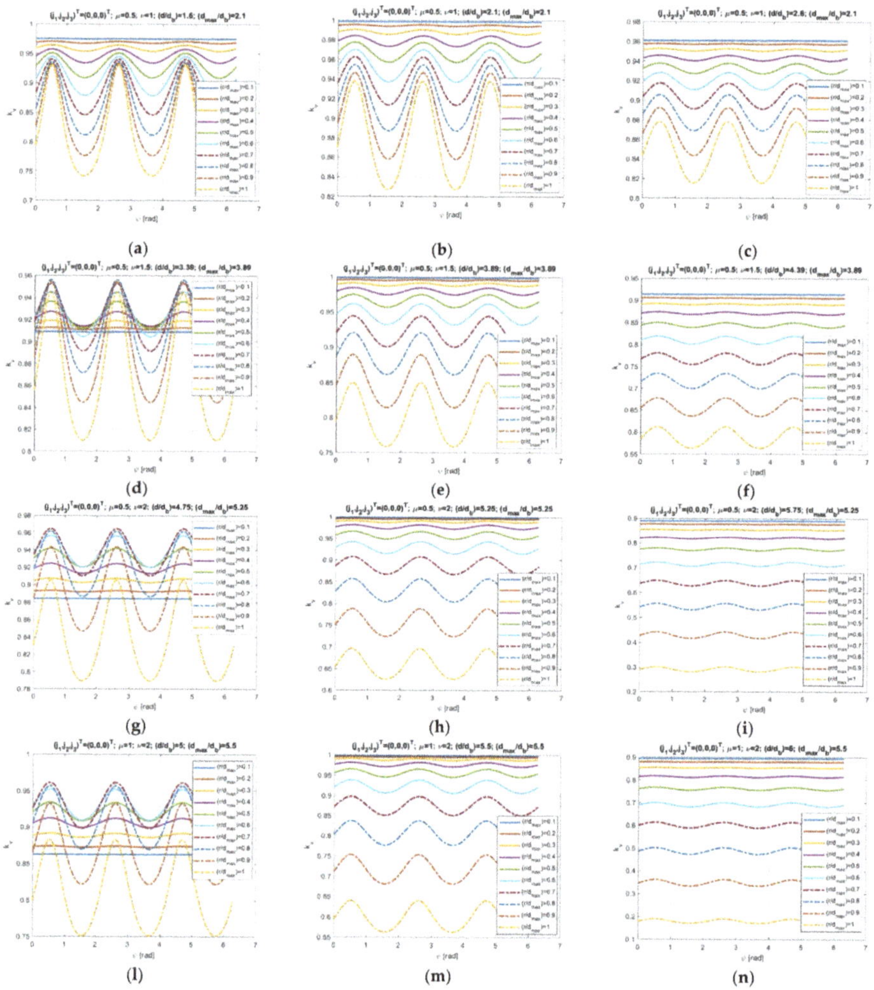

Figure 5. Values of k_v on circumferences (see Figure 2) with radius r, centered at point $D' = (-d, -d, -d)^T$ of the line x = y = z, that lie on planes perpendicular to the same line, in the case $\rho = 4d_b$ for (**a**,**d**,**g**,**l**) $(d/d_b) = (d_{max}/d_b) - 0.5$, (**b**,**e**,**h**,**m**) $(d/d_b) = (d_{max}/d_b)$, (**c**,**f**,**i**,**n**) $(d/d_b) = (d_{max}/d_b) + 0.5$, and the geometries (**a**,**b**,**c**) $\mu = 0.5$, $\nu = 1$, (**d**,**e**,**f**) $\mu = 0.5$, $\nu = 1.5$, (**g**,**h**,**i**) $\mu = 0.5$, $\nu = 2$, and (**l**,**m**,**n**) $\mu = 1$, $\nu = 2$.

3.2.3. Index k_g

Figure 2b highlights that the points O, A_i, C_i, B_i, and P lie on the same plane, which is perpendicular to the unit vector $\mathbf{g}_i$, and that the *i-th* limb, over moving together with this plane, which can rotate around the line passing through points O and A_i, performs a planar motion inside this plane. Figure 6 shows the *i-th* limb in this plane together with the force, $F_i\mathbf{v}_i$, that the limb applies to the platform and the torque, $M_i\mathbf{g}_i$, the actuator applies in the actuated-joint. The moment of equilibrium about A_i of the *i-th* limb projected along $\mathbf{g}_i$ gives the relationship $M_i = F_i s_i$ where s_i is the arm of $F_i\mathbf{v}_i$ that can be expressed as follows (see Figure 6):

$$s_i = \frac{M_i}{F_i} = f_i \sin\gamma_i = f_i \sin\theta_{i3} \tag{28}$$

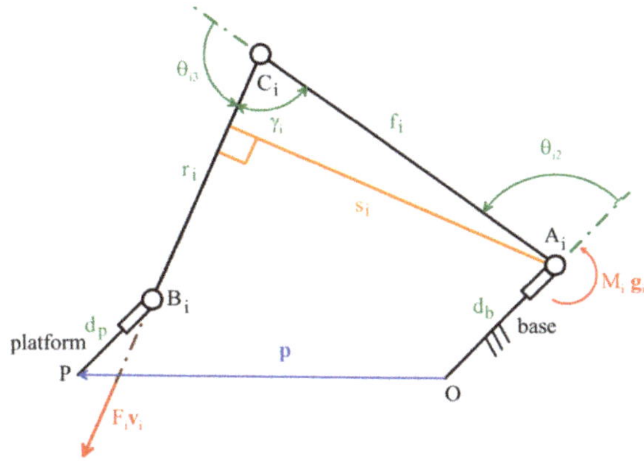

Figure 6. View of the *i-th* limb in the plane perpendicular to the unit vector $\mathbf{g}_i$ represented together with the force $F_i\mathbf{v}_i$ it applies to the platform and the torque $M_i\mathbf{g}_i$ applied by the actuator in the actuated joint.

Equation (28) reveals that s_i is a "transmission coefficient" depending on θ_{i3}, which plays the role of a "transmission angle". Such a transmission angle rates the quality of the transmission from the actuator to the platform in the *i-th* limb. Moreover, the comparison of Equation (28) with Equation (12c) provides the conclusion that k_g is just the product of the transmission coefficients, s_i for $i = 1, 2, 3$, of the three limbs. This result concurs with the general interpretation given to k_g in Section 2.

Since, in the studied TPM, the meaning of the tree factors whose product gives k_g is clear, instead of assigning a lower limit to the whole k_g (i.e., Equation (10)), a lower limit will be assigned to each factor. Such a limit can be deduced from the ample literature (see [45,46] for Refs.) on the transmission angles of planar linkages that suggest the imposition of the following limitations

$$\left|\theta_{i3} - 90^\circ\right| \leq 50^\circ \quad i = 1,\ 2,\ 3 \tag{29}$$

which yield

$$|\sin\theta_{i3}| \geq 0.643 \quad i = 1,\ 2,\ 3 \tag{30}$$

Condition (30) must be checked in the whole useful workspace by using Equation (24) every time the link lengths r_i and f_i (i.e., in our particular case, ρ and ν) are chosen. If condition (30) is satisfied for the minimum, $\theta_{i3,min}$, and the maximum, $\theta_{i3,max}$, values of θ_{i3}, it will be satisfied in the whole useful workspace. With reference to Figure 6, the minimum (maximum) value of θ_{i3} occurs when the length of the segment A_iB_i is maximum (minimum). The following relationships come out from Equation (24)

$$|\sin\theta_{i3,min}| = \sqrt{1 - \left[\frac{\left(\overline{A_iB_{i,max}}\right)^2 - f_i^2 - r_i^2}{2f_ir_i}\right]^2},\ |\sin\theta_{i3,max}| = \sqrt{1 - \left[\frac{\left(\overline{A_iB_{i,min}}\right)^2 - f_i^2 - r_i^2}{2f_ir_i}\right]^2} \tag{31}$$

3.3. Functional Parameters Determination

A useful workspace with the shape of a right circular cylinder having height equal to d_b is chosen. The above-reported results show that such a workspace shape must be always located with its axis lying on the line $x = y = z$ to get good kinetostatic performances; whereas, its position along that line and its radius must be selected by considering the chosen values of $k_{h,min}$ and $k_{v,min}$. Eventually, when the position and the sizes of the cylinder have been completely determined, condition (30) must be verified.

k_h (k_v) is a mixed product of unit vectors related to the directions of three reaction moments (forces) equilibrating external loads applied to the platform. Consequently, a reasonable minimum value of k_h (k_v) is $k_{h,min}$ = 0.5 ($k_{v,min}$ = 0.5), which, in the worst case, implies a reaction moment (force) with a magnitude 1.5 times the magnitude of the external load. Figure 3 shows that the condition $k_h \geq 0.5$ is satisfied by choosing $\lambda = (r/d) \leq 0.6$. Moreover, by choosing $\rho = 4d_b$, $\mu = 0.5$, and $\nu = 1.5$, Figure 5d–f[2] shows that the condition $k_v \geq 0.5$ is always satisfied if $(r/d_{max}) \leq 0.7$ where $d_{max} = 3.89d_b$. The conclusion is that, the choices $\rho = 4d_b$, $\mu = 0.5$, and $\nu = 1.5$ locate the right circular cylinder, used as a useful workspace, with its bases equally distant from point $D'_{max} = (-3.89d_b, -3.89d_b, -3.89d_b)^T$ and choose its radius $r = 0.6d_{max} = 2.334d_b$ (see Figure 7).

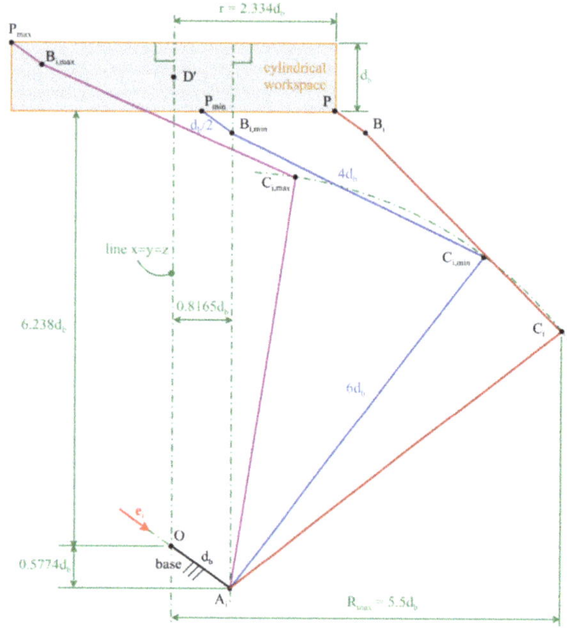

Figure 7. Cylindrical workspace: view of the *i-th* limb in the meridian plane passing through the line x = y = z and containing the coordinate axis of $Ox_by_bz_b$ that is parallel to the unit vector $\mathbf{e}_i$.

Now that the shape and the location of the useful workspace is completely defined, condition (30) must be verified. Figure 7 shows the two configurations of the *i*-th limb in which the segment A_iB_i reaches its minimum and maximum lengths. The data reported in Figure 7 make it possible to compute the following values: $\overline{A_iB_i}_{,min} = 6.5267d_b$ and $\overline{A_iB_i}_{,max} = 8.0107d_b$. The introduction of these values into Equation (31) yields $|\sin\theta_{i3,min}| = 0.9673$, which corresponds to $\theta_{i3,min} = 75^\circ$, and $|\sin\theta_{i3,max}| = 0.9806$, which corresponds to $\theta_{i3,max} = 101^\circ$. Consequently, condition (30) is verified in the whole workspace with an ample safety margin and the transmission angles are always very good.

4. Discussion

The dimensional synthesis of the LaMaViP 3-URU clearly proves that the novel method proposed for evaluating TPMs' kinetostatic performances is based on three indices whose meaning is easy to understand during the design of the machine. In particular, since two indices are dimensionless and have geometric and static meanings and the remaining third index refers to the quality of the load transmission from the actuators to the platform,

2 It is worth noting that Figure 5 holds only for the choice $\rho = 4d_b$, a different choice of ρ requires the determination of analogous diagrams through the above reported formulas.

the choice of their minimum values to satisfy inside the useful workspace comes out in a natural way from simple static and geometric considerations.

Moreover, it is worth noting that these indices provide a map of the possible design choices. Consequently, their use leaves the designer free to adopt the choices that better fit the technical requirements the machine has to satisfy.

The proposed methodology for TPMs' dimensional synthesis, even though it is based on Jacobians, does not have homogeneity issues. Such a feature makes the proposed indices suitable for comparing the kinetostatic performances of different TPMs. A possible procedure for comparing different TPMs by using these indices is to consider the minimum, the maximum, and the average values of each index in the useful workspace since they are local indices.

The proposed method requires only that the input/output instantaneous relationship of the studied LPM can be put in a canonic form in which the parallel Jacobian (i.e., the one that multiplies the platform twist) is a block diagonal Jacobian that separates platform rotations from platform translations. Therefore, over TPMs, it is certainly applicable also to spherical PMs (parallel wrists) and extendable to Schoenflies (SCARA) PMs, planar PMs, and other LPMs that can satisfy this condition.

Regarding the dimensional synthesis of the LaMaViP 3-URU, the obtained geometry and useful workspace (Figure 7) are comparable to the ones of commercial delta robots (e.g., ABB IRB360). In addition, the found limb geometry with the link adjacent to the base that is 1.5 times longer than the one adjacent to the platform suggests that the actuated-joint variable, θ_{i2}, is controllable by simply introducing a linear actuator as shown in Figure 8. Such an actuation system is easy to manufacture and makes the limb stiff enough for using the LaMaViP 3-URU in CNC machine tools.

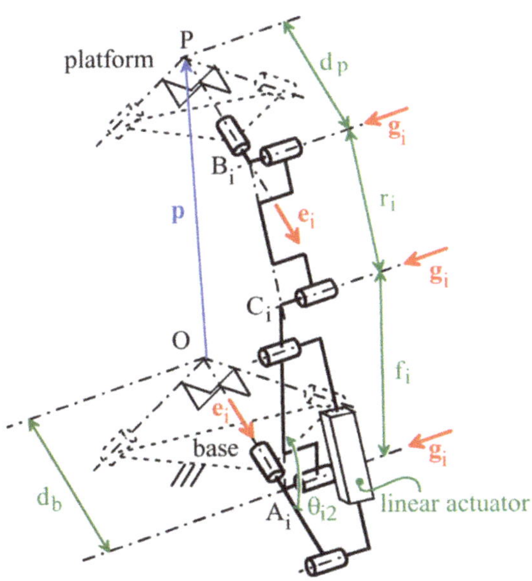

Figure 8. The *i*-th limb with a linear actuator that controls the actuated-joint variable θ_{i2}.

5. Conclusions

A novel method for the dimensional synthesis of lower-mobility PMs (LPMs) has been presented. The proposed method can be applied to all the translational PMs (TPMs) and, in general, to all the LPM types whose input/output instantaneous relationship can be put in a canonic form in which the parallel Jacobian is a block diagonal matrix that separates platform rotations from platform translations (e.g., parallel wrists).

The presented method is based on three indices, of which two are dimensionless and have clear geometric and static meanings and the remaining third rates the quality of the load transmission from the actuators to the platform. These features of the three indices make the proposed methodology not affected by homogeneity issues of the involved input/output variables. In addition, since the values of the indices are easy to relate to particular design requirements through their geometric and static meanings, the proposed technique is particularly useful for addressing the design of novel LPM architectures.

The application of the method to TPMs has been illustrated in depth and it is also illustrated by using it in the dimensional synthesis of the LaMaViP 3-URU, which is a novel TPM type recently proposed by the author. Such a synthesis has provided a clear map of the possible design choices used to determine a workspace and a machine geometry that are comparable to the ones of commercial delta robots and are suitable for industrial applications.

Future works will also present applications of the method to other types of LPMs and will try to extend the method to all non-redundant PMs. Regarding LaMaViP 3-URU, future works will address its machine-element design referring to industrial applications together with the evaluation of its dynamic performances.

6. Patents

Di Gregorio, R.: Meccanismo Parallelo Traslazionale. 23 March 2020; Italy Patent Application No. 102020000006100; published on 30 September 2021, as international PCT patent No.: WO2021/191054A1.

Funding: This research was funded by the University of Ferrara (UNIFE), FAR2020 and developed at the Laboratory of Mechatronics and Virtual Prototyping (LaMaViP), Department of Engineering, UNIFE.

Institutional Review Board Statement: Not applicable.

Informed Consent Statement: Not applicable.

Data Availability Statement: This work does not use experimental data. The data necessary to replicate the computations illustrated in the paper are included in the text of the paper.

Conflicts of Interest: The authors declare no conflict of interest. The funders had no role in the design of the study; in the collection, analyses, or interpretation of data; in the writing of the manuscript, or in the decision to publish the results.

References

1. Hartenburg, R.S.; Denavit, J. *Kinematic Synthesis of Linkages*; McGraw-Hill: New York, NY, USA, 1964; ISBN 9780070269101.
2. Tsai, L.W. *Mechanism Design: Enumeration of Kinematic Structures According to Function*; CRC Press LLC: Boca Raton, FL, USA, 2001.
3. Kong, X.; Gosselin, C.M. *Type Synthesis of Parallel Mechanisms*; Springer: Berlin/Heidelberg, Germany, 2007; ISBN 978-3-642-09118-6.
4. McCarthy, J.M.; Soh, G.S. *Geometric Design of Linkages*; Springer: Berlin/Heidelberg, Germany, 2011.
5. Bhandari, V.B. *Design of Machine Elements*, 3rd ed.; Tata McGraw-Hill: New Delhi, India, 2010.
6. Jiang, W. *Analysis and Design of Machine Elements*; Wiley: Singapore, 2019.
7. Ashby, M.F. *Materials Selection in Mechanical Design*, 5th ed.; Butterworth-Heinemann: Burlington, MA, USA, 2016.
8. Lou, Y.J.; Liu, G.F.; Li, Z.X. A general approach for optimal design of parallel manipulators. In Proceedings of the ICRA 2004, New Orleans, LA, USA, 26 April–1 May 2004.
9. de-Juan, A.; Collard, J.-F.; Fisette, P.; Garcia, P.; Sancivrian, R. Multi-objective optimization of parallel manipulators. In *New Trends in Mechanism Science, Mechanisms and Machine Science*; Pisla, D., Ceccarelli, M., Husty, M., Corves, B., Eds.; Springer: Dordrecht, The Netherlands, 2010; Volume 5. [CrossRef]
10. Angeles, J. *Fundamentals of Robotic Mechanical Systems*; Springer: Dordrecht, The Netherlands, 2014.
11. Merlet, J.P. Jacobian, Manipulability, Condition Number, and Accuracy of Parallel Robots. *ASME J. Mech. Des.* **2006**, *128*, 199–206. [CrossRef]
12. Patel, S.H.; Sobh, T. Manipulator Performance Measures—A Comprehensive Literature Survey. *J. Intell. Robot. Syst.* **2014**, *77*, 547–570. [CrossRef]
13. Rosyid, A.; El-Khasawneh, B.; Alazzam, A. Review article: Performance measures of parallel kinematics manipulators. *Mech. Sci.* **2020**, *11*, 49–73. [CrossRef]
14. Salisbury, J.K.; Craig, J.J. Articulated Hands: Force Control and Kinematic Issues. *Int. J. Robot. Res.* **1982**, *1*, 4–17. [CrossRef]

15. Gosselin, C.; Angeles, J. A Global Performance Index for the Kinematic Optimization of Robotic Manipulators. *J. Mech. Des.* **1991**, *113*, 220–226. [CrossRef]
16. Khan, W.A.; Angeles, J. The Kinetostatic Optimization of Robotic Manipulators: The Inverse and the Direct Problems. *ASME J. Mech. Des.* **2006**, *128*, 168–178. [CrossRef]
17. Yoshikawa, T. Manipulability of Robotic Mechanisms. *Int. J. Robot. Res.* **1985**, *4*, 3–9. [CrossRef]
18. Sutherland, G.; Roth, B. A transmission index for spatial mechanisms. *ASME J. Eng. Ind.* **1973**, *95*, 589–597. [CrossRef]
19. Chen, C.; Angeles, J. Generalized transmission index and transmission quality for spatial linkages. *Mech. Mach. Theory* **2007**, *42*, 1225–1237. [CrossRef]
20. Wang, J.; Wu, C.; Liu, X.J. Performance evaluation of parallel manipulators: Motion/force transmissibility and its index. *Mech. Mach. Theory* **2010**, *45*, 1462–1476. [CrossRef]
21. Liu, H.; Huang, T.; Kecskeméthy, A.; Chetwynd, D.G. A generalized approach for computing the transmission index of parallel mechanisms. *Mech. Mach. Theory* **2014**, *74*, 245–256. [CrossRef]
22. Chang, W.-T.; Lin, C.-C.; Lee, J.-J. Force Transmissibility Performance of Parallel Manipulators. *J. Robot. Syst.* **2003**, *20*, 659–670. [CrossRef]
23. Kim, H.S.; Choi, Y.J. Forward/inverse force transmission capability analyses of fully parallel manipulators. *IEEE Trans. Robot. Autom.* **2001**, *17*, 526–531. [CrossRef]
24. Ma, O.; Angeles, J. Optimum architecture design of platform manipulators. In Proceedings of the 5th International Conference on Advanced Robotics' Robots in Unstructured Environments, Pisa, Italy, 19–22 June 1991; Volume 2, pp. 1130–1135. [CrossRef]
25. Staffetti, E.; Bruyninckx, H.; De Schutter, J. On the Invariance of Manipulability Indices. In *Advances in Robot Kinematics*; Lenarčič, J., Thomas, F., Eds.; Springer: Dordrecht, The Netherlands, 2002. [CrossRef]
26. Kim, S.-G.; Ryu, J. New dimensionally homogeneous Jacobian matrix formulation by three end-effector points for optimal design of parallel manipulators. *IEEE Trans. Robot. Autom.* **2003**, *19*, 731–736. [CrossRef]
27. Cardou, P.; Bouchard, S.; Gosselin, C. Kinematic-Sensitivity Indices for Dimensionally Nonhomogeneous Jacobian Matrices. *IEEE Trans. Robot.* **2010**, *26*, 166–173. [CrossRef]
28. Hain, K. *Applied Kinematics*; McGraw-Hill: New York, NY, USA, 1967.
29. Dresner, T.L.; Buffinton, K.W. Definition of pressure and transmission angles applicable to multi-input mechanisms. *ASME J. Mech. Des.* **1991**, *113*, 495–499. [CrossRef]
30. Ball, R.S. *A Treatise on the Theory of Screws*; Cambridge University Press: Cambridge, UK, 1900.
31. Gosselin, C.M.; Angeles, J. Singularity analysis of closed-loop kinematic chains. *IEEE Trans. Robot. Autom.* **1990**, *6*, 281–290. [CrossRef]
32. Zlatanov, D.; Fenton, R.G.; Benhabib, B. A unifying framework for classification and interpretation of mechanism singularities. *ASME J. Mech. Des.* **1995**, *117*, 566–572. [CrossRef]
33. Zlatanov, D.; Bonev, I.A.; Gosselin, C.M. Constraint Singularities as C-Space Singularities. In *Advances in Robot Kinematics: Theory and Applications*; Lenarčič, J., Thomas, F., Eds.; Springer: Dordrecht, The Netherlands, 2002; pp. 183–192.
34. Hunt, K.H. *Kinematic Geometry of Mechanisms*; Clarendon Press: Oxford, UK, 1990.
35. Di Gregorio, R. A Review of the Literature on the Lower-Mobility Parallel Manipulators of 3-UPU or 3-URU Type. *Robotics* **2020**, *9*, 5. [CrossRef]
36. Brinker, J.; Corves, B.; Takeda, Y. Kinematic and dynamic dimensional synthesis of extended Delta parallel robots. In *IFToMM International Symposium on Robotics & Mechatronics (ISRM2017)*; Yang, R., Takeda, Y., Zhang, C., Fang, G., Eds.; Springer: Dordrecht, The Netherlands, 2017; pp. 131–143. [CrossRef]
37. Brinker, J.; Corves, B.; Takeda, Y. Kinematic performance evaluation of high-speed Delta parallel robots based on motion/force transmission indices. *Mech. Mach. Theory* **2018**, *125*, 111–125. [CrossRef]
38. Di Gregorio, R. A Novel 3-URU Architecture with Actuators on the Base: Kinematics and Singularity Analysis. *Robotics* **2020**, *9*, 60. [CrossRef]
39. Meyer, C.D. *Matrix Analysis and Applied Linear Algebra*; SIAM: Philadelphia, PA, USA, 2000.
40. Hervé, J.M. The Mathematical Group Structure of the Set of Displacements. *Mech. Mach. Theory* **1994**, *29*, 73–81. [CrossRef]
41. Hervé, J.M. The Lie Group of Rigid Body Displacements, a Fundamental Tool for Mechanism Design. *Mech. Mach. Theory* **1999**, *34*, 719–730. [CrossRef]
42. Di Gregorio, R. Direct Position Analysis of a Particular Translational 3-URU Manipulator. *ASME J. Mech. Robot.* **2021**, *13*, 061007. [CrossRef]
43. Di Gregorio, R.; Parenti-Castelli, V. A Translational 3-DOF Parallel Manipulator. In *Advances in Robot Kinematics: Analysis and Control*; Lenarcic, J., Husty, M.L., Eds.; Kluwer: Norwell, MA, USA, 1998; pp. 49–58.
44. Di Gregorio, R. Position Analysis of a Novel Translational 3-URU with Actuators on the Base. In *New Advances in Mechanisms, Mechanical Transmissions and Robotics. MTM & Robotics 2020*; Lovasz, E.C., Maniu, I., Doroftei, I., Ivanescu, M., Gruescu, C.-M., Eds.; Springer: Dordrecht, The Netherlands, 2020; pp. 80–90. [CrossRef]
45. Balli, S.S.; Chand, S. Transmission angle in mechanisms (Triangle in mech). *Mech. Mach. Theory* **2002**, *37*, 175–195. [CrossRef]
46. Pennestrì, E.; Valentini, P.P. A review of simple analytical methods for the kinematic synthesis of four-bar and slider-crank function generators for two and three prescribed finite positions. In *Buletin Stiintific Seria Mecanica Aplicata*; University of Pitesti: Pitesti, Romania, 2009; pp. 128–143. [CrossRef]

robotics

MDPI

Article

Kinematic Model Pruning: A Design Optimization Technique for Simultaneous Optimization of Topology and Geometry

Hannes Gamper [1,2,*], Adrien Luthi [1], Hubert Gattringer [2], Andreas Mueller [2] and Mario Di Castro [1]

[1] CERN-European Organization for Nuclear Research, Espl. des Particules 1, 1211 Meyrin, Switzerland; adrien.luthi@cern.ch (A.L.); Mario.di.castro@cern.ch (M.D.C.)

[2] Institute of Robotics, Johannes Kepler University Linz, Altenbergerstraße 69, 4040 Linz, Austria; hubert.gattringer@jku.at (H.G.); a.mueller@jku.at (A.M.)

* Correspondence: hannes.gamper@cern.ch; Tel.: +41-227-6656557

Abstract: This paper presents a method of optimizing the design of robotic manipulators using a novel kinematic model pruning technique. The optimization departs from an predefined candidate linkage consisting of a initial topology and geometry. It allows simultaneously optimizing the degree of freedom, the link lengths and other kinematic or dynamic performance criteria, while enabling the manipulator to follow the desired end-effector position and avoid collisions with the environment or itself. Current methods for design optimization rely on dedicated and complex frameworks, and solve the design optimization only as decoupled from each other in separate optimization problems. The proposed method only requires the introduction of a simple function, called a pruning function, as an objective function of an optimization problem. The introduced pruning function transforms a discrete topology optimization problem into a continuous problem that then can be solved simultaneously with other continuous objectives, using readily available optimization schemes. Two applications are presented: the optimization of a manipulator for the inspection of radio frequency cavities and a manipulator for maintenance within the future circular collider (FCC).

Keywords: kinematics; topology; design optimization; dexterity; robotics; inspection

Citation: Gamper, H.; Luthi, A.; Gattringer, H.; Mueller, A.; Di Castro, M. Kinematic Model Pruning: A Design Optimization Technique for Simultaneous Optimization of Topology and Geometry. *Robotics* **2022**, *11*, 31. https://doi.org/10.3390/robotics11020031

Academic Editor: Raffaele Di Gregorio

Received: 17 December 2021
Accepted: 1 February 2022
Published: 4 March 2022

1. Introduction

Power plants and big industrial or scientific facilities like the European Organization for Nuclear Research (CERN), are often confronted with very special automation problems in complex environments for their laboratories, experiments or test rigs, see e.g., [1,2]. These frequently lead to specific requirements that do not allow the usage of standard industrial robots. Thus, a robotic design problem with few restrictions on the actual robot design (topology and geometry), but with very hard requirements concerning other parameters, such as workspace, allowed robot space, dexterity and accuracy, has to be solved.

Design optimization in robotics is a recurrent topic and has been discussed in the literature from many different perspectives. Several publications address the continuous optimization problem of minimizing certain deterministic (mainly kinematic or dynamic) performance criteria for a given topology, as shown in [3–7]. All of the above mentioned articles use different performance criteria, optimization techniques and solvers or approaches for collision avoidance, but non of them minimize the degrees of freedom (DoF) of the mechanical structure. The discrete problem of minimizing the DoF/topology is solved, in the literature, by always decoupling it from the previously mentioned continuous optimization tasks. Extensive frameworks for structural synthesis are presented in [8,9], which start by exploring all possible combinations of joints and links. Then, the these topologies can be optimized with respect to deterministic performance criteria. Ref. [10] proposes a smart framework to optimize the link lengths and other performance criteria, but mentions that discrete criteria, such as the number of actuators, cannot be handled, since the optimization strategy requires the first and second derivatives. Based on these findings [11] extends the framework in order to enable the use of discrete criteria, but the optimization of discrete and continuous criteria are still decoupled into two separate optimization problems. Thus, complete design optimizations, in terms of topology and geometry, currently requires dedicated and complex frameworks that optimize both objectives only as decoupled from each other into two separate optimization problems.

In order to alleviate the complexity problem, in the following, a synthesis approach is presented that allows simultaneously optimizing the link length as well as the DoF. The approach is based on a candidate linkage and a pruning method that is used to optimize the candidate linkages. In other words, this work proposes a *kinematic model pruning* technique to simultaneously optimize discrete (topology/minimizing the DoF) and continuous criteria (link lengths, kinematic and dynamic performance criteria). The focus of this paper is the formulation of the optimization problem, but not its numerical solution (to this end, standard numerical solvers are applied). To achieve the simultaneous optimization of topology and geometric parameters, a certain type of function (hereafter called the *pruning function*) that facilitates the transformation from a discrete to a continuous problem, is defined. Thus, the presented kinematic model pruning technique allows optimizing the degrees of freedom, the robot link lengths and other kinematic or dynamic criteria, while ensuring that the mechanical structure reaches the desired end-effector position, avoids self collisions and collisions with its surrounding. This will be demonstrated with two applications:

- The design optimization of a surface inspection robot for radio frequency (RF) cavities, as used in the Large Hardron Collider (LHC), the Linear Accelerator (LINAC) or the Future Circular Collider (FCC).
- The design optimization of a manipulator for the 100-km-long FCC.

Section 2 summarizes the goals and limitations of the design optimization. In Section 3 the methods and work flow of the proposed algorithm are presented, and Section 4 describes the generic formulation of the design optimization algorithm in detail. In Sections 5 and 6 the proposed technique is applied to two specific problems, the cavity inspection robot and the robotic manipulator for the FCC. The last Section 7 summarizes the results and draws conclusions concerning the existing and future work.

2. Design Optimization Goals and Limitations

The proposed method is capable of modifying an initial and, thus, non-optimal candidate linkage that defines a variety of possible topologies, called *design space* (see Section 3.1). Simultaneously the geometry of the mechanical structure will be tuned such that the final result provides an optimal and practically feasible solution with respect to certain objectives:

- minimize the DoF;
- minimize the length of each robot link; and
- minimize kinematic and dynamic performance criteria of the mechanical structure.

These objectives should be optimized such that the robot is able to reach all desired positions and avoid collisions with itself as well as with the environment. The algorithm is, in general, applicable under following constraints:

- the topology can consist of arbitrary joint types, but only subsequent joints of the same type can be use for reducing the DoF; and
- the additional kinematic and dynamic performance criteria subject to optimization should, in the best case, be complementary, but never be in contradiction with each other.

3. Methodology

Figure 1 visualizes the workflow of the proposed algorithm and shows how the different methods are connected. The design optimization departs from a predefined design space (see Section 3.1), which will then be exploited by the kinematic model pruning method (see Section 3.2) in order to find the optimal design (see Section 3.3).

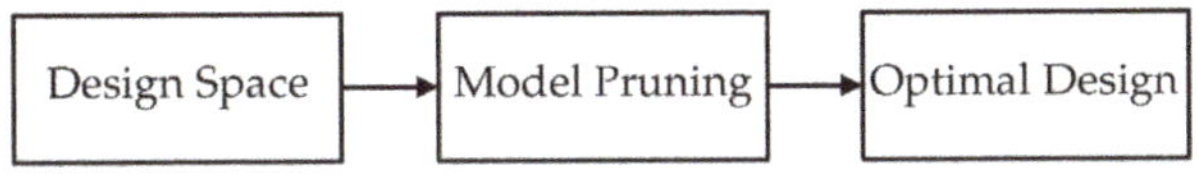

Figure 1. Workflow of the Design Optimization.

3.1. Design Space

The design space describes a variety of possible topologies based on predefined assumptions by the user. It is defined by a candidate linkage in the form of a topology and geometric parameters, see e.g., Figure 8 and Table 1. The DoF of the predefined design space needs to be greater than the expected optimal solution, since the algorithm is only able to reduce, and not increase, the DoF.

3.2. Kinematic Model Pruning

The kinematic model pruning block in Figure 1 consists of a continuous optimization problem that is to be solved using readily available optimization schemes. For detailed information about the implementation and formulation of the optimization problem see Section 4. However, the focus of this paper is the formulation of the optimization problem, but not its numerical solution (to this end standard numerical solvers are applied). As discussed in Section 1 the challenge is to combine the discrete and continuous criteria in order to be able to formulate only one optimization problem that takes all goals from Section 2 into account. At the core of the proposed kinematic model pruning technique is a certain type of function called pruning function, see Definition 1. Setting this function as the objective function enables the transformation of the discrete topology optimization problem into a continuous problem. Therefore, the topology can be optimized simultaneously with other continuous criteria or goals as presented in Section 2. The pruning function is a simple vector function with two inequality constraints on the first and second derivative.

Definition 1 (pruning function). *A vector function* $\mathbf{g} = \begin{bmatrix} g_1(l_1) & g_2(l_2) & \dots & g_N(l_N) \end{bmatrix} : \mathbb{R}^N \to \mathbb{R}^N$ *with argument* $\mathbf{p} = \begin{bmatrix} l_1 & l_2 & \dots & l_N \end{bmatrix}^T \in \mathbb{R}^N$ *that satisfies*

$$\frac{\partial \mathbf{g}}{\partial \mathbf{p}} > 0 \ \forall \ l_i > 0, \ \ i \in \{1, 2, \dots, N\} \tag{1}$$

and

$$\frac{\partial^2 \mathbf{g}}{\partial \mathbf{p}^2} < 0 \ \forall \ l_i > 0, \ \ i \in \{1, 2, \dots, N\}. \tag{2}$$

Constraint (1) ensures that the design parameters will be minimized and constraint (2) drives the design parameters to zero. In other words, constraint (2) facilitates minimization of a discrete problem like the number of DoF in a mechanical structure. In the following Example 1, the behavior of the rather abstract Definition 1 will be demonstrated with the optimization of a simple two-DoF planar robotic manipulator. Furthermore, the results of the optimization with a pruning function will be compared to the results using a quadratic function for the objective function.

Example 1. *The end effector of an $N = 2$ link planar manipulator with two DoF should be able to reach exactly one point at position $\|\mathbf{z}_d\| = 5$. Note that the orientation will not be constrained, thus creating a task redundancy. In Figure 2a the initial design, with two DoF and the link lengths l_1^a and l_2^a, is shown. Departing from this candidate linkage, the optimization should find an optimal* *design with minimal link lengths and a minimal number of DoF, while still reaching the desired position* $\|\mathbf{z}_d\|$. *A corresponding optimization problem*

$$\begin{aligned} \min_{\mathbf{x},\ \mathbf{p}} \quad & J(\mathbf{p}) = \mathbf{k}^T \mathbf{g}(\mathbf{p}) \\ s.t. \quad & \mathbf{f}(\mathbf{x},\ \mathbf{p}) - \mathbf{z}_d \ = \ \mathbf{0}, \end{aligned} \tag{3}$$

with the objective function $J(\mathbf{p})$ *can be formalized.* $\mathbf{f}(\mathbf{x},\ \mathbf{p})$ *denotes the forward kinematics, vector* $\mathbf{p} = \begin{bmatrix} l_1 & l_2 \end{bmatrix}$ *contains the two geometric parameters or link lengths, vector* $\mathbf{x} = \begin{bmatrix} q_1 & q_2 \end{bmatrix}$ *contains the two joint angles and vector* $\mathbf{k}^T = \begin{bmatrix} 1 & 1 \end{bmatrix}$ *is used as a weighting factor. For demonstration purposes, the optimization will be launched twice with different functions for* $\mathbf{g}$. *Once with a pruning function (see Definition 1)*

$$\mathbf{e}(\mathbf{p}) = \begin{bmatrix} e_1(l_1) & e_2(l_2) \end{bmatrix}^T = \begin{bmatrix} \arctan(l_1) & \arctan(l_2) \end{bmatrix}^T. \tag{4}$$

and once with a quadratic function $\mathbf{h} : \mathbb{R}^N \rightarrow \mathbb{R}^N$,

$$\mathbf{h}(\mathbf{p}) = \begin{bmatrix} h_1(l_1) & h_2(l_2) \end{bmatrix}^T = \begin{bmatrix} l_1^2 & l_2^2 \end{bmatrix}^T, \tag{5}$$

with the well-known properties

$$\frac{\partial \mathbf{h}}{\partial \mathbf{p}} > 0\,, \ \frac{\partial^2 \mathbf{h}}{\partial \mathbf{p}^2} > 0 \ \forall \ l_i > 0, \ i \in \{1,2\}. \tag{6}$$

Note that the second derivative of the quadratic function is, unlike for the pruning function, greater than zero.

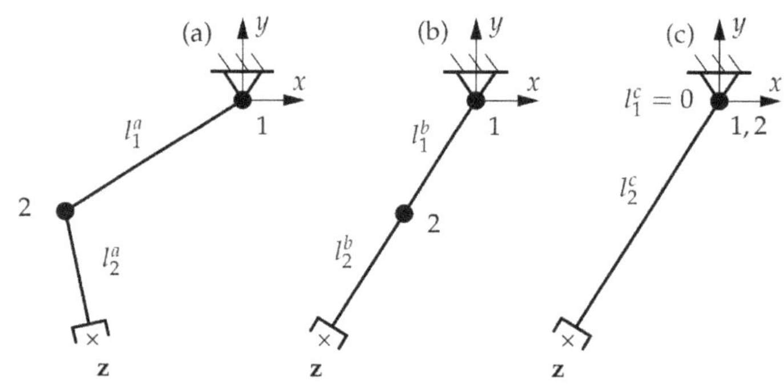

Figure 2. Optimization effects using different function types for J_1. (**a**) shows the design space. (**b**) shows the optimization results using a quadratic objective function. (**c**) shows the optimization results using kinematic model pruning.

The optimization result, when using a quadratic objective function $\mathbf{h}$, *is shown in Figure 2b. It is obvious that the optimization converges to an optimum, since the links lie on a straight line, in Euclidean space, from the base to the desired end-effector position. However, it is also clear that only one DoF would be sufficient to reach this position. Thus, it is easy to see that the quadratic objective function minimizes the total length* $l_1 + l_2$, *but splits up this length equally over both links, such that* $l_1 = l_2$ *and, hence, does not minimize the DoF.*

This behavior becomes more clear when looking at the surface and contour plots of the objective function $J(\mathbf{p})$ *over the variables* l_1 *and* l_2 *as shown in Figures 3 and 4, respectively.*

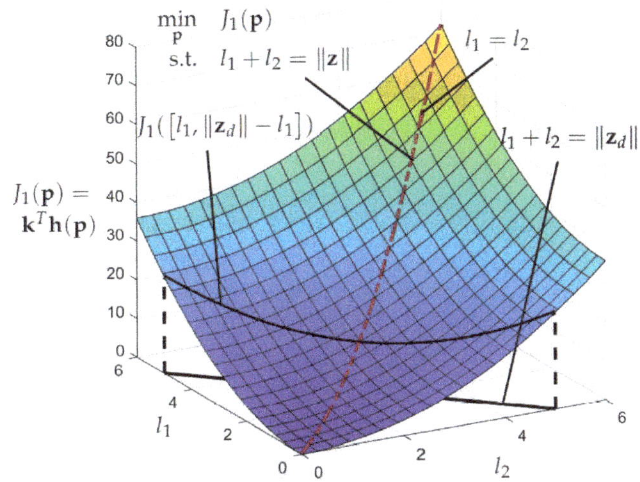

Figure 3. Surface plot of objective function $J(\mathbf{p}) = \mathbf{k}^T\mathbf{h}(\mathbf{p})$.

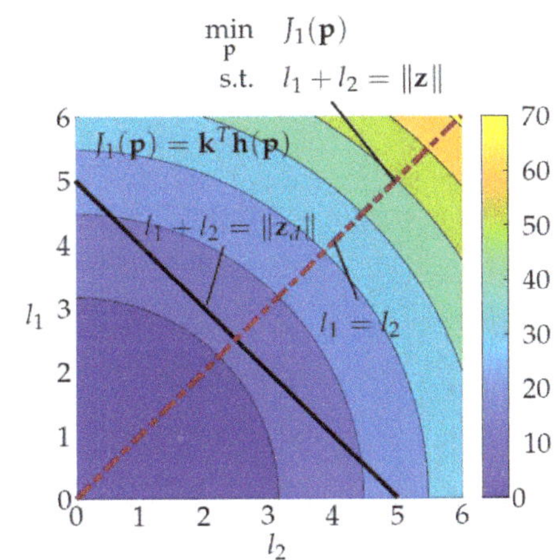

Figure 4. Contour plot of objective function $J(\mathbf{p}) = \mathbf{k}^T\mathbf{h}(\mathbf{p})$.

The black line in the $l_1 l_2$ plane represents the possible combinations of l_1 and l_2 with which the desired end-effector position can be reached and $l_1 + l_2 = \|\mathbf{z}_d\| = 5$ holds. The projection of this line on the surface of $J(\mathbf{p})$ leads to the curved line $J([l_1, \|\mathbf{z}_d\| - l_1])$ from which the optimal combination of l_1 and l_2 has to be chosen. The dashed red line visualizes the space of optimal combinations of l_1 and l_2 for arbitrary $\mathbf{z}$, which is identical to the space of combinations wherein $l_1 = l_2$. This becomes even more clear when looking at the contour plot in Figure 4. The black line $l_1 + l_2 = \|\mathbf{z}_d\|$ shows a -45° slope and the contour lines are circles of different diameter centered at the origin. Thus, the optimal solution can be found where the line $l_1 + l_2 = \|\mathbf{z}_d\|$ intersects exactly once with a contour line or in other words, where the line $l_1 + l_2 = \|\mathbf{z}_d\|$ is a tangent to the contour line. For arbitrary $\mathbf{z}$, this leads to optimal solution $l_1 = l_2$.

In order to minimize the DoF the distance should not be split up equally, but assigned to only one link, while the other link length will be set to zero. The corresponding joint to this link can then be removed, which means decreasing the DoF by one. This behavior can be achieved by using a pruning function (see Definition 1), specifically (4) for $\mathbf{g}$ in (3). Running the same optimization problem again with the new objective function, the total link length $l_1 + l_2$ is still a minimum *(straight line in Euclidean space), but is now assigned to only one link. This indicates that not more than one DoF is necessary to reach the position* $\mathbf{z}$ *(see Figure 2c). Thus, the corresponding joint i with* $l_i = 0$ *can be removed and the link lengths, and the DoF is minimized.*

Again, this behavior becomes more clear when looking at the surface and contour plots of the objective function $J(\mathbf{p})$ *over the variables* l_1 *and* l_2*, as shown in Figures 5 and 6, respectively.*

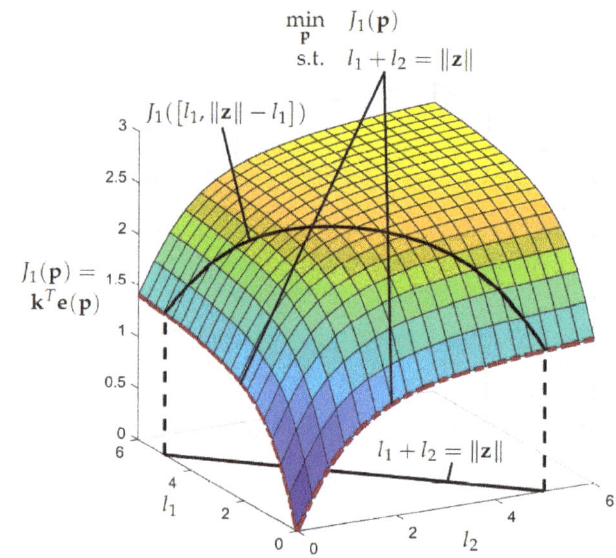

Figure 5. Surface plot of the objective function $J(\mathbf{p}) = \mathbf{k}^T\mathbf{e}(\mathbf{p})$.

The black line in the $l_1 l_2$ *plane represents the possible combinations of* l_1 *and* l_2 *with which the desired end-effector position can be reached, and* $l_1 + l_2 = \|\mathbf{z}_d\| = 5$ *holds. The projection of this line on the surface of* $J(\mathbf{p})$ *leads to the curved line* $J([l_1, \|\mathbf{z}_d\| - l_1])$ *from which the optimal combination of* l_1 *and* l_2 *has to be chosen. The dashed red lines visualize the space of optimal combinations of* l_1 *and* l_2 *for arbitrary* $\mathbf{z}$*. Compared with Figure 3 it can now be seen that for every* $\mathbf{z}_d$ *two solutions* $(l_1 = 0,\ l_2 = \|\mathbf{z}_d\|)$ *and* $(l_1 = \|\mathbf{z}_d\|,\ l_2 = 0)$ *with the same priority (for the special case* $\mathbf{k}^T = \begin{bmatrix} 1 & 1 \end{bmatrix}$*) exist. Both solution lie at the lower boundaries of the parameters* l_1 *and* l_2*. This becomes even more clear when looking at the contour plot in Figure 6. The black line* $l_1 + l_2 = \|\mathbf{z}_d\|$ *shows again a* -45° *slope, but now the contour lines are shaped in way such that the optimal solutions lie at the boundaries of* $l_1 + l_2 = \|\mathbf{z}_d\|$*. The optimal solution can be found by restricting the parameters with upper bounds in the optimization problem leading to only one solution e.g.,* $(l_1 = \|\mathbf{z}_d\|,\ l_2 = 0)$*.*

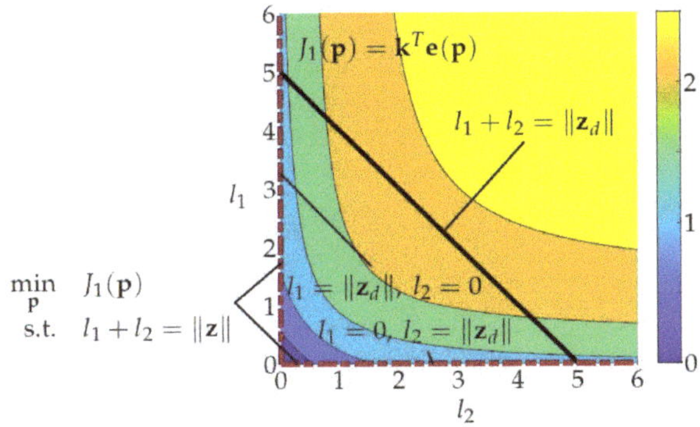

Figure 6. Contour plot of objective function $J(\mathbf{p}) = \mathbf{k}^T\mathbf{e}(\mathbf{p})$.

3.3. Optimal Design

The *optimal design* consists of the original topology, as defined in the design space, plus a set of optimal design parameters or geometric parameters. If one of the geometric parameters is driven to zero by the kinematic model pruning method, then this indicates a possible reduction of DoF and the corresponding joint can be removed.

4. Formulation of the Design Optimization Problem

In the following, the implementation and formulation of the optimization problem will be shown in more detail. In this section the problem is described in a generic manner in order to allow an easy extension of the problem with e.g., performance criteria that were not considered in this work. Specific implementations for certain problems are presented in Sections 5 and 6. In Section 4.1 a parameterized model of the initial mechanical structure, including kinematics and dynamics, are defined. General assumptions on the collision avoidance are discussed in Section 4.2. The formulation of the optimization problem is shown in Section 4.3 and the applied objective function is described in Section 4.4.

4.1. Kinematic and Dynamic Model

The forward kinematics $\mathbf{f} : \mathbb{R}^{n\mu} \to \mathbb{R}^{6\mu}$, with degrees of freedom n and μ in Cartesian positions, can be written in the form

$$\mathbf{z} = \mathbf{f}(\mathbf{x}, \mathbf{p}), \tag{7}$$

with the Cartesian positions and orientations $\mathbf{z} \in \mathbb{R}^{6\mu}$, the N geometric parameters $\mathbf{p} \in \mathbb{R}^N$ and the generalized joint coordinates $\mathbf{q} \in \mathbb{R}^n$ for every Cartesian position written in vector

$$\mathbf{x} = \begin{bmatrix} \mathbf{q}_1^T & \mathbf{q}_2^T & \dots & \mathbf{q}_\mu^T \end{bmatrix}^T \in \mathbb{R}^{n\mu}. \tag{8}$$

An explicit solution for the inverse kinematics is not computed, since it will be taken into account by non-linear equality constraints in the optimization problem.

The forward dynamics or the equation of movement for the entire robot can be written as

$$\mathbf{M}(\mathbf{p}, \mathbf{q})\ddot{\mathbf{q}} + \mathbf{g}(\mathbf{p}, \mathbf{q}, \dot{\mathbf{q}}) = \mathbf{Q}, \tag{9}$$

with the mass matrix $\mathbf{M}(\mathbf{p}, \mathbf{q})$, the non-linear term $\mathbf{g}(\mathbf{p}, \mathbf{q}, \dot{\mathbf{q}})$ containing gravitational, centrifugal and Coriolis terms and the actuator and external forces and torques $\mathbf{Q}$.

4.2. Collision Avoidance

To reduce the computational cost of the simulation, it was assumed that the mechanical design either prevents two consecutive links to collide or the collision avoidance is handled by bounds on the corresponding joint angles. Thus, for an n_R serial-link robot

$$c_{RR} = \frac{1}{2}\frac{n_R!}{(n_R-2)!} \tag{10}$$

self collisions and, in general,

$$c_{RE} = n_R n_E \tag{11}$$

collisions with the n_E environment have to be monitored.

4.3. Problem Formulation

The optimization was set up as a non-linear global optimization problem with non-linear equality and inequality constraints

$$\begin{aligned} \min_{\mathbf{x},\,\mathbf{p}} \quad & J(\mathbf{x},\,\mathbf{p}) \\ \text{s.t.} \quad & \mathbf{f}(\mathbf{x},\,\mathbf{p}) - \mathbf{z}_d = \mathbf{0} \\ & -\mathbf{c}(\mathbf{x},\,\mathbf{p}) \le \mathbf{0} \\ & \mathbf{ub}(\mathbf{x},\,\mathbf{p}) \le \mathbf{0} \\ & \mathbf{lb}(\mathbf{x},\,\mathbf{p}) \le \mathbf{0} \end{aligned} \tag{12}$$

with the objective function $J(\mathbf{x},\,\mathbf{p})$ and the N geometric parameters or link lengths of the candidate linkage

$$\mathbf{p} = \begin{bmatrix} l_1 & l_2 & \dots & l_N \end{bmatrix}^T \in \mathbb{R}^N. \tag{13}$$

The vector $\mathbf{x}$, see (8), contains the generalized coordinates in joint space for μ different desired Cartesian positions

$$\mathbf{z}_d = \begin{bmatrix} \mathbf{z}_{d,1}^T & \mathbf{z}_{d,2}^T & \cdots & \mathbf{z}_{d,\mu}^T \end{bmatrix}^T \in \mathbb{R}^{6\mu}. \tag{14}$$

The inverse kinematics is included with the equality constraint

$$\mathbf{f}(\mathbf{x},\mathbf{p}) - \mathbf{z}_d = \mathbf{0}. \tag{15}$$

The vector function

$$\mathbf{c}(\mathbf{x},\,\mathbf{p}) \in \mathbb{R}^{c_{RR}+c_{RE}} \tag{16}$$

contains the minimal distances according to self-collisions and collisions with the environment. The vector functions

$$\mathbf{ub}(\mathbf{x},\,\mathbf{p}),\ \mathbf{lb}(\mathbf{x},\,\mathbf{p}) \in \mathbb{R}^{n\mu+N} \tag{17}$$

are upper and lower bounds on the joint angles and link lengths.

4.4. Objective Function

As already discussed in Section 2, the desired objective function should minimize the DoF, the robot link lengths and other kinematic and dynamic performance criteria. This is expressed as linear combination of the objectives

$$J(\mathbf{x},\,\mathbf{p}) = \underbrace{\mathbf{k}^T\mathbf{g}(\mathbf{p})}_{J_1(\mathbf{p})} + \underbrace{\mathbf{\Gamma}^T(\mathbf{x},\,\mathbf{p})\mathbf{K}\mathbf{\Gamma}(\mathbf{x},\,\mathbf{p})}_{J_2(\mathbf{x},\,\mathbf{p})}. \tag{18}$$

In the following Sections 4.4.1 and 4.4.2 the intended effects of J_1 and J_2 on the optimization problem are discussed. The term J_1 penalizes the length of the robot links with a mapping $\mathbf{g} : \mathbb{R}^N \rightarrow \mathbb{R}^N$ and the weighting factor $\mathbf{k} \in \mathbb{R}^N$. The term J_2 represents ν kinematic and dynamic criteria summarized in $\mathbf{\Gamma}(\mathbf{x},\ \mathbf{p}) \in \mathbb{R}^{n\mu\nu}$, which is weighted with the diagonal matrix $\mathbf{K} \in \mathbb{R}^{(n\mu\nu)\times(n\mu\nu)}$.

4.4.1. Minimizing the DoF and Link Lengths

The term J_1 penalizes the robot link lengths in a way such that both the link lengths and the DoF of the robot are minimized. This requires a specific type of function, called pruning function (see Definition 1), for $\mathbf{g}$ in (18). A more detailed demonstration of the behavior of pruning functions is shown in Example 1. As a result, the total link length of the mechanical structure is minimized and, if possible (with respect to all constraints), geometric parameters are driven to zero and, thus, indicate a possible reduction of DoF.

4.4.2. Minimizing Kinematic and Dynamic Performance Criteria

The term J_2 of (18) accounts for arbitrary kinematic or dynamic criteria summarized in vector $\mathbf{\Gamma}(\mathbf{x},\ \mathbf{p})$ and multiplied with a weighting matrix $\mathbf{K}$. It is important that multiple criteria do not contradict each other or, in the best case, are complementary, in order to avoid ill-conditioned optimization problems. Possible criteria are the motor torque, distance from singularities, error propagation through the mechanical structure or kinematic and dynamic manipulabilities. Examples for such criteria are shown in Sections 5.3 and 6.2.

5. Application: Cavity Inspection Robot

Radio frequency cavities (see Figure 7) perform the linear acceleration of charged particles in straight sections of accelerator machines and, thus, make up one of the key elements in a collider complex [12]. The cavities structure and geometry define their specific radio-frequency at which the strong electromagnetic field, created inside the tubes, oscillates to accelerate each particle passing through. The inner surface quality of the cavities is critical for withstanding high energy densities, since every scratch or crack leads to higher local resistance and, thus, a rapid increase in temperature during operation and, in the end, to the failure of the system. Therefore, some kind of automated, mechanical structure has to follow the complex cavity geometry and take records of the surface quality after full assembly of the cavities. Finding the optimal topology of such a mechanical structure with respect to certain constraints, such as collision avoidance for different cavity types and minimal error propagation in direction perpendicular to the cavity surface, is a perfect example of the generic problem described in Section 1. Currently, several different system have been developed and are able to partially scan cavities (see [13–16]). However, those previously developed cavity inspection systems were extensively tested at CERN but did not satisfy the specific requirements concerning the level of automation, accuracy, repeatability and how much of the inner cavity surface could be inspected and mapped. The first prototype of the system developed at CERN and presented in [1] was not able to scan all three cavities with one robotic arm, but had to use two different arms. The aim of this design optimization is to find one topology and geometry that can handle all three cavity types and thus increase robustness and level of automation while decreasing the cost of such a system.

The main challenge for a robotic system is the complex workspace and, especially, the difference in diameter of the entrance of the smallest cavity (FCC) and the point with maximum diameter of the biggest cavity (LHC). Furthermore, the system has to detect surface anomalies of only 10 μm. A 18MP camera with liquid lens, allowing it to focus between 20 to 25 mm, is used. In order to provide one full image of the inner surface, the cavities are rotated around their axes of symmetry, while robotic manipulators are inserted along these axes. The pictures are stitched together after the inspection. Thus, the accuracy error of the end-effector position tangential to cavity surface should be not more than 1.2 mm to obtain only 10% overlapping error and not more than 1mm in the direction perpendicular to the cavity surface, which otherwise changes the contained surface area in the image.

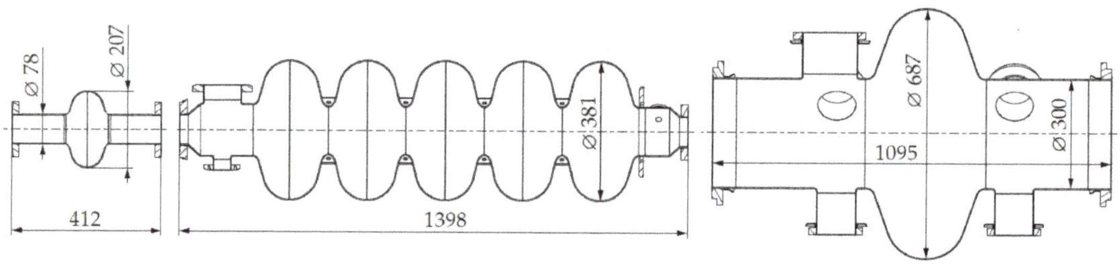

Figure 7. Cavity types for FCC, LINAC and LHC (left to right, all units in mm).

In Section 5.1 the initial model and geometry and the projection equation [17] to derive the equation of motion is described. Section 5.2 illustrates the collision-checking procedure and Section 5.3 presents the applied kinematic criteria for this example. Then, the initial states and results of the optimization are shown in Sections 5.4 and 5.5, respectively.

5.1. Model

The surrogate model defining the design space has been set up with one translational and four rotational joints with parallel rotation axes, summing up to five DoF, as shown in Figure 8. Table 1 lists the initial link lengths, where $l_{i,j}$ describes the lengths from joint i to joint j. These lengths are evaluated by launching an optimization problem, as described in Section 5.4, that returns feasible initial states. This initial set up has been used as the starting point for the actual optimization problem in (12).

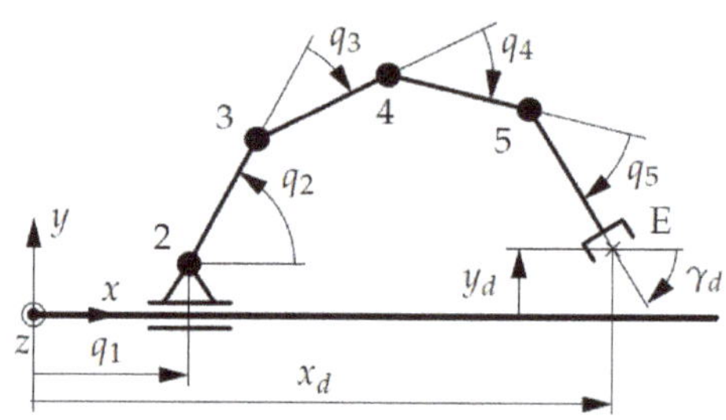

Figure 8. Design space.

Table 1. Initial geometry.

Length	[mm]
$l_{2,3}$	144.6
$l_{3,4}$	80.3
$l_{4,5}$	80.3
$l_{5,E}$	16.0

The generalized joint coordinates are set to

$$\mathbf{q} = \begin{bmatrix} q_1 & q_2 & q_3 & q_4 & q_5 \end{bmatrix}^T \in \mathbb{R}^{5\times 1} \tag{19}$$

and the desired Cartesian position and orientation are

$$\mathbf{z}_d = \begin{bmatrix} x_d & y_d & \gamma_d \end{bmatrix}^T \in \mathbb{R}^{3\times 1}, \tag{20}$$

with the position x_d and y_d and the orientation around the z-axis γ_d always keeping the end-effector orientation perpendicular to the cavity surface. The optimization parameter is set to

$$\mathbf{x} = \begin{bmatrix} \mathbf{q}_{1,1}^T & \cdots & \mathbf{q}_{1,\mu}^T & \mathbf{q}_{2,1}^T & \cdots & \mathbf{q}_{2,\mu}^T & \mathbf{q}_{n_{cav},1}^T & \cdots & \mathbf{q}_{n_{cav},\mu}^T \end{bmatrix}^T, \tag{21}$$

with the $n_{cav} = 3$ different cavities and μ different positions leading to $\mathbf{x} \in \mathbb{R}^{(n\mu n_{cav})\times 1}$. The parameter vector is defined as

$$\mathbf{p} = \begin{bmatrix} l_{2,3} & l_{3,4} & l_{4,5} & l_{5,E} \end{bmatrix}^T. \tag{22}$$

5.2. Collision Avoidance

The collisions between robot and environment are calculated by discretizing the cavity surfaces and checking the minimal distances between robot links and points on the cavity surface. First, all points on the cavity surface are transformed into the body fixed coordinate frames for each robot link and then, to decrease the computation time, only points that can possibly collide and, thus, lie in the selected points area (see Figure 9) are considered. The minimal value of the projection of all possible collision points onto the y axis of the body-fixed coordinate frame is considered the minimal distance. Collisions are taken into account in the function $\mathbf{c}(\mathbf{x},\ \mathbf{p})$ in (12). Self collisions between robot links are not considered, since it is assumed that this is mechanically impossible, as known from Scara robots.

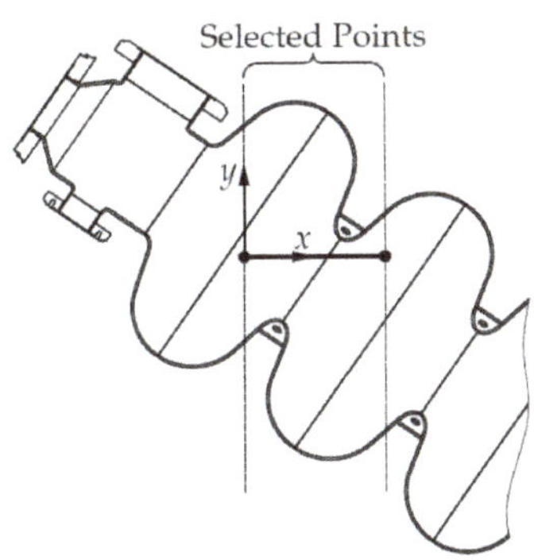

Figure 9. Collision detection for one robot link.

5.3. Kinematic and Dynamic Performance Criteria

It is crucial for the cavity inspection robot to provide a very stable base for the camera that is being used for the inspection process, since the errors that should be detected in the surface can be only micro fractures. Thus, the term J_2 of (18) is set up to optimize the error propagation through the mechanical structure in a certain direction of interest. Error propagation describes how errors that originate in joint space are being forwarded to the end-effector, such as, e.g., gear elasticity, backlash or control oscillations. This can be quantified using the directional kinematic manipulability [18]

$$w_j = \sum_{i=1}^{3} \left| \mathbf{n}_j^T \mathbf{u}_{j,i} \sigma_{j,i} \right| \tag{23}$$

with the unit vector $\mathbf{n}_j$ representing the direction of interest (perpendicular to the cavity surface) and the major and minor axes of the manipulability ellipsoid $\sigma_{j,i}\mathbf{u}_{j,i}$ obtained from the singular value decomposition of the geometric Jacobian

$$\mathbf{J}(\mathbf{q}, \mathbf{p}) = \left[\left(\frac{\partial \mathbf{v}_E}{\partial \dot{\mathbf{q}}} \right)^T \left(\frac{\partial \omega_E}{\partial \dot{\mathbf{q}}} \right)^T \right]^T, \tag{24}$$

with the linear and angular end-effector velocities $\mathbf{v}_E$ and ω_E, respectively. Looking at the mapping from joint to Cartesian space via the Jacobian and replacing the small changes in joint angles $\Delta\mathbf{q}$ with an error $\mathbf{e}$, the error in Cartesian space is

$$\Delta\mathbf{z}_e = \mathbf{J}(\mathbf{q})\mathbf{e}. \tag{25}$$

Thus, for a robot in a singular configuration such as the two-link arm in Figure 2b, the error propagation in direction of the eigenvector corresponding to the smallest singular value is zero and, thus, the repeatability only depends on manufacturing tolerances of the mechanical parts of the robot. This means that the optimization algorithm prefers configurations for which the repeatability is less dependent on the quality of gears or control.

Finally, the directional kinematic manipulability measure can be written in vector form according to (18) as

$$\boldsymbol{\Gamma}(\mathbf{x},\ \mathbf{p}) = \begin{bmatrix} w_1 & w_2 & \dots & w_\mu \end{bmatrix}^T, \tag{26}$$

with the weighting martix $\mathbf{K} \in \mathbb{R}^{\mu\times\mu}$.

5.4. Initial Configuration

The initial states, such as joint angles and link lengths, heavily influence the performance of optimization algorithms. It is important to provide feasible (in terms of the given constraints) initial states as a starting point for the optimization solvers. Starting points can be generated by either an inverse kinematics method or, as is done here, by running the optimization with the objective function (18) set to $J(\mathbf{x},\ \mathbf{p}) = 0$. Thus, the initial states are feasible with respect to all constraints, but non-optimal. The initial configurations and link lengths used as a starting point for the optimization solvers are shown for the three cavities in Figures 10–12.

The black curves represent the cavities and the colored lines represent the robot links. The gray circles indicate the desired end-effector position (x_d, y_d) from vector $\mathbf{z}_d = \begin{bmatrix} x_d & y_d & \gamma_d \end{bmatrix}^T$.

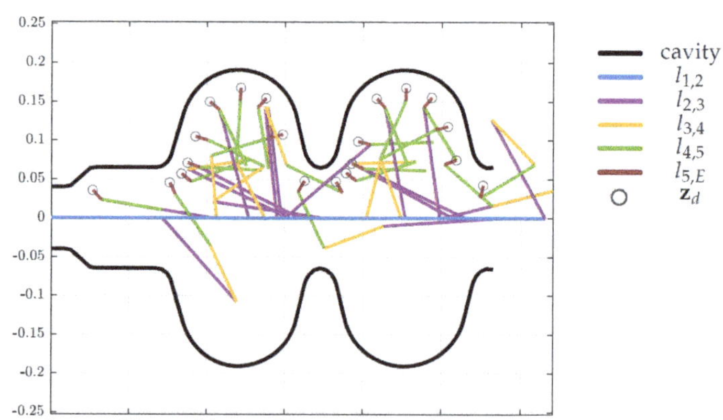

Figure 10. Initial states for the LINAC cavity (units in m).

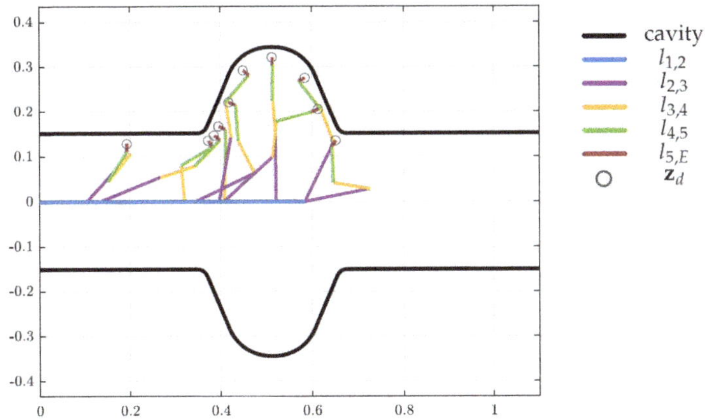

Figure 11. Initial states for the LHC cavity (units in m).

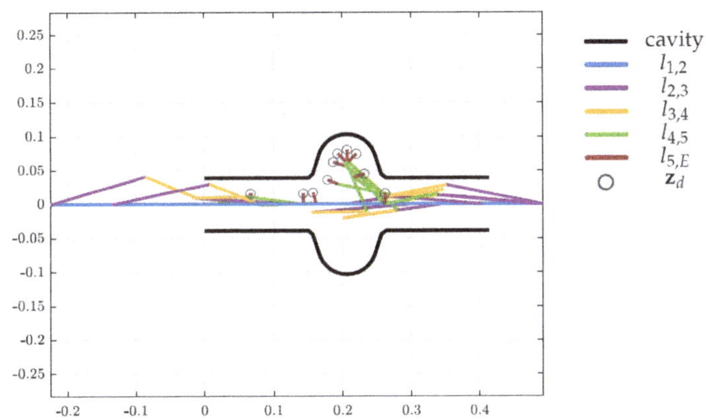

Figure 12. Initial states for the FCC cavity (units in m).

5.5. Optimization Results

The optimization depart from starting points defined for every desired end-effector position and cavity. The starting points have been found by an optimization (described in Section 5.4) and include an initial topology and geometry, as well as the configuration of the arm. Matlab's *fmincon* function is internally used as a local optimization solver, in this case, applying the interior-point algorithm [19]. The *MultiStart* and *GlobalSearch* methods are applied to solve the global optimization problem [20]. In a comparison with evolutionary algorithms, the *GlobalSearch* and *MultiStart* methods leads to better results.

In Section 5.5.1 the design optimization is launched using only the LINAC environment for demonstration purposes and in Section 5.5.2 the robotic arm is optimized to operate in all three cavities.

5.5.1. LINAC Cavity

Here, the design optimization is done only for the LINAC cavity in order to demonstrate the behaviour of the algorithm with a simpler, and, hence, more intuitive example. The initial states match the ones presented in Figure 10 and the corresponding optimized design is visualized in Figure 13.

As mentioned in Section 5.1 the desired Cartesian position and orientation $\mathbf{z} \in \mathbb{R}^3$ is of dimension three and, thus, a mechanical structure with exactly three DoF is able to reach all desired end-effector positions $\mathbf{z}$ in an environment without obstacles or other constraints. As shown in Figure 13 the restriction caused by the LINAC cavity allows a three-DoF robotic arm to reach all positions. Thus, the optimization reduces the tentative topology by two DoF, which is visible in Figure 14, but more clear in Table 2, where the lengths set to zero correspond to the removed DoF.

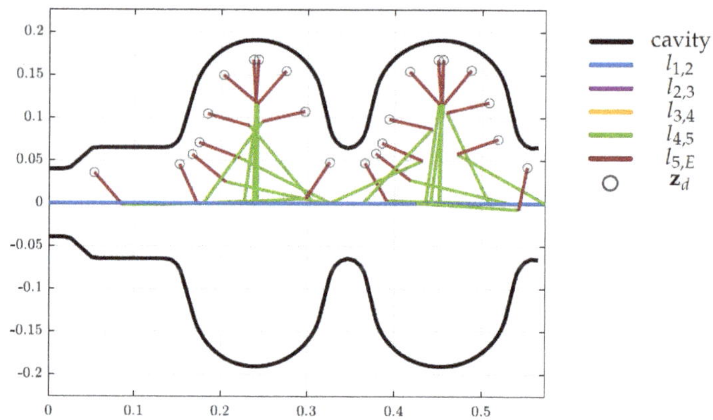

Figure 13. Optimized design for the LINAC cavity (units in m).

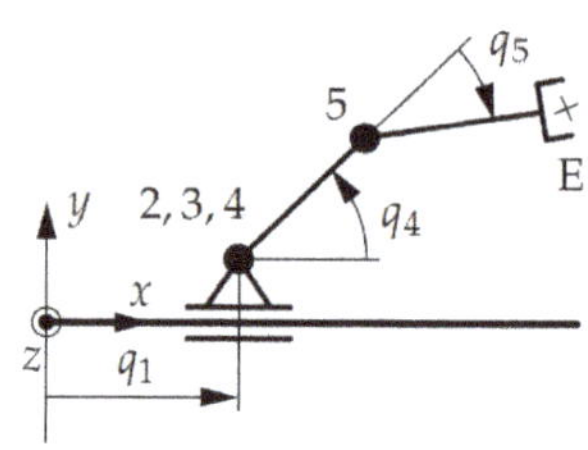

Figure 14. Optimized topology-LINAC.

Table 2. Optimized geometry-LINAC.

Length	[mm]
$l_{1,2}$	q_1
$l_{2,3}$	0
$l_{3,4}$	0
$l_{4,5}$	117.3
$l_{5,E}$	50.5

Furthermore, the length of the robotic arm has been minimized, as becomes clear when looking at the furthest point and observing that all links lie on a straight line from joint 2 to the desired end-effector position, and that this line is perpendicular to the x axis. In Figure 14 the schematic drawing of the optimal topology is shown with some joints coinciding with others and, thus, illustrating the reduction of DoF.

5.5.2. Full System

The design optimization for the full system is performed by considering all three cavities as collision objects in the non-linear inequality constraints of the optimization problem. Again, the results of the optimization are shown in Figures 15–17 for the LINAC, LHC and FCC respectively, where the colored lines represent the robot links, the black curves indicate the surface of the cavities and the gray circles show the desired end-effector positions. The resulting topology is shown in Figure 18 and the corresponding link lengths are listed in Table 3.

As is clearly visible in Table 3, at least one link length is set to zero, which means that our requirement for the tentative topology of starting the design optimization with at least one DoF higher than the expected optimal solution holds and, thus, an optimized design for the robotic arm has been found.

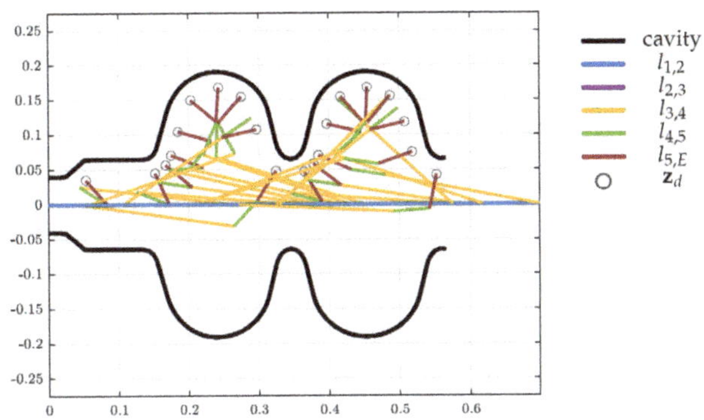

Figure 15. Optimized design for the LINAC cavity (units in m).

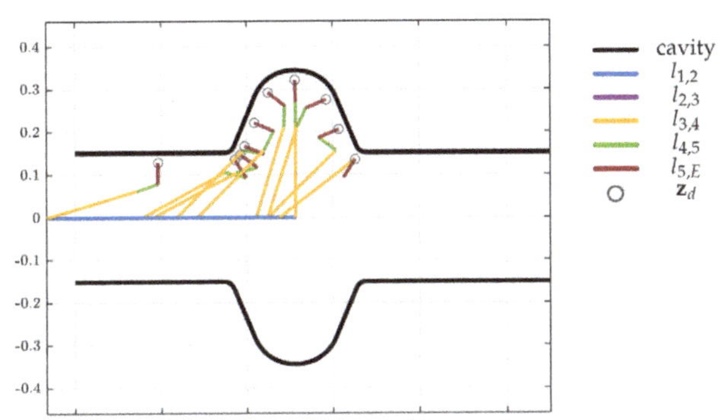

Figure 16. Optimized design for the LHC cavity (units in m).

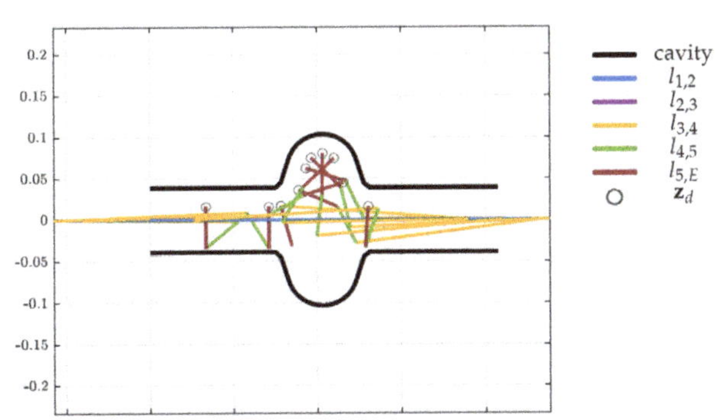

Figure 17. Optimized design for the FCC cavity (units in m).

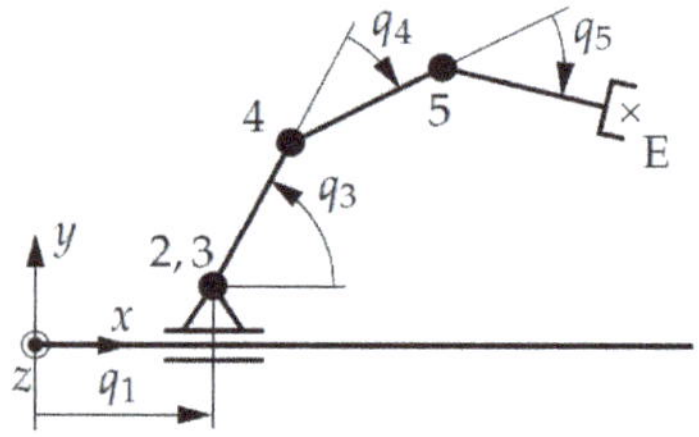

Figure 18. Optimized topology.

Table 3. Optimized geometry.

Length	[mm]
$l_{1,2}$	q_1
$l_{2,3}$	0
$l_{3,4}$	221.0
$l_{4,5}$	51.0
$l_{5,E}$	50.0

6. Application: FCC Manipulator

A detailed report on the design optimization of the FCC Manipulator (a robotic system for CERN's Future Circular Collider) has already been published in [2]. Here, only a brief summary of the assumptions and results are presented to demonstrate the algorithms capabilities.

The Future Circular Collider (FCC) is suggested to unlock observations in higher energy ranges than it is possible, now, with the current accelerator machines at CERN [21]. This particle accelerator is able to generate a center-of-mass energy of 100 TeV and has a planned circumference of 100 km (see [22,23]). The 2020 Update of the European Strategy for Particle Physics has listed the further investigation of the FCC as one of three main priorities and, thus, have launched a Technical Design Report (TDR). One of the studies that has been launched within the TDR concerns the automation of maintenance, inspection and emergency handling along the 100-km-long FCC tunnel. The automation of these tasks plays a significant role for downtime, reliability and safety of particle accelerators and decreases the radiation exposure of workers.

The tasks such an automated system has to handle and the environment it must operate in are well defined, but no restrictions on the actual design of the manipulator in terms of topology and geometry are given. Thus, the presented algorithm has been applied to optimize the tentative design of the FCC robot. In Section 6.1 the model and some assumptions are described. In Section 6.2 the applied objective function, in terms of kinematic and dynamic criteria, is analyzed and finally some results are shown in Section 6.3.

6.1. Manipulator and Environment

The process of defining the surrogate model, as shown in Figure 19, is described in detail in [2]. Here, it should just be mentioned that joints 1 and 2 are translational joints, 3–6 and 7–10 form two planar mechanisms in order to fold the arm. Joints 11–13 represent a robotic wrist and, thus, the solution for its position and orientation can be decoupled in point 12, which simplifies the optimization problem. The planar chains allow for minimizing the link lengths between joints with parallel axes, and, thus, possibly eliminating joints connected by links with zero length, hence, eventually reducing the DOF. The initial link lengths in Table 4 have been found by the same means as described in Section 5.4.

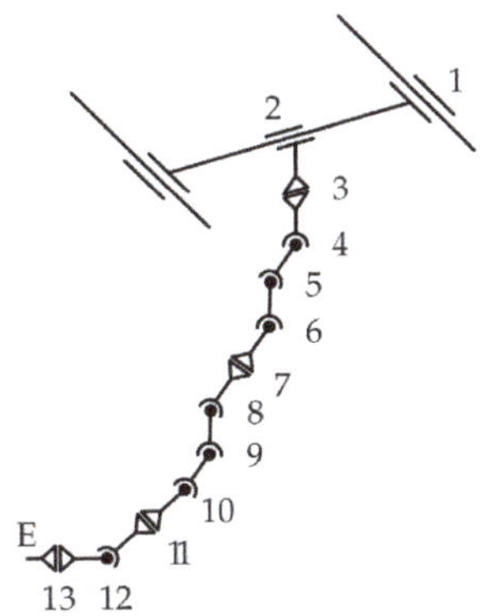

Figure 19. Design space using VDI2861 (from [2] under CC BY-NC-ND 4.0).

Table 4. Initial geometry.

Length	[mm]
$l_{1,2}$	q_2
$l_{2,4}$	288
$l_{4,5}$	500
$l_{5,6}$	500
$l_{6,8}$	400
$l_{8,9}$	400
$l_{9,10}$	400
$l_{10,12}$	200
$l_{12,E}$	100

The environment of the FCC tunnel and the robot have been approximated by convex geometric primitives; here, specifically, by Matlab's AlphaShapes [24], which can easily be passed to a function to calculate the minimal distance between two AlphaShapes. The approximation of the FCC environment with cylinders and boxes is indicated by red, dashed lines in Figures 20 and 21 for the FCC-ee and FCC-hh machines, respectively. The manipulator should be able to reach all points of interest (I–V), which represents the most diverse remote maintenance tasks at the current Large Hadron Collider—LHC, see [25–28].

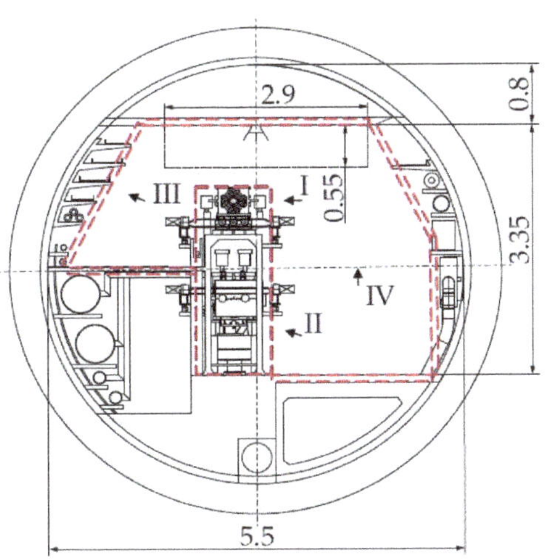

Figure 20. FCC-ee Cross Section Layout (units in m, from [2] under CC BY-NC-ND 4.0).

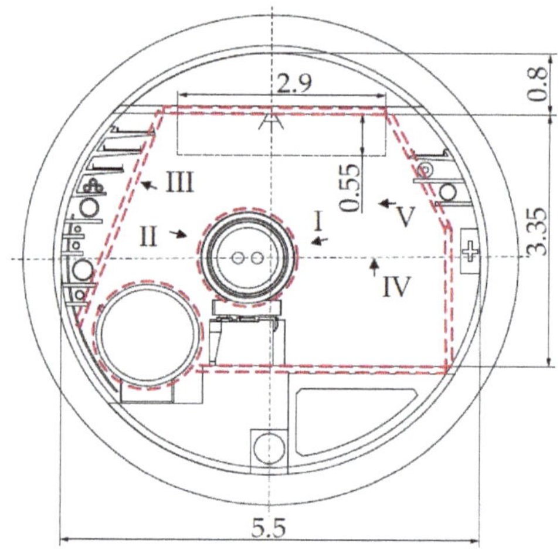

Figure 21. FCC-hh Cross Section Layout (units in m, from [2] under CC BY-NC-ND 4.0).

6.2. Kinematic and Dynamic Performance Criteria

A main objective for the design of the manipulator is to reduce the motors' torques, since the workspace, shown in Figures 20 and 21, requires a relatively long robotic arm, compared with the desired payload and weight of the robot. Thus, the dynamic measure applied in term J_2 in the objective function (18) is the motor torque of each joint. A dynamic robot model in the form (9) is used and included in the objective function (18) with

$$\mathbf{\Gamma}(\mathbf{x},\ \mathbf{p}) = \mathbf{Q}(\mathbf{q}, \dot{\mathbf{q}}, \ddot{\mathbf{q}}, \mathbf{p}) \tag{27}$$

and the weighting matrix $\mathbf{K} \in \mathbb{R}^{n \times n}$.

6.3. Optimization Results

The final results of the design optimization shows a reduced topology by two DoF. This is shown in Figure 22 by the coinciding joints 5/6 and 8/9, which correspond to the lengths $l_{5,6} = l_{8,9} = 0$ in Table 5 of the optimized geometry. In Figures 23 and 24 the optimization results are visualized using the collision objects (AlphaShapes) for the two different accelerator machines, FCC-ee and FCC-hh.

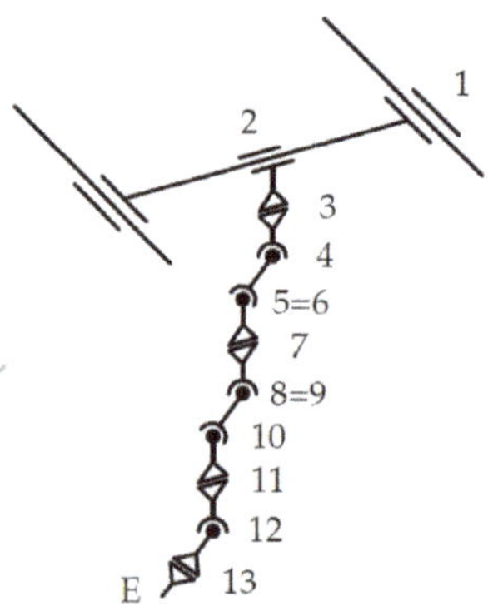

Figure 22. Optimized topology using VDI2861 (from [2] under CC BY-NC-ND 4.0).

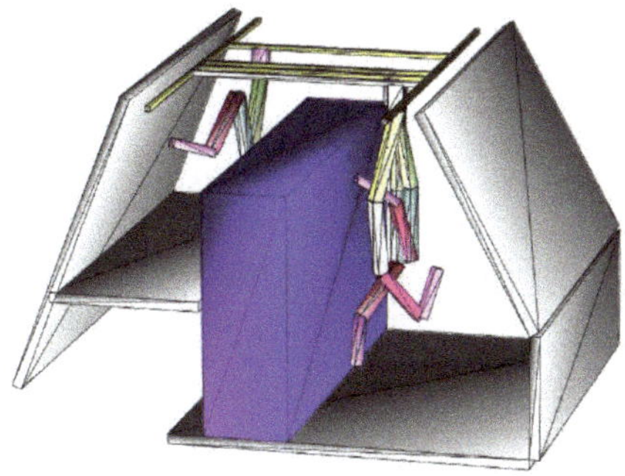

Figure 23. Optimization results FCC-ee (from [2] under CC BY-NC-ND 4.0).

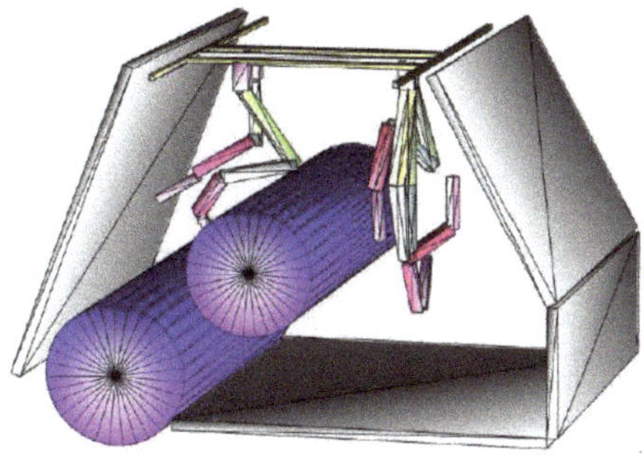

Figure 24. Optimization results FCC-hh (from [2] under CC BY-NC-ND 4.0).

Table 5. Optimized geometry.

Length	[mm]
$l_{1,2}$	q_2
$l_{2,4}$	288
$l_{4,5}$	927
$l_{5,6}$	0
$l_{6,8}$	754
$l_{8,9}$	0
$l_{9,10}$	635
$l_{10,12}$	518
$l_{12,E}$	100

7. Discussion and Conclusions

A kinematic model pruning method has been presented that allows transforming discrete optimization criteria into a continuous representation and, thus, enables simultaneous optimization of the topology and geometric parameters. The method has been demonstrated for serial mechanical structures and, therefore, covers a wide variety of robotic manipulators. The algorithm uses well-known and readily available optimization schemes without any additional, complex frameworks. The simplicity of the presented kinematic model pruning method is surely one of its main advantages, next to the simultaneous optimization of all additional criteria. It is especially simple, since no inverse kinematics are computed and, for the case without additional dynamic performance criteria, no dynamic model is needed. Thus, this method can be performed providing only the forward kinematics and (if necessary) collision detection. This allows for very quick prototyping and one does not have to rely on either complex frameworks or an "educated guess" for a new manipulator design, but, instead, can quickly produce quantified results based on deterministic performance criteria. As demonstrated with examples in Sections 5 and 6, the algorithm has led to good results for both the cavity inspection arm and the FCC Manipulator.

The choice of additional kinematic and dynamic criteria needs to be made carefully in order to not generate ill-conditioned optimization problems or even contradictions with the requirements. The weighting matrices in (18) have a major impact on the final results and can, as well, lead to infeasible solutions, in certain cases. Finding guidelines for these optimization parameters based on the mechanical structure and environment would simplify this heuristic process. The current implementation only allows for DoF reduction on two subsequent joints of the same type. An extension of the presented method to a more general use case, including closed kinematic chains, will be the subject of future work. Furthermore, a more general study on stability, convergence rates and the effects of different optimization schemes applying the proposed pruning function will be investigated.

Author Contributions: Formal analysis, H.G. (Hannes Gamper); Methodology, H.G. (Hannes Gamper); Supervision, H.G. (Hubert Gattringer), A.M. and M.D.C.; Validation, A.L.; Writing—original draft, H.G. (Hannes Gamper) and A.M. All authors have read and agreed to the published version of the manuscript.

Funding: This research received no external funding.

Institutional Review Board Statement: Not applicable.

Informed Consent Statement: Not applicable.

Conflicts of Interest: The authors declare no conflict of interest.

Abbreviations

The following abbreviations are used in this manuscript:

DoF	degrees of freedom
LHC	Large Hadron Collider
LINAC	linear accelerator
RF	radio frequency
FCC	Future Circular Collider
FCC-ee/-hh	Future Circular Collider-lepton/-hadron machine layout

References

1. Luthi, A.; Macpherson, A.; Rosario Buonocore L. ; Gamper H. ; DiCastro, M. Camera Placement in a Short Working Distance Optical Inspection System for RF Cavities. In Proceedings of the International Conference on RF Superconductivity, Virtual Conference, 28 June–2 July 2021.
2. Gamper, H.; Gattringer, H.; Müller, A.; Di Castro, M. Design Optimization of a Manipulator for CERN's Future Circular Collider (FCC). In Proceedings of the 18th International Conference on Informatics in Control, Automation and Robotics—ICINCO 2021, Online, 6–8 July 2021; pp. 320–329. [CrossRef]
3. Kelaiaia, R.; Company, O.; Zaatri, A. Multiobjective optimization of a linear Delta parallel robot. *Mech. Mach. Theory* **2012**, *50*, 159–178. [CrossRef]
4. Bi, Z.M.; Zhang, W.J. Concurrent optimal design of modular robotic configuration. *J. Robot. Syst.* **2001**, *18*, 77–87. [CrossRef]
5. Henry, R.; Chablat, D.; Porez, M.; Boyer, F.; Kanaan, D. Multi-Objective Design Optimization of the Leg Mechanism for a Piping Inspection Robot. In Proceedings of the 38th Mechanisms and Robotics Conference, Buffalo, NY, USA, 17–20 August 2014; Volume 5A. [CrossRef]
6. Van Henten, E.; Van't Slot, D.; Hol, C.; Van Willigenburg, L. Optimal manipulator design for a cucumber harvesting robot. *Comput. Electron. Agric.* **2009**, *65*, 247–257. [CrossRef]
7. Lum, M.J.; Rosen, J.; Sinanan, M.N.; Hannaford, B. Kinematic optimization of a spherical mechanism for a minimally invasive surgical robot. In Proceedings of the IEEE International Conference on Robotics and Automation, New Orleans, LA, USA; 26 April–1 May 2004; Volume 2004, pp. 829–834. [CrossRef]
8. Schappler, M.; Ortmaier, T. Dimensional Synthesis of Parallel Robots: Unified Kinematics and Dynamics using Full Kinematic Constraints. In Proceedings of the Sechste IFToMM D-A-CH Konferenz 2020, Lienz, Austria, 27–28 February 2020; Volume 2020. [CrossRef]

9. Rodriguez, D.A.R. *Automatic Generation of Task-Specific Serial Mechanisms Using Combined Structural and Dimensional Synthesis*; Institutionelles Repositorium der Leibniz Universität Hannover: Hannover, Germany; 2018. [CrossRef]
10. Ha, S.; Coros, S.; Alspach, A.; Kim, J.; Yamane, K. Computational co-optimization of design parameters and motion trajectories for robotic systems. *Int. J. Robot. Res.* **2018**, *37*, 1521–1536. [CrossRef]
11. Whitman, J.; Choset, H. Task-Specific Manipulator Design and Trajectory Synthesis. *IEEE Robot. Autom. Lett.* **2019**, *4*, 301–308. [CrossRef]
12. CERN. Accelerating: Radiofrequency Cavities. Available online: https://home.cern/science/engineering/accelerating-radiofrequency-cavities (accessed on 2 December 2021).
13. Watanabe, K. Review of optical inspection methods. In Proceedings of the 14th International Conference on RF Superconductivity, Berlin-Dresden, Germany, 20–25 September 2009.
14. Wenskat, M. Automated optical inspection and image analysis of superconducting radio-frequency cavities. *J. Instrum.* **2017**, *12, 1–10.* . [CrossRef]
15. Lemke, M.; Elsen, E.; Aderhold, S.; Cornett, U.; Falley, G.; Karstensen, S.; Külper, T.; Navitski, A.; Schaffran, J.; Schlander, F.; et al. Optical Bench for Automated Cavity Inspection with High Resolution on Short Time Scales. ILC-HiGrade-Reports 2013. Available online: https://bib-pubdb1.desy.de/record/166454/files/ILC-HiGrade-2013-001-1.pdf?version=1 (accessed on 14 December 2021).
16. Iwashita, Y.; Tajima, Y.; Hayano, H. Development of High Resolution Camera for Observations of Superconducting Cavities. *Rev. Mod. Phys. Sept.* **2008**. [CrossRef]
17. Gattringer, H. *Starr-Elastische Robotersysteme*; Springer: Berlin/Heidelberg, Germany, 2011.
18. Nait-Chabane, K.; Hoppenot, P.; Colle, E. Directional Manipulability for Motion Coordination of an Assistive Mobile Arm. In Proceedings of the 4th International Conference on Informatics in Contronl, Automation and Robotics, Angers, France, 9–12 May 2007.
19. Byrd, R.H.; Hribar, M.E.; Nocedal, J. An Interior Point Algorithm for Large-Scale Nonlinear Programming. *SIAM J. Optim.* **1999**, *9*, 877–900. [CrossRef]
20. Ugray, Z.; Lasdon, L.; Plummer, J.; Glover, F.; Kelly, J.; Marti, R. Scatter Search and Local NLP Solvers: A Multistart Framework for Global Optimization. *INFORMS J. Comput.* **2007**, *19*, 328–340. [CrossRef]
21. CERN. Future Circular Collider (FCC). Available online: https://home.cern/science/accelerators/future-circular-collider (accessed on 8 December 2021).
22. CDR. FCC-ee: The Lepton Collider. *Eur. Phys. J. Spec. Top.* **2019**, *228*, 261–623. [CrossRef]
23. CDR. FCC-hh: The Hadron Collider. *Eur. Phys. J. Spec. Top.* **2019**, *228*, 755–1107. [CrossRef]
24. The MathWorks Inc., Matlab Version 2019, Computer software. Available online: https://ch.mathworks.com/ (accessed on 10 July 2021).
25. Bestmann, P.; CERN The Control of the LHC Alignment Using a Robot, CERN-TS-NOTE-2008-020. May 2008. Available online: https://cds.cern.ch/record/1119514 (accessed on 2 December 2021)
26. Missiaen, D.; Steinhagen, R.; Quesnel, J. The Alignment of the LHC. In proceedings of the 23rd Particle Accelerator Conference, Vancouver, BC, Canada, 4–8 May 2009.
27. Valentino, G.; Aßmann, R.; Bruce, R.; Redaelli, S.; Rossi, A.; Sammut, N.; Wollmann, D. Semiautomatic beam-based LHC collimator alignment. *Am. Phys. Soc.* **2012**, *15*, 051002. [CrossRef]
28. Bajko, F.B.M. *Report of the Task Force on the Incident of 19 September 2008 at the LHC*; LHC Project Report 1168; 2009. Available online: https://inspirehep.net/literature/824814 (accessed on 2 December 2021).

robotics

MDPI

Article

Kinematic Synthesis and Analysis of the RoboMech Class Parallel Manipulator with Two Grippers

Zhumadil Baigunchekov [1,2], Med Amine Laribi [3,*], Azamat Mustafa [2,*] and Abzal Kassinov [1]

1 Department of Mechanics, Al–Farabi Kazakh National University, 71 Al–Farabi Ave., Almaty 050040, Kazakhstan; bzh47@mail.ru (Z.B.); kasinov07@gmail.com (A.K.)
2 Department of Mechanical Engineering and Modeling, Satbayev University, 22a Satpaev Str., Almaty 050013, Kazakhstan
3 Department GMSC, Institut PPRIME, CNRS, Université de Poitiers, ENSMA, UPR 3346, 86962 Poitiers, France
* Correspondence: med.amine.laribi@univ-poitiers.fr (M.A.L.); mustafa_azamat@mail.ru (A.M.)

Abstract: In this paper, methods of kinematic synthesis and analysis of the RoboMech class parallel manipulator (PM) with two grippers (end effectors) are presented. This PM is formed by connecting two output objects (grippers) with a base using two passive and one negative closing kinematic chains (CKCs). A PM with two end effectors can be used for reloading operations of stamped products between two adjacent main technologies in a cold stamping line. Passive CKCs represent two serial manipulators with two degrees of freedom, and negative CKC is a three-joined link with three negative degrees of freedom. A negative CKC imposes three geometric constraints on the movements of the two output objects. Geometric parameters of the negative CKC are determined on the basis of the problems of the Chebyshev and least-square approximations. Problems of positions and analogues of velocities and accelerations of the PM with two end effectors have been solved.

Keywords: parallel manipulator; RoboMech; kinematic synthesis and analysis; Chebyshev and least-square approximations

Citation: Baigunchekov, Z.; Laribi, M.A.; Mustafa, A.; Kassinov, A. Kinematic Synthesis and Analysis of the RoboMech Class Parallel Manipulator with Two Grippers. *Robotics* **2021**, *10*, 99. https://doi.org/10.3390/robotics10030099

Academic Editor: Raffaele Di Gregorio

Received: 12 June 2021
Accepted: 30 July 2021
Published: 3 August 2021

1. Introduction

There are technological processes in industry where it is necessary to perform several operations simultaneously or sequentially, for example, in stamping production, in loading and unloading operations. For the simultaneous or sequential execution of several operations, it is advisable to use manipulation robots with many end effectors.

In this paper, a PM with two end effectors is synthesized that can be used to perform reloading operations from one technological equipment to another. This PM with two end effectors replaces two industrial serial robots in the existing production line of cold stamping and it belongs to the RoboMech class PM. The PM, simultaneously setting the laws of motions of the end effectors and actuators, is called the RoboMech class PM [1]. Setting the laws of motion of the actuators monotonously and uniformly but not defining by solving the inverse kinematics problem simplifies the control system and improves dynamics. Replacing two industrial robots with one RoboMech class PM with two end effectors simplifies the control system and increases the productivity and reliability of the technological line.

Since in the RoboMech class PMs simultaneously set the laws of motion of the end effectors and actuators, they work with certain structural schemes and geometric parameters of their links. The existing methods of kinematic analysis and synthesis of mechanisms and manipulators are based on the derivation of loop-closure equations and their study: in kinematic analysis, using known constant geometric parameters of links and variable generalized coordinates, variable parameters characterizing the relative movements of elements of kinematic pairs are determined, and in kinematic synthesis (dimensional or parametric synthesis) for the given positions of the input and output links, constant geometric parameters of the links are determined. Loop-closure equations are derived on the base of vector and matrix methods [2–10], and the theory of screws [11–13], which are leads to polynomials of higher degrees. Then, examining the resulting polynomials using computers, depending on the assigned tasks, the kinematic analysis or synthesis is performed. McCarthy in his papers [14,15] shows the close relationship between the kinematics, synthesis, polynomials, and computations in the 21st century. In the considered approach of kinematic analysis and synthesis of mechanisms and manipulators, it is rather difficult to obtain the polynomials; moreover, with the complication of the structures of mechanisms and manipulators, the formation of polynomials becomes more complicated and their degree increases. Performance analysis and applications of the PMs and robots are also presented in [16–21].

In this paper, kinematic synthesis of the PM with two end effectors is carried out on the basis of a modular approach [22,23], according to which PMs, regardless of their complexity, are formed by connecting the output objects (end effectors) with a base using closing kinematic chains (CKCs), which are structural modules. CKCs can be active, passive, and negative, which have positive, zero, and negative DOFs, respectively. The active and negative CKCs impose geometric constraints on the motions of the output objects, and passive CKCs do not impose geometric constraints. The representation of PMs from separate structural modules simplifies the methods of their investigation.

2. Kinematic Synthesis of the PM with Two Grippers

A PM with two end effectors can be used in a cold stamping technological line for reloading operations between two hydraulic presses [24].

Figure 1 shows a structural scheme of the PM with two end effectors in two positions.

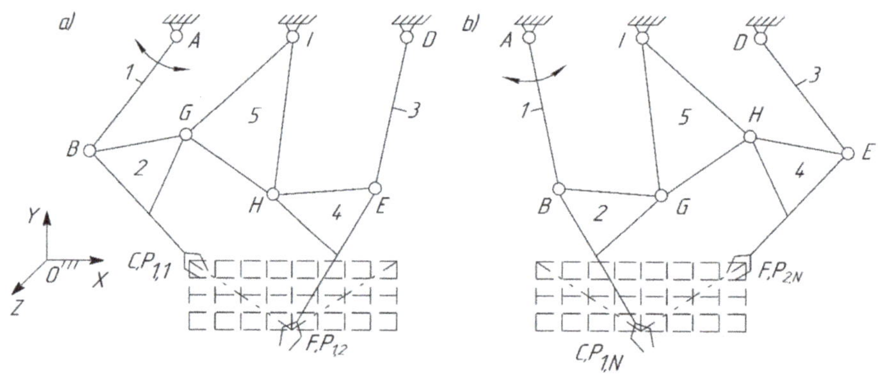

Figure 1. PM with two end effectors in two positions: (**a**) the first position (**b**) the second position.

In the first position (Figure 1a), the first gripper P_1 in position $P_{1,1}$ takes the workpiece after processing in the first hydraulic press for delivery to the store. At this time, the second gripper P_2 in position $P_{2,1}$ takes the previous workpiece processed in the first hydraulic press for delivery to the second hydraulic press for further processing.

In the second position (Figure 1b), the first gripper P_1 in position $P_{1,N}$ delivers the workpiece to the store and the second gripper P_2 in position $P_{2,N}$ delivers the previous workpiece to the second hydraulic press. The cycle is then repeated.

The considered positioning PM with two end effectors is formed by connecting two output objects (grippers P_1 and P_2) with a base using two passive and one negative CKC in the following sequence. First, the grippers P_1 and P_2 are connected to the base using passive CKCs ABC and DEF with revolute kinematic pairs, respectively, which have two degrees of freedom. Since passive CKCs ABC and DEF have two degrees of freedom, they can reproduce the given laws of motion of the output points P_1 and P_2. Then, to form a single movable PM with two end effectors, we connect the links *BC* and *EF* of the passive CKCs *ABC* and *DEF* with the base using a negative CKC *GHI* with three negative degrees of freedom. Figure 2 shows a block structure of the formed PM with two end effectors.

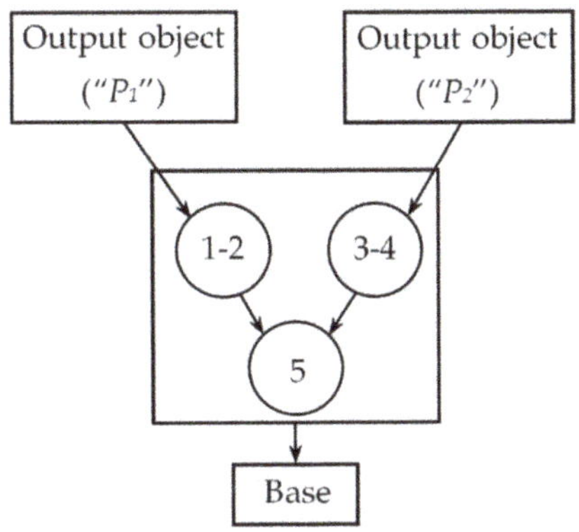

Figure 2. Block structure of the PM with two end effectors.

According to the block structure (Figure 2), the parametric synthesis of the PM with two end effectors (Figure 3) consists of the parametric synthesis of two passive CKCs—*ABC* and *DEF*—and one negative CKC—*GHI*. The parameters of the synthesis of two passive CKCs—*ABC* and *DEF*—are X_A, Z_A, l_{AB}, l_{BC} and X_D, Z_D, l_{DE}, l_{EF}, respectively, where X_A, Z_A and X_D, Z_D are the coordinates of the fixed joints *A* and *D* in the absolute coordinate system *OXYZ*; $l_{AB}, l_{BC}, l_{DE}, l_{EF}$ are the length of the links *AB*, *BC*, *DE*, *EF*. Let denote these parameters by the vectors $\mathbf{p}_1$ and $\mathbf{p}_2$, where $\mathbf{p}_1 = [X_A, Z_A, l_{AB}, l_{BC}]^T$ and $\mathbf{p}_2 = [X_D, Z_D, l_{DE}, l_{EF}]^T$.

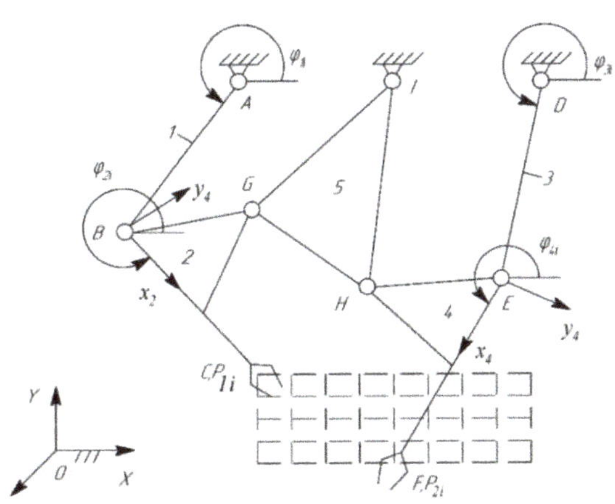

Figure 3. PM with two end effectors in the first position.

Since the passive CKCs do not impose geometric constraints on the movements of the output points *C* and *F*, the vectors of the synthesis parameters $\mathbf{p}_1$ and $\mathbf{p}_2$ are varied by the generator of LP_τ sequence [25] to satisfy the constraints of the negative CKC *IDH*. In this case, the following conditions should be fulfilled:

$$|l_{AB} - l_{BC}| \leq \rho_{1\min},\ \ l_{AB} + l_{BC} \geq \rho_{1\max}, \tag{1}$$

$$|l_{DE} - l_{EF}| \leq \rho_{2\min},\ \ l_{DE} + l_{EF} \geq \rho_{2\max}, \tag{2}$$

where:

$$\rho_{1\min} = \min_{X_{P_1}, Y_{P_1}} |l_{AP_{1i}}|,\ \rho_{1\max} = \max_{X_{P_1}, Y_{P_1}} |l_{AP_{1i}}|, \tag{3}$$

$$\rho_{2\min} = \min_{X_{P_2}, Y_{P_2}} |l_{DP_{2i}}|,\ \ \rho_{2\max} = \max_{X_{P_2}, Y_{P_2}} |l_{DP_{2i}}|. \tag{4}$$

The variable distances $l_{AP_{1i}}$ and $l_{CP_{2i}}$ in the Expressions (1)–(4) are determined by the equations:

$$l_{AP_{1i}} = \left[\left(X_{P_{1i}} - X_A \right)^2 + \left(Y_{P_{1i}} - Y_A \right)^2 \right]^{\frac{1}{2}}, (i = 1, 2, \ldots, N), \tag{5}$$

$$l_{DP_{2i}} = \left[\left(X_{P_{2i}} - X_D \right)^2 + \left(Y_{P_{2i}} - Y_D \right)^2 \right]^{\frac{1}{2}}. \tag{6}$$

Let us consider the parametric synthesis of the negative CKC *GHI* with three negative degrees of freedom, determined by the Chebyshev formula [26]:

$$W = 3n - 2p_5 = 3 \cdot 1 - 2 \cdot 3 = -3, \tag{7}$$

where n is number of links, p_5 is the kinematic pairs of the fifth class.

To do this, we preliminarily determine the positions of links 2 and 4 of the passive CKCs *ABC* and *DEF* by the equations:

$$\varphi_{2i} = \mathrm{tg}^{-1} \frac{Z_{P_{1i}} - Z_{B_i}}{X_{P_{1i}} - X_{B_i}}, \tag{8}$$

$$\varphi_{4i} = \mathrm{tg}^{-1} \frac{Z_{P_{2i}} - Z_{E_i}}{X_{P_{2i}} - X_{E_i}}, \tag{9}$$

where:

$$\begin{bmatrix} X_{Bi} \\ Z_{Bi} \end{bmatrix} = \begin{bmatrix} X_A \\ Z_A \end{bmatrix} + l_{AB} \begin{bmatrix} \cos \varphi_{1i} \\ \sin \varphi_{1i} \end{bmatrix}, \tag{10}$$

$$\begin{bmatrix} X_{Ei} \\ Z_{Ei} \end{bmatrix} = \begin{bmatrix} X_D \\ Z_D \end{bmatrix} + l_{DE} \begin{bmatrix} \cos \varphi_{3i} \\ \sin \varphi_{3i} \end{bmatrix}, \tag{11}$$

$$\varphi_{1i} = \varphi_{AP_{1i}} - \cos^{-1} \frac{l_{AB}^2 + l_{AP_{1i}}^2 - l_{BC}^2}{2 l_{AB} l_{AP_{1i}}}, \tag{12}$$

$$\varphi_{3i} = \varphi_{DP_{2i}} + \cos^{-1} \frac{l_{DE}^2 + l_{DP_{2i}}^2 - l_{EF}^2}{2 l_{DE} l_{DP_{2i}}}, \tag{13}$$

$$\varphi_{AP_{1i}} = \mathrm{tg}^{-1} \frac{Z_{P_{1i}} - Z_A}{X_{P_{1i}} - X_A}, \tag{14}$$

$$\varphi_{DP_{2i}} = \mathrm{tg}^{-1} \frac{Z_{P_{2i}} - Z_D}{X_{P_{2i}} - X_D}. \tag{15}$$

Let us attach the coordinate systems Bx_2z_2 and Ex_4z_4 with the links *BC* and *EF* of the passive CKCs *ABC* and *DEF*, where the axes Bx_2 and Ex_4 are directed along the links *BC* and *EF*, respectively (Figure 3). Then, the synthesis parameters of the negative CKC *GHI* are $x_G^{(2)}, z_G^{(2)}, x_H^{(4)}, z_H^{(4)}$, $l_{GH}, X_I, Z_I, l_{GI}, l_{HI}$ where $x_G^{(2)}, z_G^{(2)}, x_H^{(4)}, z_H^{(4)}$, X_I, Z_I are the coordinates of the joints *G*, *H*, *I* in the coordinate systems Bx_2z_2, Dx_4z_4, *OXYZ*, respectively;

l_{GH}, l_{GI}, l_{HI} are the lengths of the sides *GH, GI, HI* of the link *GHI*. Let us denote these synthesis parameters through the vector $\mathbf{p}_3 = [x_G^{(2)}, z_G^{(2)}, x_H^{(4)}, z_H^{(4)}, l_{GH}, X_I, Z_I, l_{GI}, l_{HI}]$.

Since the three-joined link *GHI* imposes three geometric constraints on the movements of the links of two passive CKCs *ABC* and *DEF*, we derive three functions of weighted differences:

$$\Delta q_{1i} = (x_{Hi}^{(2)} - x_G^{(2)})^2 + (z_{Hi}^{(2)} - z_G^{(2)})^2 - l_{HG}^2, \tag{16}$$

$$\Delta q_{2i} = (X_{Gi} - X_I)^2 + (Z_{Gi} - Z_I)^2 - l_{GI}^2, \tag{17}$$

$$\Delta q_{3i} = (X_{Hi} - X_I)^2 + (Z_{Hi} - Z_I)^2 - l_{HI}^2, \tag{18}$$

where $x_{Hi}^{(2)}$ and $z_{Hi}^{(2)}$ are the coordinates of the joint *H* in the local coordinate system Bx_2z_2; Z_{Gi}, X_{Gi} and X_{Hi}, Z_{Hi} are the coordinates of the joints *G* and *H* in the absolute coordinate system *OXYZ*, which are determined by the equations:

$$\begin{bmatrix} x_{Hi}^{(2)} \\ z_{Hi}^{(2)} \end{bmatrix} = \begin{bmatrix} \cos\varphi_{2i} & \sin\varphi_{2i} \\ -\sin\varphi_{2i} & \cos\varphi_{2i} \end{bmatrix} \cdot \begin{bmatrix} X_{Hi} - X_{Bi} \\ Z_{Hi} - Z_{Bi} \end{bmatrix}, \tag{19}$$

$$\begin{bmatrix} X_{Gi} \\ Z_{Gi} \end{bmatrix} = \begin{bmatrix} X_{Bi} \\ Z_{Bi} \end{bmatrix} + \begin{bmatrix} \cos\varphi_{2i} & -\sin\varphi_{2i} \\ \sin\varphi_{2i} & \cos\varphi_{2i} \end{bmatrix} \cdot \begin{bmatrix} x_G^{(2)} \\ z_G^{(2)} \end{bmatrix}, \tag{20}$$

$$\begin{bmatrix} X_{Hi} \\ Z_{Hi} \end{bmatrix} = \begin{bmatrix} X_{Ei} \\ Z_{Ei} \end{bmatrix} + \begin{bmatrix} \cos\varphi_{4i} & -\sin\varphi_{4i} \\ \sin\varphi_{4i} & \cos\varphi_{4i} \end{bmatrix} \cdot \begin{bmatrix} x_H^{(4)} \\ z_H^{(4)} \end{bmatrix}, \tag{21}$$

where:

$$\begin{bmatrix} X_{Bi} \\ Z_{Bi} \end{bmatrix} = \begin{bmatrix} X_A \\ Z_A \end{bmatrix} + l_{AB} \begin{bmatrix} \cos\varphi_{1i} \\ \sin\varphi_{1i} \end{bmatrix}, \tag{22}$$

$$\begin{bmatrix} X_{Ei} \\ Z_{Ei} \end{bmatrix} = \begin{bmatrix} X_D \\ Z_D \end{bmatrix} + l_{DE} \begin{bmatrix} \cos\varphi_{3i} \\ \sin\varphi_{3i} \end{bmatrix}. \tag{23}$$

The geometric meanings of Functions (16)–(18) are the deviations of the coordinates of the joints *H* and *G* from circles with radiuses l_{HG}, l_{GI}, l_{HI} in the relative motion of the plane Ex_4z_4 and in the absolute motion of link 5.

After replacing the synthesis parameters of the form:

$$\begin{bmatrix} p_1 \\ p_2 \end{bmatrix} = \begin{bmatrix} x_G^{(2)} \\ y_G^{(2)} \end{bmatrix}, \begin{bmatrix} p_4 \\ p_5 \end{bmatrix} = \begin{bmatrix} x_H^{(4)} \\ y_H^{(4)} \end{bmatrix},$$

$$p_3 = \frac{1}{2}(x_G^{(2)^2} + z_G^{(2)^2} + x_H^{(4)^2} + z_H^{(4)^2} - l_{GH}^{(2)}),$$

$$\begin{bmatrix} p_6 \\ p_7 \end{bmatrix} = \begin{bmatrix} X_I \\ Z_I \end{bmatrix}, \; p_8 = \frac{1}{2}(x_G^{(2)^2} + z_G^{(2)^2} + X_I^2 + Z_I^2 - l_{GI}^2),$$

$$p_9 = \frac{1}{2}(x_H^{(4)^2} + z_H^{(4)^2} + X_I^2 + Z_I^2 - l_{HI}^2) \tag{24}$$

Functions (16)–(18) are expressed linearly in the following vectors of synthesis parameters $\mathbf{p}_3^{(1)} = [p_1, p_2, p_3]^T, \mathbf{p}_3^{(2)} = [p_4, p_5, p_3]^T, \mathbf{p}_3^{(3)} = [p_6, p_7, p_8]^T, \mathbf{p}_3^{(4)} = [p_1, p_2, p_8]^T$, $\mathbf{p}_3^{(5)} = [p_6, p_7, p_9]^T, \mathbf{p}_3^{(6)} = [p_4, p_5, p_9]^T$ as:

$$\Delta q_{1i}^{(k)} = 2(\mathbf{g}_{1i}^{(k)^T} \cdot \mathbf{p}_3^{(k)} - g_{0i}^{(k)}), \; k = 1, 2, \tag{25}$$

$$\Delta q_{2i}^{(k)} = 2(\mathbf{g}_{2i}^{(k)^T} \cdot \mathbf{p}_3^{(k)} - g_{0i}^{(k)}), \; k = 3, 4, \tag{26}$$

$$\Delta q_{3i}^{(k)} = 2(\mathbf{g}_{3i}^{(k)^T} \cdot \mathbf{p}_3^{(k)} - g_{0i}^{(k)}), \; k = 5, 6, \tag{27}$$

where:

$$\mathbf{g}_{1i}^{(1)} = -\begin{bmatrix} \mathbf{\Gamma}^{-1}(\varphi_{2i}) & & 0 \\ & & 0 \\ 0 & 0 & 1 \end{bmatrix} \cdot \begin{bmatrix} X_{Ei} - X_{Bi} \\ Z_{Ei} - Z_{Bi} \\ 1 \end{bmatrix} - \begin{bmatrix} \mathbf{\Gamma}(\varphi_{4i} - \varphi_{2i}) & & 0 \\ & & 0 \\ 0 & 0 & 1 \end{bmatrix} \cdot \begin{bmatrix} p_4 \\ p_5 \\ 0 \end{bmatrix}, \tag{28}$$

$$\mathbf{g}_{1i}^{(2)} = \begin{bmatrix} \mathbf{\Gamma}^{-1}(\varphi_{4i}) & & 0 \\ & & 0 \\ 0 & 0 & 1 \end{bmatrix} \cdot \begin{bmatrix} X_{Ei} - X_{Bi} \\ Z_{Ei} - Z_{Bi} \\ 1 \end{bmatrix} - \begin{bmatrix} \mathbf{\Gamma}^{-1}(\varphi_{4i} - \varphi_{2i}) & & 0 \\ & & 0 \\ 0 & 0 & 0 \end{bmatrix} \cdot \begin{bmatrix} p_1 \\ p_2 \\ 0 \end{bmatrix}, \tag{29}$$

$$g_{0i}^{(1)} = -\frac{1}{2}[(X_{Ei} - X_{Bi})^2 + (Z_{Ei} - Z_{Bi})^2] + [X_{Ei} - X_{Bi}, Z_{Ei} - Z_{Bi}] \cdot \mathbf{\Gamma}(\varphi_{4i}) \cdot \begin{bmatrix} p_4 \\ p_5 \end{bmatrix}, \tag{30}$$

$$g_{0i}^{(2)} = -\frac{1}{2}[(X_{Ei} - X_{Bi})^2 + (Z_{Ei} - Z_{Bi})^2] - [X_{Ei} - X_{Bi}, Z_{Ei} - Z_{Bi}] \cdot \mathbf{\Gamma}(\varphi_{2i}) \cdot \begin{bmatrix} p_1 \\ p_2 \end{bmatrix}, \tag{31}$$

$$\mathbf{g}_{2i}^{(3)} = -\begin{bmatrix} X_{Bi} \\ Z_{Bi} \\ 1 \end{bmatrix} - \begin{bmatrix} \mathbf{\Gamma}(\varphi_{2i}) & & 0 \\ & & 0 \\ 0 & 0 & 1 \end{bmatrix} \cdot \begin{bmatrix} p_1 \\ p_2 \\ 0 \end{bmatrix}, \tag{32}$$

$$\mathbf{g}_{2i}^{(4)} = \begin{bmatrix} \mathbf{\Gamma}^{-1}(\varphi_{2i}) & & 0 \\ & & 0 \\ 0 & 0 & 1 \end{bmatrix} \cdot \begin{bmatrix} X_{Bi} \\ Z_{Bi} \\ 1 \end{bmatrix} - \begin{bmatrix} \mathbf{\Gamma}(\varphi_{2i}) & & 0 \\ & & 0 \\ 0 & 0 & 1 \end{bmatrix} \cdot \begin{bmatrix} p_6 \\ p_7 \\ 0 \end{bmatrix}, \tag{33}$$

$$g_{0i}^{(3)} = -\frac{1}{2}[X_{Bi}{}^2 + Z_{Bi}{}^2] + [X_{Bi}, Z_{Bi}] \cdot \mathbf{\Gamma}(\varphi_{2i}) \cdot \begin{bmatrix} p_1 \\ p_2 \end{bmatrix}, \tag{34}$$

$$g_{0i}^{(4)} = -\frac{1}{2}[X_{Bi}{}^2 + Z_{Bi}{}^2] - [X_{Bi}, Z_{Bi}] \cdot \begin{bmatrix} p_6 \\ p_7 \end{bmatrix}, \tag{35}$$

$$\mathbf{g}_{3i}^{(5)} = -\begin{bmatrix} X_{Ei} \\ Z_{Ei} \\ 1 \end{bmatrix} - \begin{bmatrix} \mathbf{\Gamma}(\varphi_{4i}) & & 0 \\ & & 0 \\ 0 & 0 & 1 \end{bmatrix} \cdot \begin{bmatrix} p_4 \\ p_5 \\ 0 \end{bmatrix}, \tag{36}$$

$$\mathbf{g}_{3i}^{(6)} = \begin{bmatrix} \mathbf{\Gamma}^{-1}(\varphi_{4i}) & & 0 \\ & & 0 \\ 0 & 0 & 1 \end{bmatrix} \cdot \begin{bmatrix} X_{Ei} \\ Z_{Ei} \\ 1 \end{bmatrix} + \begin{bmatrix} \mathbf{\Gamma}(\varphi_{4i}) & & 0 \\ & & 0 \\ 0 & 0 & 1 \end{bmatrix} \cdot \begin{bmatrix} p_6 \\ p_7 \\ 0 \end{bmatrix}. \tag{37}$$

$$g_{0i}^{(5)} = -\frac{1}{2}[X_{Ei}{}^2 + Z_{Ei}{}^2] + [X_{Ei}, Z_{Ei}] \cdot \mathbf{\Gamma}(\varphi_{4i}) \cdot \begin{bmatrix} p_4 \\ p_5 \end{bmatrix}, \tag{38}$$

$$g_{0i}^{(6)} = -\frac{1}{2}[X_{Ei}{}^2 + Z_{Ei}{}^2] - [X_{Ei}, Z_{Ei}] \cdot \begin{bmatrix} p_6 \\ p_7 \end{bmatrix}. \tag{39}$$

Furthermore, the synthesis parameters of the negative CKC *GHI* are determined on the basis of the problems of Chebyshev and least-square approximations [16,17].

3. Kinematic Analysis of the PM with Two Grippers

In the kinematic analysis of the PM with two end effectors (Figure 4) for the given geometric parameters of the links and the input angle φ_{1i}, it is necessary to determine the positions and analogues of the velocities and accelerations of the links, including the output points *C* and *F*.

The considered PM with two end effectors has the structural formula:

$$\mathrm{I}(1) \to \mathrm{II}(2,5), \to \mathrm{II}(3,4), \tag{40}$$

i.e., it contains two dyads $\mathrm{II}(2,5)$ and $\mathrm{II}(3,4)$.

According to the structural Formula (40), first, a kinematic analysis of the dyad $\mathrm{II}(2,5)$ is carried out, and then of the dyad $\mathrm{II}(3,4)$.

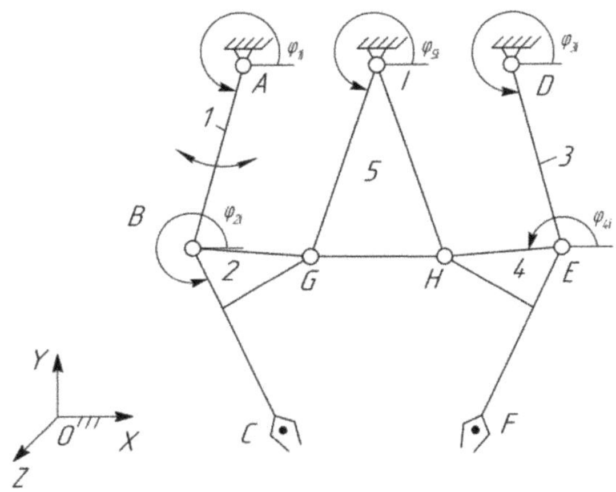

Figure 4. PM with two end effectors.

3.1. *Kinematic Analysis of the PM with Two Grippers*

Let us derive a vector *BGI* loop-closure equation:

$$l_{BG}\mathbf{e}_{2i} - l_{IG}\mathbf{e}_{5i} + l_{(IB)i}\mathbf{e}_{(IB)i} = 0, \tag{41}$$

where:

$$l_{(IB)i} = \left[(X_{Bi} - X_I)^2 + (Z_{Bi} - Z_I)^2\right]^{\frac{1}{2}}, \tag{42}$$

$$\varphi_{(IB)i} = \mathrm{tg}^{-1}\frac{Z_{Bi} - Z_I}{X_{Bi} - X_I}, \tag{43}$$

$$\begin{bmatrix} X_{Bi} \\ Z_{Bi} \end{bmatrix} = \begin{bmatrix} X_A \\ Z_A \end{bmatrix} + l_{AB}\begin{bmatrix} \cos\varphi_{1i} \\ \sin\varphi_{1i} \end{bmatrix}. \tag{44}$$

Transfer $l_{BG}\mathbf{e}_{2i}$ to the right side of Equation (41) and square both sides. As a result, we obtain:

$$\varphi_{5i} = \varphi_{(IB)i} + \cos^{-1}\frac{l_{(IB)i}^2 + l_{IG}^2 - l_{BG}^2}{2l_{(IB)i}l_{IG}}. \tag{45}$$

Next, we define:

$$\begin{bmatrix} X_{Gi} \\ Z_{Gi} \end{bmatrix} = \begin{bmatrix} X_I \\ Z_I \end{bmatrix} + l_{IG}\begin{bmatrix} \cos\varphi_{5i} \\ \sin\varphi_{5i} \end{bmatrix}, \tag{46}$$

$$\varphi_{2i} = \mathrm{tg}^{-1}\frac{Z_{Gi} - Z_{Bi}}{X_{Gi} - X_{Bi}}. \tag{47}$$

To solve the problem of the positions of the dyad II (3.4), we derive a vector *DEH* loop-closure equation:

$$l_{DE}\mathbf{e}_{3i} + l_{EH}\mathbf{e}_{4i} - l_{(DH)i}\mathbf{e}_{(DH)i} = 0, \tag{48}$$

where:

$$l_{(DH)i} = \left[(X_{Hi} - X_D)^2 + (Z_{Hi} - Z_D)^2\right]^{\frac{1}{2}}, \tag{49}$$

$$\varphi_{(DH)i} = \mathrm{tg}^{-1}\frac{Z_{Hi} - Z_D}{X_{Hi} - X_D}, \tag{50}$$

$$\begin{bmatrix} X_{Hi} \\ Z_{Hi} \end{bmatrix} = \begin{bmatrix} X_I \\ Z_I \end{bmatrix} + \begin{bmatrix} \cos\varphi_{5i} & -\sin\varphi_{5i} \\ \sin\varphi_{5i} & \cos\varphi_{5i} \end{bmatrix} \cdot \begin{bmatrix} x_G^{(5)} \\ y_G^{(5)} \end{bmatrix}. \tag{51}$$

Transfer $l_{EH}\mathbf{e}_{4i}$ to the right side of Equation (48) and square both sides. As a result, we obtain:

$$\varphi_{3i} = \varphi_{(DH)i} + \cos^{-1}\frac{l_{DE}^2 + l_{(DH)i}^2 - l_{EH}^2}{2l_{DE} \cdot l_{(DH)i}}. \tag{52}$$

Next, we define:

$$\begin{bmatrix} X_{Ei} \\ Z_{Ei} \end{bmatrix} = \begin{bmatrix} X_D \\ Z_D \end{bmatrix} + l_{DE}\begin{bmatrix} \cos\varphi_{3i} \\ \sin\varphi_{3i} \end{bmatrix}, \tag{53}$$

$$\varphi_{4i} = \mathrm{tg}^{-1}\frac{Z_{Hi} - Z_{Ei}}{X_{Hi} - X_{Ei}}. \tag{54}$$

Coordinates of the output points C and F in the absolute coordinate system $OXYZ$ are determined by the equations:

$$\begin{bmatrix} X_{Ci} \\ Z_{Ci} \end{bmatrix} = \begin{bmatrix} X_{Bi} \\ Z_{Bi} \end{bmatrix} + \begin{bmatrix} \cos\varphi_{2i} & -\sin\varphi_{2i} \\ \sin\varphi_{2i} & \cos\varphi_{2i} \end{bmatrix} \cdot \begin{bmatrix} x_C^{(2)} \\ y_C^{(2)} \end{bmatrix}, \tag{55}$$

$$\begin{bmatrix} X_{Fi} \\ Z_{Fi} \end{bmatrix} = \begin{bmatrix} X_{Ei} \\ Z_{Ei} \end{bmatrix} + \begin{bmatrix} \cos\varphi_{4i} & -\sin\varphi_{4i} \\ \sin\varphi_{4i} & \cos\varphi_{4i} \end{bmatrix} \cdot \begin{bmatrix} x_F^{(4)} \\ y_F^{(4)} \end{bmatrix}. \tag{56}$$

3.2. Analogues of Velocities and Accelerations

To determine the analogues of the angular velocities of the PM with two end effectors, we derive the vector $ABGI$ and $IHED$ loop-closure equations:

$$l_{AB}\mathbf{e}_{1i} + l_{BG}\mathbf{e}_{2i} - l_{IG}\mathbf{e}_{5i} - l_{AI}\mathbf{e}_{AI} = 0 \tag{57}$$

and

$$l_{IH}\mathbf{e}_{(IH)i} - l_{EH}\mathbf{e}_{4i} - l_{DE}\mathbf{e}_{3i} - l_{ID}\mathbf{e}_{ID} = 0 \tag{58}$$

and project them on the axes OX and OZ of the absolute coordinate system $OXYZ$

$$\left.\begin{aligned} l_{AB}\cos\varphi_{1i} + l_{BG}\cos\varphi_{2i} - l_{IG}\cos\varphi_{5i} - l_{AI}\cos\varphi_{AI} = 0 \\ l_{AB}\sin\varphi_{1i} + l_{BG}\sin\varphi_{2i} - l_{IG}\sin\varphi_{5i} - l_{AI}\sin\varphi_{AI} = 0 \end{aligned}\right\} \tag{59}$$

and

$$\left.\begin{aligned} l_{IH}\cos(\varphi_{5i} + \alpha_5) - l_{EH}\cos\varphi_{4i} - l_{DE}\cos\varphi_{3i} - l_{ID}\cos\varphi_{ID} = 0 \\ l_{IH}\sin(\varphi_{5i} + \alpha_5) - l_{EH}\sin\varphi_{4i} - l_{DE}\sin\varphi_{3i} - l_{ID}\sin\varphi_{ID} = 0 \end{aligned}\right\} \tag{60}$$

differentiate the systems of Equations (59) and (60) with respect to the generalized coordinate φ_{1i}

$$\left.\begin{aligned} -l_{AB}\sin\varphi_{1i} - l_{BG}\sin\varphi_{2i} \cdot \varphi'_{2i} + l_{IG}\sin\varphi_{5i} \cdot \varphi'_{5i} = 0 \\ l_{AB}\cos\varphi_{1i} + l_{BG}\cos\varphi_{2i} \cdot \varphi'_{2i} - l_{IG}\cos\varphi_{5i} \cdot \varphi'_{5i} = 0 \end{aligned}\right\} \tag{61}$$

and

$$\left.\begin{aligned} -l_{IH}\sin(\varphi_{5i} + \alpha_5) \cdot \varphi'_{5i} + l_{EH}\sin\varphi_{4i} \cdot \varphi'_{4i} + l_{DE}\sin\varphi_{3i} \cdot \varphi'_{3i} = 0 \\ l_{IH}\cos(\varphi_{5i} + \alpha_5) \cdot \varphi'_{5i} - l_{EH}\cos\varphi_{4i} \cdot \varphi'_{4i} - l_{DE}\cos\varphi_{3i} \cdot \varphi'_{3i} = 0 \end{aligned}\right\} \tag{62}$$

From the system of Equation (61), we determine the analogues of the angular velocities φ'_{2i} and φ'_{5i}

$$\mathbf{A}_1^{-1} \cdot \mathbf{u}_1 = \mathbf{b}_1 \tag{63}$$

where:

$$\mathbf{A}_1 = \begin{bmatrix} Z_{Bi} - Z_{Gi} & Z_{Gi} - Z_I \\ X_{Gi} - X_{Bi} & X_I - X_{Gi} \end{bmatrix},$$

$$\mathbf{u}_1 = \begin{bmatrix} \varphi'_{2i} \\ \varphi'_{5i} \end{bmatrix}, \quad \mathbf{b}_1 = \begin{bmatrix} Z_{Bi} - Z_A \\ X_A - X_{Bi} \end{bmatrix}.$$

Substituting the obtained values of the angular velocity analogue φ'_{5i} into the system of Equation (62), from this system, we determine the angular velocities analogues φ'_{3i} and φ'_{4i}

$$\mathbf{A}_2^{-1} \cdot \mathbf{u}_2 = \mathbf{b}_2, \tag{64}$$

where:

$$\mathbf{A}_2 = \begin{bmatrix} Z_{Ei} - Z_D & Z_{Hi} - Z_{Ei} \\ X_D - X_{Ei} & X_{Ei} - X_{Hi} \end{bmatrix},$$

$$\mathbf{u}_2 = \begin{bmatrix} \varphi'_{3i} \\ \varphi'_{4i} \end{bmatrix}, \qquad \mathbf{b}_2 = \begin{bmatrix} (Z_{Hi} - Z_I) \cdot \varphi'_{5i} \\ (X_I - X_{Hi}) \cdot \varphi'_{5i} \end{bmatrix}.$$

Projections of the linear velocities analogues of the output points C and F on the axis of the absolute coordinate system $OXYZ$ are determined by differentiating Equations (55) and (56) with respect to the generalized coordinate φ_{1i}

$$\begin{bmatrix} u^X_{Ci} \\ u^Z_{Ci} \end{bmatrix} = \begin{bmatrix} u^X_{Bi} \\ u^Z_{Bi} \end{bmatrix} + \begin{bmatrix} -\sin\varphi_{2i} & -\cos\varphi_{2i} \\ \cos\varphi_{2i} & -\sin\varphi_{2i} \end{bmatrix} \cdot \begin{bmatrix} x^{(2)}_C \\ y^{(2)}_C \end{bmatrix} \cdot \varphi'_{2i}, \tag{65}$$

$$\begin{bmatrix} u^X_{Fi} \\ u^Z_{Fi} \end{bmatrix} = \begin{bmatrix} u^X_{Ei} \\ u^Z_{Ei} \end{bmatrix} + \begin{bmatrix} -\sin\varphi_{4i} & -\cos\varphi_{4i} \\ \cos\varphi_{4i} & -\sin\varphi_{4i} \end{bmatrix} \cdot \begin{bmatrix} x^{(4)}_F \\ y^{(4)}_F \end{bmatrix} \cdot \varphi'_{4i}, \tag{66}$$

where the projections of the linear velocity analogues of the joints B and E are determined by differentiating Equations (44) and (53) with respect to the generalized coordinate φ_{1i}

$$\begin{bmatrix} u^X_{Bi} \\ u^Z_{Bi} \end{bmatrix} = l_{AB} \begin{bmatrix} -\sin\varphi_{1i} \\ \cos\varphi_{1i} \end{bmatrix}, \tag{67}$$

$$\begin{bmatrix} u^X_{Ei} \\ u^Z_{Ei} \end{bmatrix} = l_{DE} \begin{bmatrix} -\sin\varphi_{3i} \\ \cos\varphi_{3i} \end{bmatrix} \cdot \varphi'_{3i}. \tag{68}$$

To determine the angular acceleration analogues of the links, we differentiate the systems of Equations (61) and (62) with respect to the generalized coordinate φ_{1i}

$$\left.\begin{array}{c} -l_{AB}\cos\varphi_{1i} - l_{BG}\cos\varphi_{2i} \cdot \varphi'^2_{2i} - l_{BG}\sin\varphi_{2i} \cdot \varphi''_{2i} + \\ l_{IG}\cos\varphi_{5i} \cdot \varphi'^2_{5i} + l_{IG}\sin\varphi_{5i} \cdot \varphi''_{5i} = 0 \\ -l_{AB}\sin\varphi_{1i} - l_{BG}\sin\varphi_{2i} \cdot \varphi'^2_{2i} + l_{BG}\cos\varphi_{2i} \cdot \varphi''_{2i} + \\ l_{IG}\sin\varphi_{5i} \cdot \varphi'^2_{5i} - l_{IG}\cos\varphi_{5i} \cdot \varphi''_{5i} = 0 \end{array}\right\} \tag{69}$$

and

$$\left.\begin{array}{l} -l_{IH}\cos(\varphi_{5i} + \alpha_5)\cdot\varphi'^2_{5i} - l_{IH}\sin(\varphi_{5i} + \alpha_5)\cdot\varphi''_{5i} + l_{EH}\cos\varphi_{4i}\cdot\varphi'^2_{4i} \\ +l_{EH}\sin\varphi_{4i}\cdot\varphi''_{4i} + l_{DE}\cos\varphi_{3i}\cdot\varphi'^2_{3i} + l_{DE}\sin\varphi_{3i}\cdot\varphi''_{3i} = 0 \\ -l_{IH}\sin(\varphi_{5i} + \alpha_5)\cdot\varphi'^2_{5i} + l_{IH}\cos(\varphi_{5i} + \alpha_5)\cdot\varphi''_{5i} + l_{EH}\sin\varphi_{4i}\cdot\varphi'^2_{4i} \\ -l_{EH}\cos\varphi_{4i}\cdot\varphi''_{4i} + l_{DE}\sin\varphi_{3i}\cdot\varphi'^2_{3i} - l_{DE}\cos\varphi_{3i}\cdot\varphi''_{3i} = 0 \end{array}\right\} \tag{70}$$

From the systems of Equation (69), we determine the angular accelerations analogues φ''_{2i} and φ''_{5i}

$$\mathbf{A}_1^{-1} \cdot \mathbf{w}_1 = \mathbf{b}_3, \tag{71}$$

where:

$$\mathbf{w}_1 = \begin{bmatrix} \varphi''_{2i} \\ \varphi''_{5i} \end{bmatrix}, \quad \mathbf{b}_3 = \begin{bmatrix} (X_{Bi} - X_A) + (X_{Gi} - X_{Bi}) \cdot \varphi'^2_{2i} + (X_I - X_{Gi}) \cdot \varphi'^2_{5i} \\ (Z_{Bi} - Z_A) + (Z_{Gi} - Z_A) \cdot \varphi'^2_{2i} + (Z_I - Z_{Gi}) \cdot \varphi'^2_{5i} \end{bmatrix}.$$

Substituting the obtained values of the angular velocity analogues φ''_{2i} and φ''_{5i} into the system of Equation (70), from this system, we determine the angular velocities analogues φ''_{3i} and φ''_{4i}

$$\mathbf{A}_2^{-1} \cdot \mathbf{w}_2 = \mathbf{b}_4, \tag{72}$$

where:

$$\mathbf{w}_2 = \begin{bmatrix} \varphi''_{3i} \\ \varphi''_{4i} \end{bmatrix}, \mathbf{b}_4 = \begin{bmatrix} (X_{Hi} - X_I) \cdot \varphi'^2_{5i} + (X_{Ei} - X_{Hi}) \cdot \varphi'^2_{4i} + (X_D - X_{Ei}) \cdot \varphi'^2_{3i} \\ (Z_{Hi} - Z_I) \cdot \varphi'^2_{5i} + (Z_{Ei} - Z_{Hi}) \cdot \varphi'^2_{4i} + (Z_D - Z_{Ei}) \cdot \varphi'^2_{3i} \end{bmatrix}. \tag{73}$$

Projections of the linear velocities analogues of the output points C and F on the axis of the absolute coordinate system $OXYZ$ are determined by differentiating Equations (65) and (66) with respect to the generalized coordinate φ_{1i}

$$\begin{bmatrix} w^X_{Ci} \\ w^Z_{Ci} \end{bmatrix} = \begin{bmatrix} w^X_{Bi} \\ w^Z_{Bi} \end{bmatrix} + \begin{bmatrix} -\cos\varphi_{2i} & \sin\varphi_{2i} \\ -\sin\varphi_{2i} & -\cos\varphi_{2i} \end{bmatrix} \cdot \begin{bmatrix} x^{(2)}_C \\ z^{(2)}_C \end{bmatrix} \cdot \varphi'^2_{2i} + \begin{bmatrix} -\sin\varphi_{2i} & -\cos\varphi_{2i} \\ \cos\varphi_{2i} & -\sin\varphi_{2i} \end{bmatrix} \cdot \begin{bmatrix} x^{(2)}_C \\ z^{(2)}_C \end{bmatrix} \cdot \varphi''_{2i}, \tag{74}$$

$$\begin{bmatrix} w^X_{Fi} \\ w^Z_{Fi} \end{bmatrix} = \begin{bmatrix} w^X_{Ei} \\ w^Z_{Ei} \end{bmatrix} + \begin{bmatrix} -\cos\varphi_{4i} & \sin\varphi_{4i} \\ -\sin\varphi_{4i} & -\cos\varphi_{4i} \end{bmatrix} \cdot \begin{bmatrix} x^{(4)}_F \\ z^{(4)}_F \end{bmatrix} \cdot \varphi'^2_{4i} + \begin{bmatrix} -\sin\varphi_{4i} & -\cos\varphi_{4i} \\ \cos\varphi_{4i} & -\sin\varphi_{4i} \end{bmatrix} \cdot \begin{bmatrix} x^{(2)}_F \\ z^{(2)}_F \end{bmatrix} \cdot \varphi''_{4i}, \tag{75}$$

where the projections of the linear velocity analogues of the joints B and E are determined by differentiating Equations (67) and (68) with respect to the generalized coordinate φ_{1i}

$$\begin{bmatrix} w^X_{Bi} \\ w^Z_{Bi} \end{bmatrix} = l_{AB} \begin{bmatrix} -\cos\varphi_{1i} \\ -\sin\varphi_{1i} \end{bmatrix}, \tag{76}$$

$$\begin{bmatrix} w^X_{Ei} \\ w^Z_{Ei} \end{bmatrix} = l_{DE} \begin{bmatrix} -\cos\varphi_{3i} \\ -\sin\varphi_{3i} \end{bmatrix} \cdot \varphi'^2_{3i} + \begin{bmatrix} -\sin\varphi_{3i} \\ \cos\varphi_{3i} \end{bmatrix} \cdot \varphi''_{3i}. \tag{77}$$

4. Numerical Results

Table 1 shows N = 11 positions of the grippers P_1 and P_2 of the PM with two end effectors.

Table 1. Positions of the grippers P_1 and P_2.

i	1	2	3	4	5	6	7	8	9	10	11
$X_{P_{1,i}}$, mm	0	4.6	9.2	13.8	18.40	23.0	27.60	32.20	36.80	41.40	46.3882
$Y_{P_{1,i}}$, mm	0	−6.2512	−11.0478	−14.8398	−17.8092	−20.1433	−21.8922	−23.1166	−23.8528	−24.1065	−23.8384
$X_{P_{2,i}}$, mm	46.3882	53.5938	59.8359	65.3367	70.4344	75.0223	79.1896	82.9453	86.4738	89.6476	92.7766
$Y_{P_{2,i}}$, mm	−23.8384	−24.0439	−23.2688	−21.8398	−19.8480	−17.4431	−14.6785	−11.6242	−8.1590	−4.4228	0

Tables 2–4 show the obtained values of the synthesis parameters of the two passive CKCs ABC, DEF and negative CKC GHI, respectively.

Table 2. Synthesis parameters of the CKC ABC.

X_A, mm	Z_A, mm	l_{AB}, mm	l_{BC}, mm
14.04	−57.11	46.5309	43.0

Table 3. Synthesis parameters of the CKC DEF.

X_D, mm	Z_D, mm	l_{DE}, mm	l_{EF}, mm
78.74	−57.11	46.5309	43.0

Table 4. Synthesis parameters of the CKC *GHI*.

$x_G^{(2)}$, mm	$y_G^{(2)}$, mm	$x_H^{(4)}$, mm	$y_H^{(4)}$, mm	l_{GH}, mm	X_I, mm	Z_I, mm	l_{GI}, mm	l_{HI}, mm
6.6072	9.9607	−6.6072	−9.9607	45.0	46.39	57.11	50.0	50.0

3D CAD model of the synthesized PM with two grippers is shown in Figure 5.

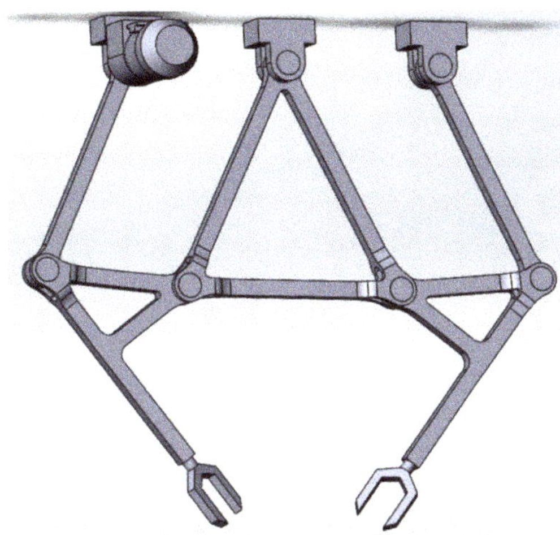

Figure 5. 3D CAD model of the PM with two grippers.

Table 5 shows the obtained values of the coordinates $X_{P_{1,i}}, Y_{P_{1,i}}, X_{P_{2,i}}, Y_{P_{2,i}}$ and projections of analogues of linear velocities $u^X_{P_{1,i}}, u^Z_{P_{1,i}}, u^X_{P_{2,i}}, u^Z_{P_{2,i}}$ and linear accelerations $w^X_{P_{1,i}}, w^Z_{P_{1,i}}, w^X_{P_{2,i}}, w^Z_{P_{2,i}}$ of the grippers of the PM with two end effectors.

Table 5. Positions and analogues of linear velocities and accelerations of the grippers.

i	1	2	3	4	5	6	7	8	9	10	11
$X_{P_{1,i}}$, mm	0	4.6	9.2	13.8	18.40	23.0	27.60	32.20	36.80	41.40	46.3882
$Z_{P_{1,i}}$, mm	0	−6.2512	−11.0478	−14.8398	−17.8092	−20.1433	−21.8922	−23.1166	−23.8528	−24.1065	−23.8384
$X_{P_{2,i}}$, mm	46.3882	53.5938	59.8359	65.3367	70.4344	75.0223	79.1896	82.9453	86.4738	89.6476	92.7766
$Z_{P_{2,i}}$, mm	−23.8384	−24.0439	−23.2688	−21.8398	−19.8480	−17.4431	−14.6785	−11.6242	−8.1590	−4.4228	0
$u^X_{P_{1,i}}$, mm	0.8440	0.5204	0.2432	−0.026	−0.276	−0.495	−0.693	−0.8729	−1.036	−1.184	−1.325
$u^Z_{P_{1,i}}$, mm	0.9830	0.9627	0.9489	0.9362	0.9236	0.9110	0.8977	0.8829	0.8662	0.8465	0.8211
$u^X_{P_{2,i}}$, mm	1.0733	1.0380	1.0173	0.9998	0.9845	0.9710	0.9583	0.9459	0.9335	0.9212	0.9094
$u^Z_{P_{2,i}}$, mm	−0.1071	−0.0733	−0.0650	−0.0515	−0.0399	−0.0294	−0.0189	−0.0075	0.0062	0.0240	0.0515
$w^X_{P_{1,i}}$, mm	−0.1096	−0.1066	−0.0868	−0.1056	−0.1964	−0.3570	−0.5768	−0.8396	−1.1265	−1.4157	−1.6956
$w^Z_{P_{1,i}}$, mm	−0.8688	−0.2504	−0.0283	0.0221	−0.0380	−0.1620	−0.3271	−0.5223	−0.7421	−0.9843	−1.2642
$w^X_{P_{2,i}}$, mm	−0.1116	−0.0885	−0.0968	−0.1112	−0.1353	−0.1676	−0.2100	−0.2661	−0.3420	−0.4486	−0.6182
$w^Z_{P_{2,i}}$, mm	0.6401	0.4710	0.3803	0.3069	0.2416	0.1771	0.1038	0.01319	−0.1059	−0.2713	−0.5320

Positions and modules of the velocities and acceleration analogues of the synthesized PM grippers P_1 and P_2 are also presented with the graphical plots in Figures 6–8.

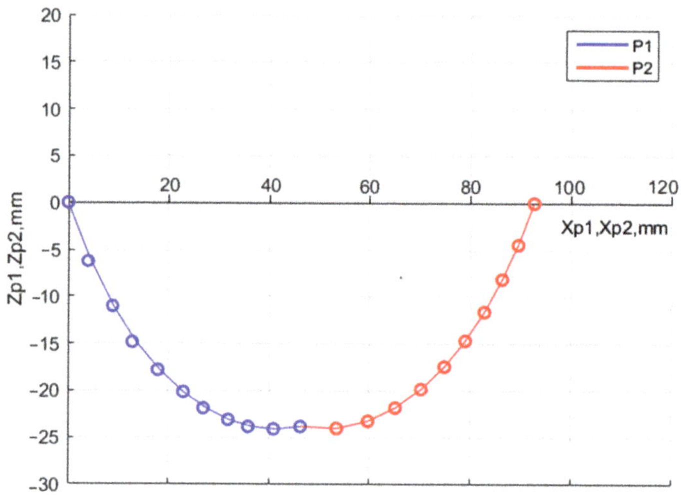

Figure 6. Graphics of the grippers P_1 and P_2 positions.

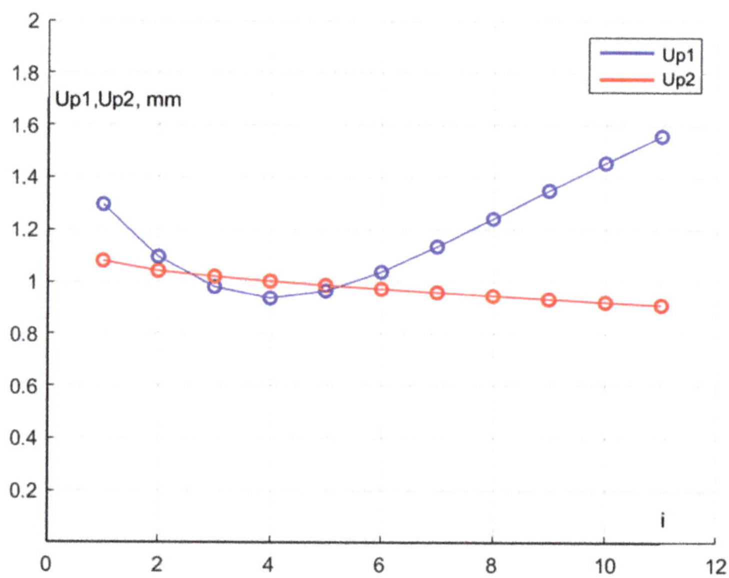

Figure 7. Graphics of the grippers P_1 and P_2 linear velocities analogues.

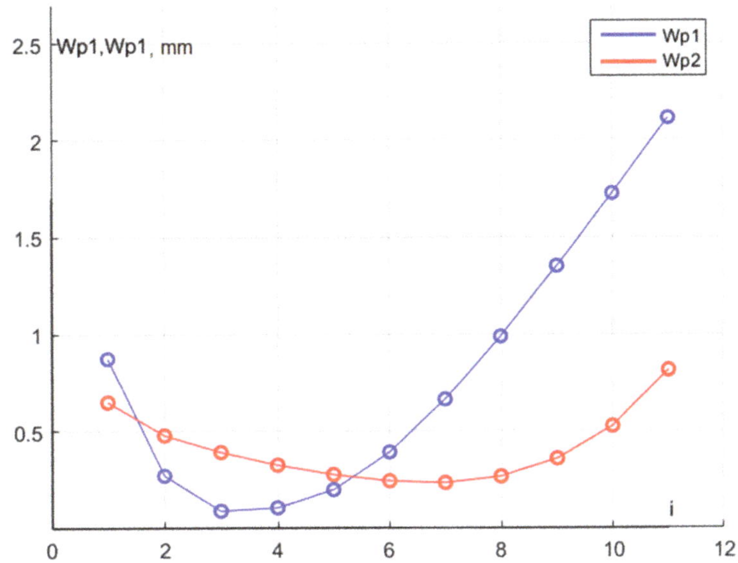

Figure 8. Graphics of the grippers P_1 and P_2 linear accelerations analogues.

5. Conclusions

Kinematic synthesis and analysis of the PM with two end effectors have been carried out. In the kinematic synthesis according to the given laws of motions (or positions) of two end effectors, the structural scheme, and geometric parameters of links of the synthesized PM are determined. The structural scheme of this PM is formed by connecting two output objects (end effectors) and a base using three CKCs: two passive and one negative CKC. Passive and negative CKCs are structural modules from which the PM is formed. Passive CKCs are two movable serial manipulators, and the negative CKCs is a three-jointed link. Serial manipulators (passive CKCs) do not impose geometric constraints on the movement of the output objects, and the three-jointed link (negative CKC) imposes three geometric constraints. Therefore, the geometric parameters of the links of the negative CKCs are determined, and the geometric parameters of the links of the passive CKCs are varied depending on the imposed geometric constraints of the negative CKC. Kinematic synthesis of the negative CKC was carried out on the basis of the Chebyshev and least-square approximations. Since the structure of the synthesized PM consists of two dyads, the position analysis is solved analytically. Analogues of angular velocities and accelerations are determined from two systems of linear equations obtained by differentiating the loop-closure equations with respect to the generalized coordinate.

Author Contributions: Z.B. established the methods of structural and kinematic synthesis of the RoboMech class parallel manipulator (PM) with two grippers. M.A.L. established the methods of kinematic analysis of this PM. A.M. created the programs in the MatLab for kinematic synthesis of the PM and obtained numerical results. A.K. created the programs in the MatLab for kinematic analysis of the PM and obtained numerical results. All authors contributed equally to the evaluation of the results and writing of this paper. All authors have read and agreed to the published version of the manuscript.

Funding: This research is founded by the Science Committee of the Ministry of Education and Science of the Republic of Kazakhstan (Grant No. AP08857522).

Institutional Review Board Statement: Not applicable.

Informed Consent Statement: Not applicable.

Data Availability Statement: The data presented in this study are available on request from the corresponding author.

Conflicts of Interest: The authors declare no conflict of interest.

References

1. Baigunchekov, Z.; Kalimoldayev, M.; Ibrayev, S.; Izmambetov, M.; Baigunchekov, T.; Naurushev, B.; Aisa, N. Parallel Manipulator of a Class RoboMech. *Lect. Notes Electr. Eng.* **2017**, *408*, 547–557.
2. Paul, B. *Kinematics and Dynamics of Planar Machinery*; Prentice-Hall: Hoboken, NJ, USA, 1979.
3. Hartenberg, R.S.; Denavit, J. *Kinematic Synthesis of Linkages*; McGraw-Hill: New York, NY, USA, 1964.
4. Rodhavan, M.; Roth, B. Solving Polynomial Systems for the Kinematic Analysis and Synthesis of Mechanisms and Robot Manipulators. *J. Mech. Des.* **1995**, *117*, 71–79.
5. Erdman, A.G.; Sandor, G.N.; Kota, S. *Mechanism Design: Analysis and Synthesis Lecture Notes*; McGill University: Montreal, QC, Canada, 2001.
6. Uicker, J.J.; Pennack, G.R.; Shigley, J.E. *Theory of Machines and Mechanisms*, 4th ed.; Oxford University Press: New York, NY, USA, 2001.
7. Hernandez, A.; Petya, V. Position Analysis of Planar Mechanisms with R-pairs Using Geometrical-Iterative Method. *Mech. Mach. Theory* **2004**, *39*, 50–61. [CrossRef]
8. Pennock, G.R.; Israr, A. Kinematic Analysis and Synthesis of an Adjustable Six-Bar Linkage. *Mech. Mach. Theory* **2009**, *44*, 306–323. [CrossRef]
9. McCarthy, J.M.; Soh, G.M. *Geometric Design of Linkages*, 2nd ed.; Springer Science & Business Media: Berlin/Heidelberg, Germany, 2010.
10. Angeles, J.; Bai, S.H. *Kinematic Synthesis, Lecture Notes*; McGill University: Montreal, QC, Canada, 2016.
11. Duffy, J. *Analysis of Mechanisms and Robot Manipulator*; Edward Arnold: London, UK, 1980.
12. Tsai, L.W. *Robot Analysis: The Mechanics of Serial and Parallel Manipulators*; John Wiley and Sons, Inc.: New York, NY, USA, 1999.
13. Glazunov, V. Design of Decoupled Parallel Manipulators by Means of the Theory of Screws. *Mech. Mach. Theory* **2010**, *45*, 209–230. [CrossRef]
14. McCarthy, J.M. Kinematics, Polynomials, and Computers-A Brief History. *J. Mech. Robot.* **2011**, *3*, 010201-1. [CrossRef]
15. McCarthy, J.M. 21st Kinematics Synthesis, Compliance, and Tensegrity. *J. Mech. Robot.* **2011**, *3*, 020202-1. [CrossRef]
16. Wang, D.; Wang, L.; Wu, J. Physics-Based Mechatronics Modeling and Application of an Industrial-Grade Parallel Tool Head. *Mech. Syst. Signal Process.* **2021**, *148*, 107158. [CrossRef]
17. Bourbonnais, F.; Bigras, P.; Bonev, I.A. Minimum-Time Trajectory Planning and Control of a Pick-and-Place Five-Bar Parallel Robot. *IEEE/ASME Trans. Mechatron.* **2014**, *20*, 740–749. [CrossRef]
18. Zhang, Z.; Wang, L.; Shao, Z. Improving the Kinematic Performance of a Planar 3-RRR Parallel Manipulator Through Actuation Mode Conversion. *Mech. Mach. Theory* **2018**, *130*, 86–108. [CrossRef]
19. Carabin, G.; Scalera, L.; Wongratanaphisan, T.; Vidoni, R. An Energy-Efficient Approach for 3D Printing with a Linear Delta Robot Equipped with Optimal Springs. *Robot. Comput. Integr. Manuf.* **2021**, *67*, 102045. [CrossRef]
20. Scalera, L.; Carabin, G.; Vidoni, R.; Wongratanaphisan, T. Energy Efficiency in a 4-DOF Parallel Robot Featuring Compliant Elements . *Int. J. Mech. Control* **2019**, *20*, 49–57.
21. Hu, B.; Shi, D.; Xie, T.; Hu, B.; Ye, N. Kinematically Identical Manipulators Derivation for the 2-RPU+ UPR Parallel Manipulator and Their Constraint Performance Comparison. *J. Mech. Robot.* **2020**, *6*, 061011. [CrossRef]
22. Baigunchekov, Z.; Ibrayev, S.; Izmambetov, M.; Mustafa, A.; Sagyntay, M. Synthesis of Reconfigurable Positioning Parallel Manipulator of a Class RoboMech. In Proceedings of the 2018 International Conference on Reconfigurable Mechanisms and Robots, ReMAR, Delft, The Netherlands, 20–22 June 2018.
23. Baigunchekov, Z.; Izmambetov, M.; Zhumasheva, Z.; Baigunchekov, T.; Mustafa, A. Parallel Manipulator of a Class RoboMech for Generation of Horizontal Trajectories Family. *Mech. Mach. Sci.* **2019**, *73*, 1395–1402.
24. Grabchenko, I.P.; Klepnikov, V.B.; Dobroskok, V.L.; Kryzhni, G.K. *Introduction to Mechatronics*; HPI: Harkov, Ukraine, 2017.
25. Sobol, I.M.; Statnikov, R.B. *Choice of Optimal Parameters in Problems with Many Criteria*; Nauka: Moscow, Russian, 1981.
26. Artobolevskiy, I.I. *Theory of Mechanisms and Machines*; Nauka: Moscow, Russian, 1988.

robotics

MDPI

Article

Bio-Inspired Conceptual Mechanical Design and Control of a New Human Upper Limb Exoskeleton

Narek Zakaryan *, Mikayel Harutyunyan and Yuri Sargsyan

Department of Mechanics and Machine Science, National Polytechnic University of Armenia, Yerevan 0009, Armenia; mharut@seua.am (M.H.); yusarg@seua.am (Y.S.)
* Correspondence: n.zaqaryan@polytechnic.am; Tel.: +37-494-355-234

Abstract: Safe operation, energy efficiency, versatility and kinematic compatibility are the most important aspects in the design of rehabilitation exoskeletons. This paper focuses on the conceptual bio-inspired mechanical design and equilibrium point control (EP) of a new human upper limb exoskeleton. Considering the upper limb as a multi-muscle redundant system, a similar over-actuated but cable-driven mechatronic system is developed to imitate upper limb motor functions. Additional torque adjusting systems at the joints allow users to lift light weights necessary for activities of daily living (ADL) without increasing electric motor powers of the device. A theoretical model of the "ideal" artificial muscle exoskeleton is also developed using Hill's natural muscle model. Optimal design parameters of the exoskeleton are defined using the differential evolution (DE) method as a technique of a multi-objective optimization. The proposed cable-driven exoskeleton was then fabricated and tested on a healthy subject. Results showed that the proposed system fulfils the desired aim properly, so that it can be utilized in the design of rehabilitation robots. Further studies may include a spatial mechanism design, which is especially important for the shoulder rehabilitation, and development of reinforcement learning control algorithms to provide more efficient rehabilitation treatment.

Keywords: upper limb rehabilitation; bio-inspired exoskeleton; cable-driven system; over-actuation; Hill's model; EP control; torque adjusting mechanism; differential evolution

Citation: Zakaryan, N.; Harutyunyan, M.; Sargsyan, Y. Bio-Inspired Conceptual Mechanical Design and Control of a New Human Upper Limb Exoskeleton. *Robotics* **2021**, *10*, 123. https://doi.org/10.3390/robotics10040123

Academic Editor: Raffaele Di Gregorio

Received: 13 October 2021
Accepted: 9 November 2021
Published: 12 November 2021

1. Introduction

The nowadays fast developing robotics technology promises to improve human upper-limb functionalities required for performing ADL [1–5]. The following technical challenges urgently need to be studied:

- Kinematic compatibility,
- Safety,
- Control strategy.

The bio-inspired design (bio-inspiration) can offer the best solutions in rehabilitation robotics, particularly in wearable robotics. They lead to a similarly efficient and effective robotic system design. The main characteristics of a bio-inspired design include the redundant actuation, flexible actuators, flexible links and joints, and an appropriate bio-inspired control. Successful examples of upper limb and hand bio-inspired exoskeletons have been presented recently by Ning Li et al. and Ong et al. [6,7]. They used tension lines, cables and flexible bands to simulate muscles and muscle tendons of the human upper limb. Test results confirmed that a natural-like joint motion range and a trajectory curve are provided; thus, the effectiveness of the bio-inspired design is proven. These studies are devoted to bio-inspired mechanical design, i.e., ensuring kinematic compatibility, but no appropriate control systems have been developed, and the capabilities of these devices have not been evaluated when used in human ADL.

The well-known studies of bio-inspired exoskeletons mainly refer to cable-driven exoskeletons, where Bowden cables are widely used [8–11]. Many leading scientists have studied cable-driven exoskeletons for stroke patients [12–15]. This innovative design increases the robot safety and effectiveness. It is shown that the remote control of arm movements helps to achieve a light weight and low inertia design properties. Besides, the cable-driven architecture can eliminate possible joint misalignment between the human upper limb and exoskeleton, thus reducing the chances of injuries for the patient during robotic rehabilitation [16,17]. However, bio-inspired control methods are proposed in those studies, although they are not derived from the control principles of natural biological systems, and they do not take into account that the bio-inspired control mostly depends on the nonlinear properties of natural muscles.

The basic notions of the EP control theory are related to the multi-muscle and multi-degrees of freedom redundancy [18–22]. As is known, the shift of equilibrium positions is ensured in the result of the organism–environment interaction; therefore, a new generation of a robot–human interaction can play a crucial role. According to the investigations of Feldman et al., the notion of EP characterizes the equilibrium state in terms of equilibrium positions of body segments and the muscle torques at these positions as two-dimensional vectors [23]. However, each EP is associated with a static (steady) state of the system, the notion of EP shifts is essentially dynamical, and its presence essentially influences the dynamics of motor behavior. In fact, the EP does not conflict with dynamic systems theory; rather, it complements it. It is a specific form of dynamic systems theory.

Spiers A. et al. have studied muscle models as well, highlighting their role in bio-inspired robots [24]. A popular method of characterizing muscle motion is the Hill model with its variations [25–27]. Emerged actuator technologies such as electroactive polymers (EAPs) are promising alternatives for natural muscles, but their characteristics are not satisfactory yet, and they are in the stage of research and improvement [28].

In this paper, a conceptual model of a cable-driven over-actuated upper limb rehabilitation device with mechanical variable torque mechanisms at the joints and EP control, providing safety and human-like motions, is described. A universal artificial muscle-driven exoskeleton using Hill's natural muscle model as an actuator is also considered, and results are compared. The proposed cable-driven exoskeleton is prepared and tested, which ensured light weight, flexibility, kinematic compatibility of the device and, therefore is safe and compfortable for the user, confirming the theoretical results, and creating a basis for further development. Our eventual goal is the development of a theoretical model of an upper limb exoskeleton, which can be used as a guide to evaluate the performance of the proposed cable-driven device, as well as other similar artificial muscle-devices, and finally, the use of the architecture described herein to create robots which will be suitable for use in domestic and rehabilitation environments, where they will perform tasks that are normally performed in human ADL.

2. Conceptual Mechanical Design of the Exoskeleton

The human upper limb is a system with great mobility and muscle redundancy, which can be modeled by an approximation biomechanical model consisting of only 7 DoFs (shoulder-3 DoF, elbow-2DoF, wrist-2 DoF). The total number of upper limb muscles is about 30, among which there are biarticular muscles, complicating the system even more. Many of the studied human motion dynamic optimization methods have included models of muscle forces and dynamics [29,30]. The fact that several muscles are involved in the same joint movement was also considered [31]. The movement of a joint is a combination of these effects (synergy). For example, elbow flexion (i.e., folding of the elbow so that the angle between the arm and forearm decreases) has various levels of torque contributions from four separate muscles in the arm (the brachialis, biceps brachii, brachioradialis and pronator teres muscles).

It is proposed to simplify the system in this work, by reducing the degrees of freedom of the biomechanical model and considering its movements in only the sagittal plane. Based on the agonist–antagonist structure of the natural muscular system of human limbs, a new concept of a device, actuated by a cable-driven system, is presented (Figure 1).

Bi-articular cables (actuation of two joints with single cable) are also included, which makes the system more complex, but closer to the human muscular system. In addition, it has variable torque mechanisms at the joints, with $h_1, h_1', h_2, h_2', h_3, h_3'$ variable lever arms.

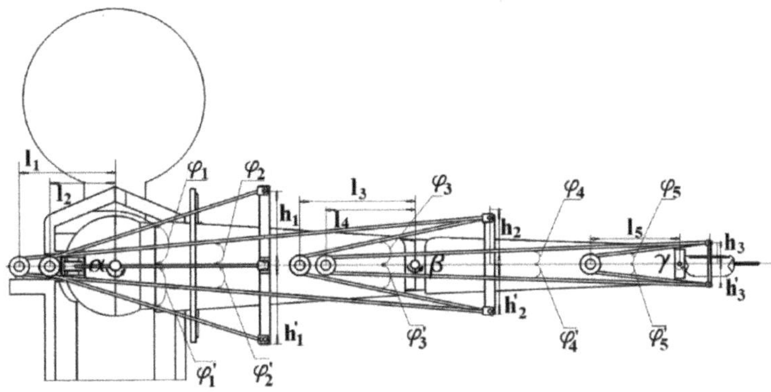

Figure 1. The upper limb bio-inspired exoskeleton (sagittal plane, conceptual design).

Cable tensions can be represented by generalized coordinates in order to use Lagrange's method. Suppose that c_j are position vectors of the connection point of the jth cable to the exoskeleton expressed in the Cartesian frame. The generalized forces (Q_i^t) of the system can be expressed in cable forces as follows:

$$Q_i^t = \sum_{j=1}^{10}\left(F_j w_j \cdot \frac{\partial c_j}{\partial q_i}\right), i = 1\ldots 3, \tag{1}$$

where w_j, c_j are unit direction vectors of the jth cable and the corresponding moment arm on the end-effector, respectively, $q_i(i = 1\ldots 3)$ or α, β, γ are generalized coordinates.

Equilibrium equations of the system can be written as follows:

$$KF = M_f \tag{2}$$

where F is the matrix of cable tensions:

$$F = \left|\, F_1 \quad F_2 \quad \ldots F_{10} \,\right|^T. \tag{3}$$

M_f includes all the external forces and represents the following matrix:

$$M_f = \left|\, \frac{d}{dt}\left(\frac{\partial L}{\partial \dot{\alpha}}\right) - \frac{\partial L}{\partial \alpha} - Q_{g_1} \quad \frac{d}{dt}\left(\frac{\partial L}{\partial \dot{\beta}}\right) - \frac{\partial L}{\partial \beta} - Q_{g_2} \quad \frac{d}{dt}\left(\frac{\partial L}{\partial \dot{\gamma}}\right) - \frac{\partial L}{\partial \gamma} - Q_{g_3} \,\right|^T \tag{4}$$

where L is the Lagrangian, $Q_{g_1}, Q_{g_2}, Q_{g_3}$ are gravity forces, K is the structural matrix, including unit direction vectors w_j, c_j, L is determined by the difference between kinetic and potential energies of the system.

The lever arms (a_i) can be defined by the following diagram (Figure 2):

$$\begin{gathered} \sin\varphi_i = \frac{a_i}{l_i + b_i} = \frac{r_i}{b_i} \Rightarrow a_i = l_i \sin\varphi_i + r_i, i = 1\ldots 5 \\ \text{or } a_i = l_i \frac{h_i}{AF} + r_i = l_i \frac{h_i}{\sqrt{h_i^2 + AG^2}} + r_i \end{gathered} \tag{5}$$

where the distances AG and AF can be chosen based on the sizes of the human upper limb.

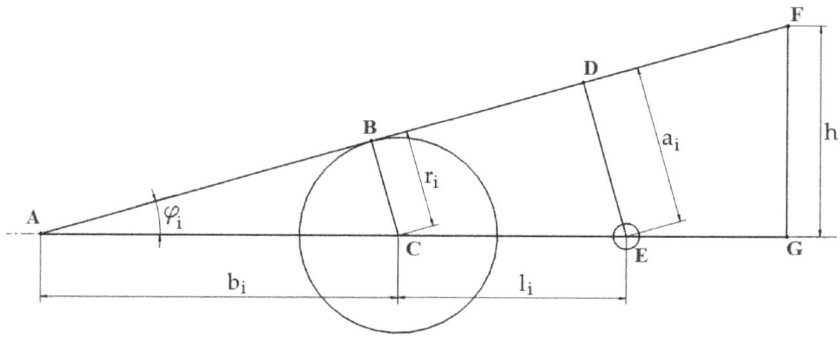

Figure 2. Geometrical parameters of the exoskeleton actuation system.

Equations (1)–(5) allow to establish the relation between cable tensions, geometrical parameters of their attachments, radiuses of pulleys and variable lever arms.

Let us now consider the structure and operation of the actuation system (Figure 3).

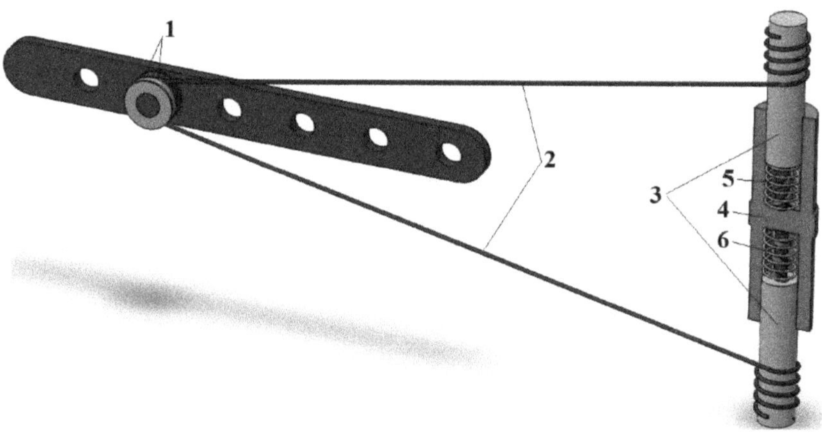

Figure 3. Conceptual design of the exoskeleton actuation module.

The device works as follows: the agonist and antagonist actuators (2) are connected to the corresponding levers (3) of the mechanism pulleys (1). The levers form screw pairs with the links (5) and are mounted in the cylinder (4) by attaching to the spring (6). When the values of tensile forces of the actuators are exceeding a certain value during the operation of the device, the levers start to rotate towards the cylinder in the anti-clockwise direction, the lever arms of the cable tensile forces are increasing, and consequently, the torques at the joints are also increasing. When the load is decreasing or partly restored muscles participate in the movement, the spring restores the lever initial length, reducing the torque.

Thus, a cable-driven bioinspired exoskeleton was developed, and a mechanism providing variable torque in the joints was additionally installed to regulate the robot's lifting capability. The next step is to determine the optimal values of the device design parameters. However, the analysis of the torque adjusting mechanism will be considered in our future studies.

3. Design Optimization Using Differential Evolution Method

Differential evolution (DE) is an evolutionary algorithm, which uses the difference of solution vectors to create new candidate solutions. DE was originally proposed by

Rainer Storn and Kenneth Price in 1997 [32]. DE is a very simple, yet very powerful and useful algorithm, and can be used to deal with a wide variety of optimization problems. In this paper, we design a bio-inspired rehabilitation exoskeleton using multi-objective DE algorithm. According to the proposed control strategy (EP hypothesis), the system's movement consists of balanced positions, which is treated as a constraint. Thus, the multi-objective optimization considers the following three objectives to minimize:

(1) The total mass of the device,
(2) The maximal magnitudes of cable tensions,
(3) The maximal difference between magnitudes of agonist-antagonist cable tensions.

The first goal is dedicated to reducing the weights of arm, forearm and hand segments of the exoskeleton (m_{up}, m_f, m_h), and the main components of the cable-driven mechanism (pulleys (m_{p_i}) and electric motors (m_{e_i})), in order to design a lightweight and wearable rehabilitation device. The total mass of the exoskeleton can be defined as follows:

$$M_{exos} = m_{up} + m_f + m_h + \sum_{i=1}^{10} (m_{p_i} + m_{e_i}), \quad (6)$$

The second objective concerns the reduction of cable tensions, which will lead to the use of smaller electric motors and, consequently, energy saving. In multi-body cable-driven mechanisms, the effect of cables is modeled as point forces applied to the links, i.e., inertia and elasticity of the cables are ignored. For simplicity, it is accepted that the radii of the agonist and antagonist pulleys are equal.

Achieving the third goal will increase the rigidity of joints, will make the device safer, increase the accuracy of upper limb movements, but as a result, energy consumption will be increased. It can be expressed as follows:

$$\min_{t} \big| |F_i| - |F_i'| \big|, i = 1 \ldots 10. \quad (7)$$

According to the design objectives, the following design variables have been selected:

- the masses of segments of the exoskeleton (m_{up}, m_f, m_h), pulleys ($m_{p_1}, m_{p_2}, m_{p_3}, m_{p_4}, m_{p_5}, m'_{p_1}, m'_{p_2}, m'_{p_3}, m'_{p_4}, m'_{p_5}$) and electric motors ($m_{e_1}, m_{e_2}, m_{e_3}, m_{e_4}, m_{e_5}, m'_{e_1}, m'_{e_2}, m'_{e_3}, m'_{e_4}, m'_{e_5}$),
- positions of pulleys installation (l_1, l_2, l_4, l_5, l_7), radii of pulleys (r_1, r_2, r_3, r_4, r_5), and cable connection angles ($\varphi_1, \varphi_2, \varphi_3, \varphi_4, \varphi_5, \varphi'_1, \varphi'_2, \varphi'_3, \varphi'_4, \varphi'_5$),
- cable tensions ($F_1, F_2, F_3, F_4, F_5, F'_1, F'_2, F'_3, F'_4, F'_5$).

Thus, the vector of design variables, R, will be defined as follows:

$$R = \left[m_{up}, m_f, m_h, m_{p_i}, m_{e_i}, r_i, \varphi_i, \varphi'_i, F_i, F'_i\right], i = 1, 2, \ldots, 5. \quad (8)$$

The next important step is to specify the range of possible values of the design parameters (Table 1) that meet the requirements of wearable rehabilitation devices.

Table 1. Bounds of design variables.

Variables	Range	Units
m_{up}	[1, 5]	kg
m_f	[0.5, 4]	kg
m_h	[0.1, 0.5]	kg
$m^j_{p_i}$	[0.1, 0.5]	kg
$m^j_{e_i}$	[0.5, 1]	kg
r^j_i	[0.01, 0.1]	m
$\varphi^j_i, \varphi^{j'}_j$	[5, 90]	deg
$F^j_i, F^{j'}_i$	[1, 50]	N

An algorithm is described to apply according to the simple evolutionary algorithm [33–35]. In simple DE, an initial random population consisting of NP vectors is randomly generated according to a uniform distribution within the lower and upper boundaries (a_{jL}, a_{jU}). After initialization, these individuals are evolved by mutation and crossover to generate a trial vector. A comparison between the parent and its trial vector is then done to select the vector which should survive to the next generation. To start the optimization process, an initial population must be created. Each jth component ($j = 1, 2, \ldots, D$) of the ith individuals ($i = 1, 2, \ldots, NP$) is obtained as follows:

$$a_{ji} = a_{jL} + \mathrm{rand}(0,1)\left(a_{jU} - a_{jL}\right), \tag{9}$$

where $\mathrm{rand}(0,1)$ returns a random number in $[0,1]$, D is the number of variables to be optimized.

A mutant vector μ_i is generated according to the following:

$$\mu_i = a_{n_1} + F(a_{n_2} - a_{n_3}), \tag{10}$$

where $n_1, n_2, n_3 \in \{1, 2, \ldots, NP\}$, are randomly chosen indices, $F = [0, 2]$ is a real number to control the amplification of the difference vector. The crossover rate (CR ∈ [0, 1]) is also important and introduced to control the number of components inherited from the mutant vector.

MATLAB codes of the described above algorithm are presented in Appendix A.

4. Control Strategy Analysis: EP Control

Wearable robots are designed to interact with humans. Consequently, they must be designed for intrinsic safety and should repeat human natural movements. The key design features, promoting intrinsic safety, are the following:

- mechanical compliance to accommodate interactions,
- light weight to minimize kinetic energy,
- bio-inspired control strategy.

Both joint stiffness and equilibrium position of each joint in the robot can be controlled using the MIMO control system [36]. The errors in joint stiffness k and equilibrium position θ_{EP}, along with the actual angular position, are transferred into a decoupling block (Figure 4), which uses the partial derivatives of joint stiffness $\partial K_i/\partial F_i$ and joint EP $\partial\theta_{EP}/\partial F_i$ ($i = 1, 2, 3$) with respect to cable tension to transform stiffness and EP to error in tension for each cable.

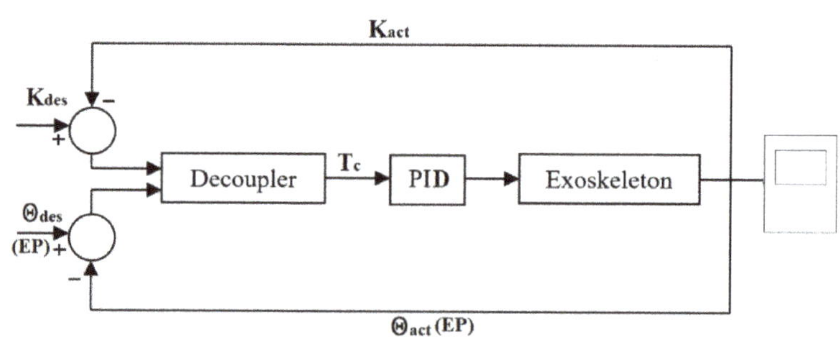

Figure 4. EP control architecture of the exoskeleton.

In order to obtain these derivatives, we note that torque in each joint is given by

$$\tau = K(\theta - \theta_{EP}) = l\sum \overline{F}. \tag{11}$$

The PID controller was tuned using MSC ADAMS control toolkit methods.

The hand position can be determined analytically, in terms of two dimensional homogeneous transformations—rotational and translational:

$$P_H = R(\alpha)T(L_1)R(\beta)T(L_2)R(\gamma)T(L_3). \tag{12}$$

By multiplying all of the matrices, we get

$$P_H = \begin{pmatrix} C_1 & -S_1 & L_2C_2 + L_1C_3 + L_3C_1 \\ S_1 & C_1 & L_2S_2 + L_1S_3 + L_3S_1 \\ 0 & 0 & 1 \end{pmatrix}, \tag{13}$$

where $C_1 = \cos(\alpha+\beta+\gamma)$, $C_2 = \cos(\alpha+\beta)$, $C_3 = \cos\alpha$, $S_1 = \sin(\alpha+\beta+\gamma)$, $S_2 = \sin(\alpha+\beta)$, $S_3 = \sin\alpha$.

The joint stiffnesses are controlled by the agonist–antagonist artificial muscular system with considerable accuracy, which is very important for the rehabilitation/assistive device; therefore, consideration of frictional forces in dynamic equations of the system is also important:

$$\frac{d}{dt}\left(\frac{\partial T}{\partial \dot{\theta}_i}\right) - \frac{\partial T}{\partial \theta_i} + \frac{\partial V}{\partial \theta_i} + \frac{\partial F_R}{\partial \dot{\theta}_i} = \tau_i, i = 1,2,3, \tag{14}$$

where $\theta = \begin{bmatrix} \alpha & \beta & \gamma \end{bmatrix}^T$, $\tau = \begin{bmatrix} \tau_1 & \tau_2 & \tau_3 \end{bmatrix}^T$, T-system kinetic energy, V-system potential energy, F_R-energy dissipation by joint frictions, which can be defined by the following expressions:

$$T = \frac{1}{2}\sum_{i=1}^{3} J_i\dot{\theta}_i^2 + m_i\left[\frac{d}{dt}\left(\sum_{k=i-2}^{i-1}(L_k \sin\theta_k + a_i \sin\theta_i)\right)\right]^2, \tag{15}$$

$$V = \sum_{i=1}^{3} m_i g\left(a_i \cos\theta_i + \sum_{k=i-2}^{i-1} L_k \cos\theta_k\right), \tag{16}$$

$$F_R = \frac{1}{2}\sum_{i=1}^{3} c_i(\theta_i - \theta_{i-1})^2, \tag{17}$$

where L_i, a_i, m_i, J_i and θ_i are link lengths, distances from the centroid to corresponding axes, masses, moments of inertia computed by means of three link centroids, and angles of deviation from vertical direction are measured clockwise, respectively, $i = 1,2,3$ are indexes of three links, while the frictional torque coefficients of three joints are denoted as c_1, c_2 and c_3.

Kinematic modeling and new control principle based on the EP hypothesis can be implemented by using MSC ADAMS software with the following assumptions: the whole range of upper limb segments motion is divided into 26 equilibrium position subranges (Figure 5a), using flexion/extension ranges of upper limb joints (shoulder: (−180 ... +80), elbow: (−10 ... +145), wrist: (−90 ... +70)), and the movement is considered in the sagittal plane, where flexion/extension rotations of the shoulder, elbow and wrist joints are only possible.

The control system works as follows: desired angles of joints rotations, and the durations of breaks or balancing are prescribed, then the algorithm finds the necessary torques, which can provide the given motion.

Computer simulations (Figure 5b) are performed, according to the positions shown in Figure 5a.

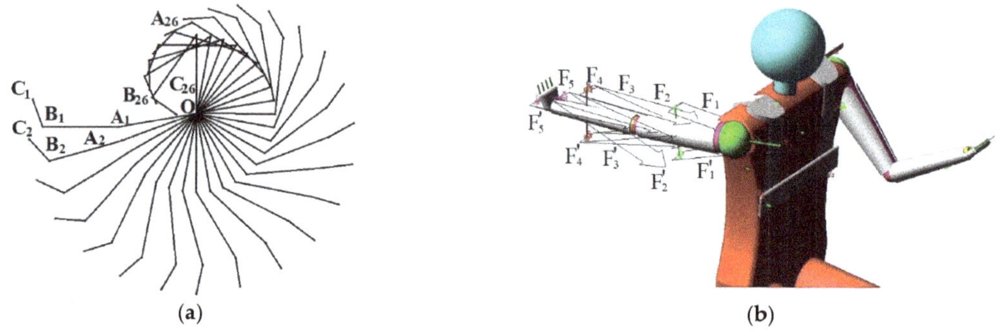

Figure 5. (**a**) 26 balancing postures of the upper limb (Arm-OA_i, Forearm-A_iB_i, Hand-$B_iC_i, i = 1,2,\ldots,26$), (**b**) MSC ADAMS simulation of the over-actuated upper limb model (Wrist: STEP (time, 0, 0, 1, −6.15 d)+ ... + (26th)STEP (time, 0, 0, 1, −6.15 d), Elbow: STEP (time, 0, 0, 1, −5.96 d)+ ... + (26th)STEP (time, 0, 0, 1, −5.96 d), Shoulder: STEP (time, 0, 0, 1, −10 d)+ ... + (26th)STEP (time, 0, 0, 1, −10 d).

5. Artificial Muscle Model and System

In this part, the cables are replaced by artificial muscles, of which the properties are similar to the properties of natural muscles. For this purpose, Hill's natural muscle model is studied, and is used to represent all necessary properties of the proposed artificial muscle [37]. The series elastic element models the behaviour of the tendon and the connective tissues. The parallel elastic element reflects the resistance of the muscle to passive stretching, while the damper models the dynamic resistance to movement, which is speed dependent. There is only one active element that models the contraction of the fibres; this is a force generator.

Based on the Hills natural muscle model, a new model of an artificial muscle is designed. For this purpose, the series elastic element is presented by two triangular components to ensure the stability of the system (Figure 6a).

Figure 6. (**a**) A 1 DOF manipulator system actuated by Hill's model artificial muscle, (**b**) the kinematic scheme.

Computer experiments of this 1 DoF manipulator system (only the elbow joint is actuated) are implemented by means of MSC ADAMS software.

Kinematics and dynamics of a single link musculoskeletal structure are presented by Shukor et al. [38], where the lengths of actuators are determined by a joint rotation angle, expressed by coordinates of actuator connection points. In this case, the total length of the actuator is the sum of current lengths of the muscle model series and parallel elastic elements. In addition, the connection to the second link is made rigidly by means of additional h_1 length link and vertically attached to it. Therefore, the total muscle length (l_m) can be determined using the kinematic scheme (Figure 6b).

According to the generalized Pythagorean theorem:

$$l_m = \sqrt{l_1^2 + h_1^2 + l_2^2 - 2l_1\sqrt{h_1^2 + l_2^2}\cos\left(\theta_1 + \arctan\frac{h_1}{l_2}\right)}, \qquad (18)$$
$$l_m = x_1 + x_2 \sin\vartheta,$$

where ϑ—connection angle of series elastic elements, x_1, x_2—deformations of parallel and series elastic elements, respectively.

Since we have a muscle model, the well-known formulas for muscle contraction can be used to get an idea of how the model works. As is known, the muscle contraction force is determined by the rate of contraction (V_m), the length of the muscle (l_m), and activation ($a(t) \in [0,1]$) [39].

$$F_c = F_{max}a(t)f(l_m)f(V_m), \tag{19}$$

where: $f(l_m)$—force-length relation of the muscle, $f(V_m)$—force-velocity relation of the muscle; F_{max}—maximal isometric force at optimum muscle fiber length and zero velocity. Muscle length and velocity can be chosen as state space variables: $X = \begin{bmatrix} l_m & \dot{l}_m \end{bmatrix}^T$.

The contraction force (Figure 7a) and the rate of contraction (Figure 7b) are obtained from the motion study of the upper limb from the horizontal position to the maximum flexed position. The motion duration is 2 s. The average speed: 0.12 m/s: The stiffness and damper coefficients of parallel elastic element and parallel damper are determined by acceptable maximum values of force generator through computer simulations: 0.5 N/mm and 0.3 N·s/mm respectively.

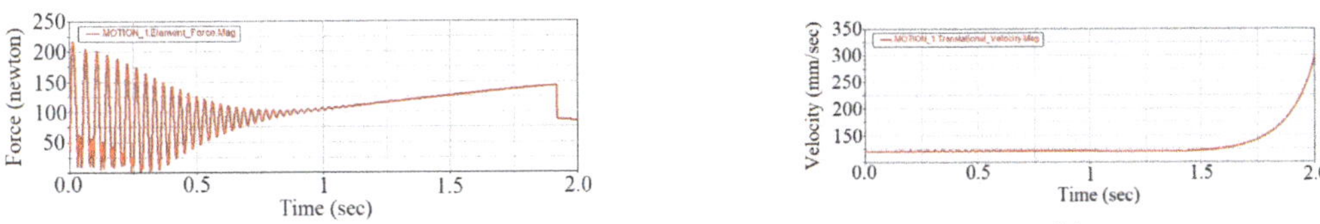

Figure 7. Motion study of the 1 DOF manipulator system: (**a**) the contraction force, (**b**) the rate of contraction.

Substituting obtained values from the computer simulation to the (19), we can calculate F_{max}, which represents the necessary preloads of springs. On the reducing of vibrations and maximal forces should be dedicated the optimal design: the determination of optimal values of elastic elements, dampers and system sizes.

There exist similar studies of generalized models with one link and multiple actuators [37], with two links and four actuators [38], with three links and nine actuators [39]. However, there is no generalized study of the model with 3 links and multiple natural muscle-like actuators.

Finally, computer simulations of the conceptual model of the upper limb-exoskeleton system actuated by artificial muscles with properties similar to natural muscles (Hill's model) can be provided (Figure 8).

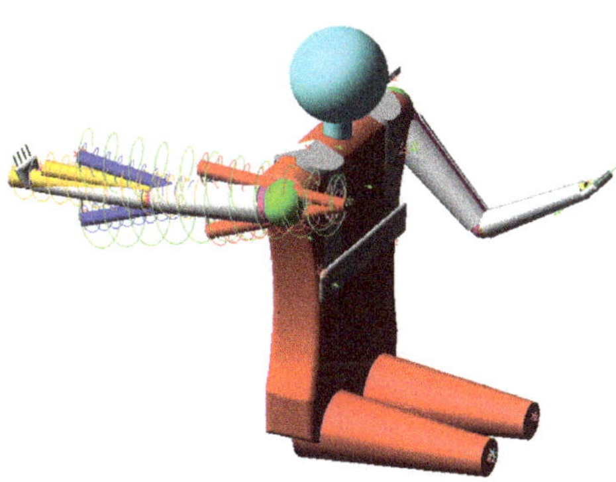

Figure 8. MSC ADAMS model of the upper-limb exoskeleton system with Hill's type artificial muscles.

Obtained values of the forces can be used to determine the stiffness and damping coefficients of the natural muscle model components. Results of the computer modelling are presented in the Table 2. For the sake of simplicity, stiffness of parallel and series elastic elements are considered equivalent.

Table 2. Mechanical properties of Hill's model components.

Stiffness,	s_1	s'_1	s_2	s'_2	s_3	s'_3	s_4	s'_4	s_5	s'_5
N/m	300	300	140	100	800	270	750	220	70	80
Damping,	d_1	d'_1	d_2	d'_2	d_3	d'_3	d_4	d'_4	d_5	d'_5
N·s/m	95	95	95	95	40	40	40	40	17	17

6. Experimental Validation

The proposed exoskeleton is fabricated (Figure 9) and tested on a healthy subject (see Video S1, Supplementary Materials).

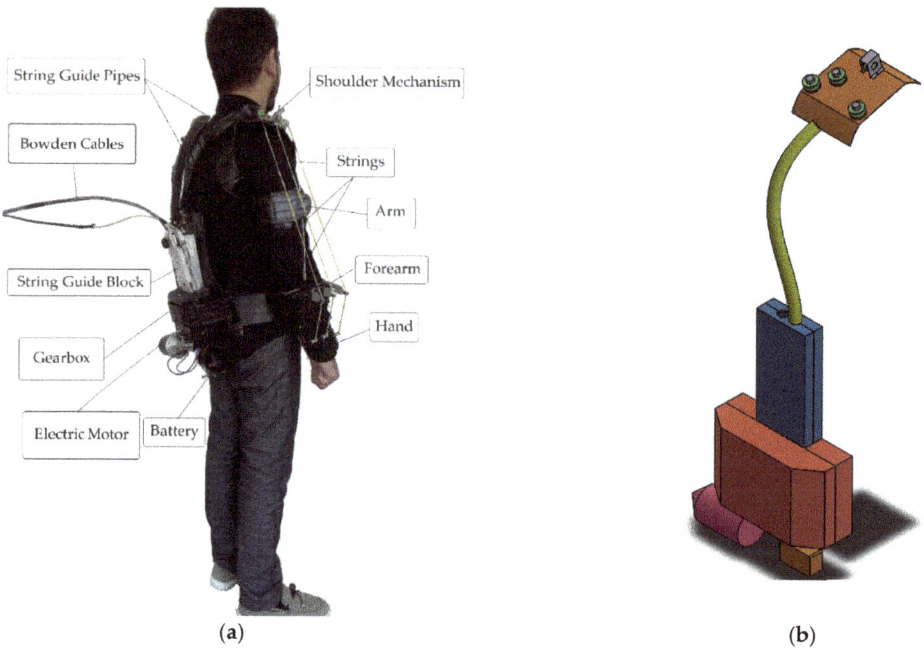

Figure 9. (**a**) The structure of the cable-driven exoskeleton of the upper limb, (**b**) CAD model.

For technical simplicity and testing convenience, instead of small electric motors actuating each string, one 30 W electric motor is used, which transmits the motion to each string via a gearbox (Figure 10a). The device is mechanically controlled (i.e., pulling the corresponding strings) by means of Bowden cables. The separation of strings and their tension process is regulated by means of pulleys placed in the string guide box. The strings are guided from the human back to the shoulder mechanism (Figure 10b) by the flexible string guide pipes. Then, they are attached to the arm, forearm and hand segments. The device works with a 12 V, 1.5 A small battery. The total weight of the device is 3.5 kg. Experiments show the effectiveness of the device and its compliance with the requirements of rehabilitation robotics.

(**a**) (**b**)

Figure 10. (**a**) Gearbox structure of the exoskeleton, (**b**) shoulder mechanism.

A more detailed description of the developed exoskeleton including the wearing and installation methods, as well as a long-term evaluation of the results during the rehabilitation treatment, are subjects of another large-scale study; we plan to present this in future studies.

7. Results

As a result of the optimization, the optimal weights of the exoskeleton segments, pulleys and electric motors, radii of pulleys, cable tensions, as well as cable connection angles or, in other words, the angles between tension lines and upper limb segments are determined (Table 3).

Table 3. The optimal weights of the exoskeleton segments, pulleys and electric motors, radii of pulleys, cable tensions, and cable connection angles.

m_{up}	m_f	m_h	m_p	m_e	r_1	r_2	r_3	r_4	r_5
2	1.1	0.2	0.1	0.4	0.07	0.05	0.02	0.02	0.03
F_1	F_1'	F_2	F_2'	F_3	F_3'	F_4	F_4'	F_5	F_5'
44.2	33.8	27.9	24.5	27.8	25.1	25.5	24.9	12.5	13.1
$\sin\varphi_1$	$\sin\varphi_2$	$\sin\varphi_3$	$\sin\varphi_4$	$\sin\varphi_5$	$\sin\varphi_6$	$\sin\varphi_7$	$\sin\varphi_8$	$\sin\varphi_9$	$\sin\varphi_{10}$
0.08	0.6	0.18	0.8	0.3	0.5	0.2	0.5	0.6	0.5

Now, let's discuss the effects of variable lever arms. For the simulation of human daily activities, we add a load of 5 kg to a user's hand. The following cable tension average values for the same lever arms were obtained (Table 4).

Table 4. Average values of cable tensions (additional loading).

F_1	F_1'	F_2	F_2'	F_3	F_3'	F_4	F_4'	F_5	F_5'
353	353	138	138	251	251	60	60	52	52

If we increase the lever arms by 30 mm, we get the following values of cable tensions (Table 5).

Table 5. Average values of cable tensions (increasing lever arms).

F_1	F_1'	F_2	F_2'	F_3	F_3'	F_4	F_4'	F_5	F_5'
150	150	67	67	75	75	41	41	32	32

Finally, the necessary forces developed by the Hill model artificial muscles and general deformations in the whole range of the simulation were obtained, which can be a guideline for other similar studies (Figure 11).

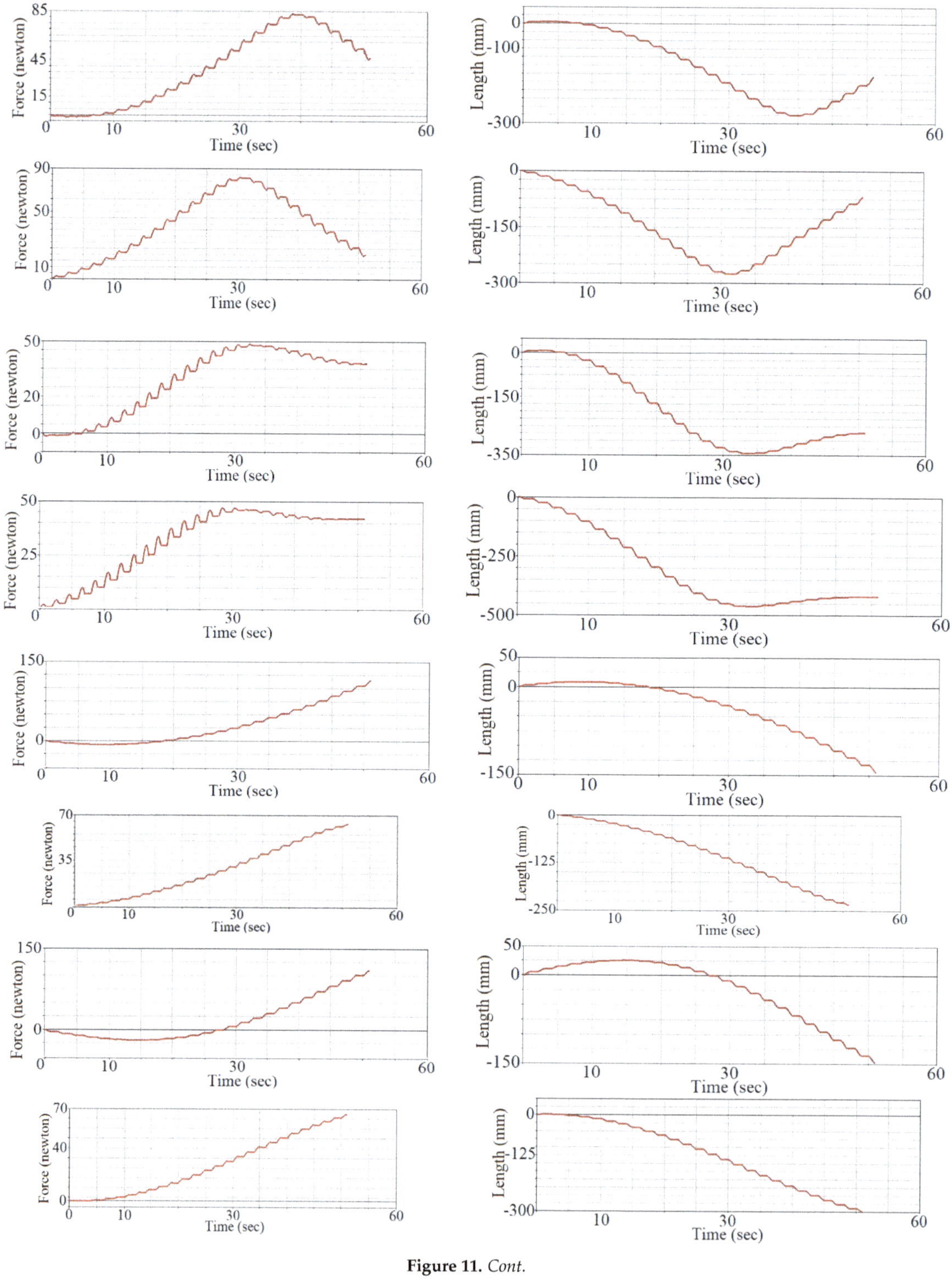

Figure 11. *Cont.*

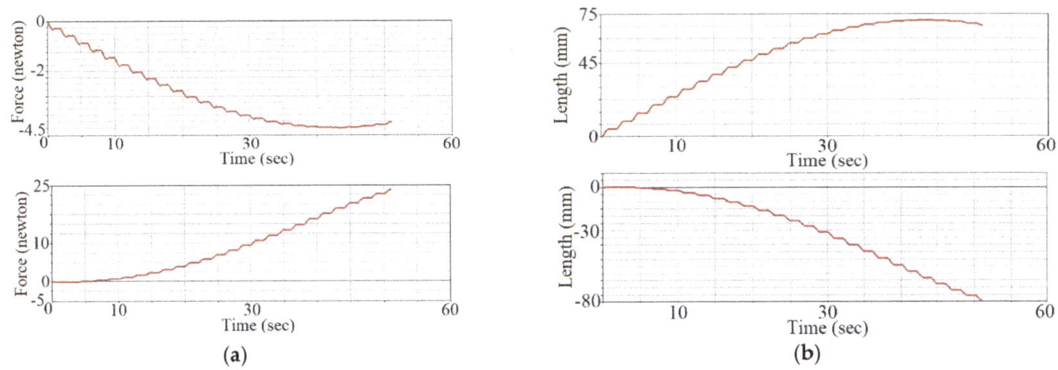

Figure 11. (**a**) The developed forces ($F_1, F_1', F_2, F_2', F_3, F_3', F_4, F_4', F_5, F_5'$), (**b**) general deformations of the Hill's model artificial muscles.

8. Discussion

In this paper, we have presented a bio-inspired mechanical design and control method for a new bio-inspired upper limb exoskeleton. Some of the main requirements to modern wearable exoskeletons: safety, stability, energy efficiency and versatility were achieved due to cable-driven over-actuation, EP control and variable torque generating mechanisms installed at the joints. Shift of the hand position during the whole motion was reached, and the relationship between the values of variable lever arms and cable tensions was revealed. The generalized exoskeleton model with artificial muscles has been proposed, taking into account Hill's model of natural muscles, which, as an "ideal" theoretical model, allows us to compare the results of the design of these types of exoskeletons and serves as a guide. The mathematical modeling and control of the proposed device were carried out by means of the MSC ADAMS system. As extensions of the present work, the following directions of further studies aimed to improve the main functional characteristics of the proposed bio-inspired exoskeleton can be outlined:

- A spatial model design, which will allow us to activate all degrees of freedom of upper limb, and consequently restore muscles functions.
- Nowadays, requirements of exoskeletons also include the ability to learn new skills, i.e., the creation of a so-called "smart" device is needed, which will greatly increase the efficiency of the device. This is again a good target for further follow-up studies.

Supplementary Materials: The following are available online at https://www.mdpi.com/article/10.3390/robotics10040123/s1, Video S1: manuscript-supplementary.MOV.

Author Contributions: N.Z. and M.H. established the methods of conceptual mechanical design of the exoskeleton, created Matlab algorithm for the design optimization and obtained numerical results, N.Z. established the methods of EP control, Y.S. established the methods of using artificial muscle model in the exoskeleton design. All authors contributed equally to the evaluation of the results and writing of this paper. All authors have read and agreed to the published version of the manuscript.

Funding: This research is funded by the Science Committee of the Ministry of Education and Science of the Republic of Armenia (Grant No. 21DP-2D006).

Institutional Review Board Statement: The study was conducted in accordance with the Declaration of Helsinki, and the protocol was approved by the Ethics Committee of robotics-1440822.

Informed Consent Statement: Informed consent was obtained from all subjects involved in the study.

Data Availability Statement: The data presented in this study are available on request from the corresponding author.

Conflicts of Interest: The Authors declare that there is no conflict of interest.

Appendix A

```
clear all; close all; clc
D = 30;
objf=inline(‘(−0.5*(x1 + 4 × x4 + 4 × x5) − x2 + 2 × x4 + 2 × x5 + x3)) × 9.8 × 0.3 × 0.17 − (x6 × x16/(0.1 × x21 + x16)) + (x7 × x16/(0.1 × x26 + x16)) − (x10 × x17/(0.15 × x23 + x17)) + (x11 × x17/(0.15 × x23 + x17))’,‘x1’,‘x2’,‘x3’,‘x4’,‘x5’,‘x6’,‘x7’,‘x8’,‘x9’,‘x10’,‘x11’,‘x12’,‘x13’, ‘x14’,‘x15’,‘x16’,‘x17’,‘x18’,‘x19’,‘x20’,‘x21’,‘x22’,‘x23’,‘x24’,‘x25’,‘x26’,‘x27’,‘x28’,‘x29’,‘x30’);
objf = inline(‘(−0.5 × (x2 + 2 × x4 + 2 × x5 − x3) × 9.8 × 0.3 × 0.34 − (x8 × x18/(0.1 × x23 + x18)) + (x9 × x18/(0.1 × x28 + x18)) − (x10 × x17/(0.1 × x22 + x17)) + (x11 × x17/(0.1 × x27 + x17)) − (x14 × x19/(0.15 + 0.3) × x24 + x19)) + (x15 × x19/(0.15 + 0.3) × x29 + x19))’,‘x1’,‘x2’,‘x3’,‘x4’,‘x5’,‘x6’,‘x7’,‘x8’,‘x9’,‘x10’,‘x11’,‘x12’,‘x13’,‘x14’,‘x15’,‘x16’,‘x17’,‘x18’, ‘x19’,‘x20’,‘x21’,‘x22’,‘x23’,‘x24’,‘x25’,‘x26’,‘x27’,‘x28’,‘x29’,‘x30’);
objf = inline(‘−0.5 × x3 × 9.8 × 0.1 × 0.5 − (x12 × x20/(0.1 × x25 + x20)) + (x13 × x20/(0.1 × x30 + x20)) − (x14 × x19/(0.4 × x24 + x19)) + (x15 × x19/(0.4 × x28 + x19))’,‘x1’,‘x2’,‘x3’,‘x4’, ‘x5’,‘x6’,‘x7’,‘x8’,‘x9’,‘x10’,‘x11’,‘x12’,‘x13’,‘x14’,‘x15’,‘x16’,‘x17’,‘x18’,‘x19’,‘x20’,‘x21’,‘x22’, ‘x23’,‘x24’,‘x25’,‘x26’,‘x27’,‘x28’,‘x29’,‘x30’);
objf = inline(‘abs(x6 − x7),abs(x8 − x9),abs(x10 − x11),abs(x12 − x13),abs(x14 − x15)’,‘x1’,‘x2’,‘x3’,‘x4’,‘x5’,‘x6’,‘x7’,‘x8’,‘x9’,‘x10’,‘x11’,‘x12’,‘x13’,‘x14’,‘x15’,‘x16’,‘x17’,‘x18’, ‘x19’,‘x20’,‘x21’,‘x22’,‘x23’,‘x24’,‘x25’,‘x26’,‘x27’,‘x28’,‘x29’,‘x30’);
objf = iline(‘x1 + x2 + x3 + 10 × x4 + 10 × x5’,‘x1’,‘x2’,‘x3’,‘x4’,‘x5’,‘x6’,‘x7’,‘x8’,‘x9’,‘x10’, ‘x11’,‘x12’,‘x13’,‘x14’,‘x15’,‘x16’,‘x17’,‘x18’,‘x19’,‘x20’,‘x21’,‘x22’,‘x23’,‘x24’,‘x25’,‘x26’,‘x27’, ‘x28’,‘x29’,‘x30’);
objf = vectorize(objf);
N = 500;
Itmax = 100;
F = 1/30; CR = 0.8;
a(1:N,1) = 2; b(1:N,1) = 5; a(1:N,2) = 1; b(1:N,2) = 4; a(1:N,3) = 0.2; b(1:N,3) = 0.6; a(1:N,4) = 0.1; b(1:N,4) = 0.5; a(1:N,5) = 0.5; b(1:N,5) = 1; a(1:N,6) = 1; b(1:N,6) = 50; a(1:N,7) = 1; b(1:N,7) = 50; a(1:N,8) = 1; b(1:N,8) = 50; a(1:N,9) = 1; b(1:N,9) = 50; a(1:N,10) = 1; b(1:N,10) = 50; a(1:N,11) = 1; b(1:N,11) = 50; a(1:N,12) = 1; b(1:N,12) = 50; a(1:N,13) = 1; b(1:N,13) = 50; a(1:N,14) = 1; b(1:N,14) = 50; a(1:N,15) = 1; b(1:N,15) = 50; a(1:N,16) = 0.01; b(1:N,16) = 0.1; a(1:N,17) = 0.01; b(1:N,17) = 0.1; a(1:N,18) = 0.01; b(1:N,18) = 0.1; a(1:N,19) = 0.01; b(1:N,19) = 0.1; a(1:N,20) = 0.01; b(1:N,20) = 0.1; a(1:N,21) = 0.08; b(1:N,21) = 1; a(1:N,22) = 0.08; b(1:N,22) = 1; a(1:N,23) = 0.08; b(1:N,23) = 1; a(1:N,24) = 0.08; b(1:N,24) = 1; a(1:N,25) = 0.08; b(1:N,25) = 1; a(1:N,26) = 0.08; b(1:N,26) = 1; a(1:N,27) = 0.08; b(1:N,27) = 1; a(1:N,28) = 0.08; b(1:N,28) = 1; a(1:N,29) = 0.08; b(1:N,29) = 1; a(1:N,30) = 0.08; b(1:N,30) = 1;
d = (b − a);
basemat = repmat(int16(linspace(1,N,N)),N,1);
basej = repmat(int16(linspace(1,D,D)),N,1);
x = a + d. × rand(N,D);
fx = objf(x(:,1),x(:,2),x(:,3),x(:,4),x(:,5),x(:,6),x(:,7),x(:,8),x(:,9),x(:,10),x(:,11),x(:,12),x(:,13), x(:,14),x(:,15),x(:,16),x(:,17),x(:,18),x(:,19),x(:,20),x(:,21),x(:,22),x(:,23),x(:,24),x(:,25),x(:,26),x(:,27), x(:,28),x(:,29),x(:,30));
[fxbest,ixbest] = min(fx);
xbest = x(ixbest,1:D);
for it = 1:itmax;
permat = bsxfun(@(x,y) x(randperm(y(1))),basemat’,N(ones(N,1)))’;
v(1:N,1:D) = repmat(xbest,N,1) + F × (x(permat(1:N,1),1:D) − x(permat(1:N,3),1:D));
r = repmat(randi([1 D],N,1),1,D);
muv = ((rand(N,D)<CR) + (basej = = r)) ~ = 0;
mux = 1 − muv;
u(1:N,1:D) = x(1:N,1:D). × mux(1:N,1:D) + v(1:N,1:D). × muv(1:N,1:D);
```

```
fu = objf(u(:,1),u(:,2),u(:,3),u(:,4),u(:,5),u(:,6),u(:,7),u(:,8),u(:,9),u(:,10),u(:,11),u(:,12),u(:,13),
u(:,14),u(:,15),u(:,16),u(:,17),u(:,18),u(:,19),u(:,20),u(:,21),u(:,22),u(:,23),u(:,24),u(:,25),u(:,26),
u(:,27),u(:,28),u(:,29),u(:,30));
    idx = fu < fx;
    fx(idx) = fu(idx);
    x(idx,1:D) = u(idx,1:D);
    [fxbest,ixbest] = min(fx);
    xbest = x(ixbest,1:D);
    end
    [xbest,fxbest]
```

References

1. Gull, M.A.; Bai, S.; Bak, T. A Review on Design of Upper Limb Exoskeletons. *Robotics* **2020**, *9*, 16. [CrossRef]
2. Maciejasz, P.; Eschweiler, J.; GerlachHahn, K.; Jansen-Troy, A.; Leonhardt, S. A survey on robotic devices for upper limb rehabilitation. *J. NeuroEngineering Rehabil.* **2014**, *11*, 1–29. [CrossRef] [PubMed]
3. Birouaș, F.I.; Țarcă, R.C.; Dzitac, S.; Dzitac, I. Preliminary Results in Testing of a Novel Asymmetric Underactuated Robotic Hand Exoskeleton for Motor Impairment Rehabilitation. *Symmetry* **2020**, *12*, 1470. [CrossRef]
4. VélezGuerrero, M.A.; CallejasCuervo, M.; Mazzoleni, S. Design, Development, and Testing of an Intelligent Wearable Robotic Exoskeleton Prototype for Upper Limb Rehabilitation. *Sensors* **2021**, *21*, 5411. [CrossRef] [PubMed]
5. Gull, M.A.; Thoegersen, M.; Bengtson, S.H.; Mohammadi, M.; Andreasen Struijk, L.N.S.; Moeslund, T.B.; Bak, T.; Bai, S. A 4−DOF Upper Limb Exoskeleton for Physical Assistance: Design, Modeling, Control and Performance Evaluation. *Appl. Sci.* **2021**, *11*, 5865. [CrossRef]
6. Li, N.; Yu, P.; Zhao, L.; Yang, T. Bioinspired wearable soft upperlimb exoskeleton robot for stroke survivors. In Proceedings of the 2017 IEEE International Conference on Robotics and Biomimetics (ROBIO), Macau, China, 5–8 December 2017; pp. 2693–2698. [CrossRef]
7. Ong, A.P.; Bugtaib, N.T. A Bioinspired Design of a Hand Robotic Exoskeleton for Rehabilitation. *AIP Conf. Proc.* **2018**, *1933*, 1–8.
8. Lenzi, T.; de Rossi, S. The neurorobotics paradigm: NEURARM, NEUROExos, HANDEXOS. In Proceedings of the 2009 Annual International Conference of the IEEE Engineering in Medicine and Biology Society, Minneapolis, MN, USA, 3–6 September 2009; pp. 2430–2433.
9. Dinha, B.; Xiloyannis, M. Adaptive backlash compensation in upper limb soft wearable exoskeletons. *Robot. Auton. Syst.* **2017**, *92*, 173–186. [CrossRef]
10. Cui, X.; Chen, W.; Jin, X.; Argawal, S.K. Design of a 7DOF CableDriven Arm Exoskeleton 11 (CAREX7) and a Controller for Dexterous Motion Training or Assistance. *IEEE/ASME Trans. Mechatron.* **2017**, *22*, 161–172. [CrossRef]
11. Mao, Y.; Agrawal, S.K. A cable driven upper arm exoskeleton for upper extremity rehabilitation. In Proceedings of the 2011 IEEE International Conference on Robotics and Automation, Shanghai, China, 9–13 May 2011; pp. 4163–4168. [CrossRef]
12. Sanjuan, J.D.; Castillo, A.D. Cable driven exoskeleton for upperlimb rehabilitation: A design review. *Robot. Auton. Syst.* **2020**, *126*, 1–25. [CrossRef]
13. Lessard, S.; Pansodtee, P.; Robbins, A.; BaltaxeAdmony, L. A compliant robotic upperextremity exosuit for lightweight, portable, multi−joint muscular augmentation. In Proceedings of the 2017 International Conference on Rehabilitation Robotics (ICORR), London, UK, 17–20 July 2017; pp. 1633–1638. [CrossRef]
14. Xiao, F.; Gao, Y.; Wang, Y.; Zhu, Y.; Zhao, J. Design and evaluation of a 7DOF cabledriven upper limb exoskeleton. *J. Mech. Sci. Technol.* **2018**, *32*, 855–864. [CrossRef]
15. Li, Z.; Li, W.; Chen, W.; Zhang, J.; Wang, J.; Fang, Z.; Yang, G. Mechatronics design and testing of a cabledriven upper limb rehabilitation exoskeleton with variable stiffness. *Rev. Sci. Instrum.* **2021**, *92*, 024101. [CrossRef]
16. Folgheraiter, M.; de Gea, J.; Bongardt, B.; Albiez, J.; Kirchner, F. Bioinspired control of an arm exoskeleton joint with active−compliant actuation system. *Appl. Bionics Biomech.* **2009**, *6*, 193–204. [CrossRef]
17. Elizabeth, A.B.; Mao, Y.; Agrawal, S.K.; Annapragada, M.; Venkatesh, N.D. Dynamics and Control of a 4dof Wearable CableDriven Upper Arm Exoskeleton. In Proceedings of the 2009 IEEE International Conference on Robotics and Automation, Kobe, Japan, 12–17 May 2009; pp. 2300–2305. [CrossRef]
18. Bizzi, E.; Hogan, N.; Mussalvaldi, F.A.; Giszter, S.F. Does the nervous system use equilibriumpoint control to guide single and multiple joint movements? *Behav. Brain. Sci.* **1992**, *15*, 603–613. [CrossRef]
19. Bernstein, N. *The CoOrdination and Regulation of Movement*, 1st ed.; Pergamon Press: Oxford, UK, 1967; pp. 159–196.
20. Clapa, D.J.; Croft, E.A.; Hodgson, A.J. Equilibrium Point Control of a 2DOF Manipulator. *ASME J. Dyn. Sys. Meas. Control* **2006**, *128*, 134–141. [CrossRef]
21. Kim, B.S.; Park, S.; Song, J.B.; Kim, B. Equilibrium point control of a robot manipulator using biologicallyinspired redundant actuation system. *Adv. Robot.* **2013**, *27*, 567–579. [CrossRef]
22. Mukaibo, Y.; Park, S.; Maeno, T. Equilibrium Point Control of a Robot Arm with a Double Actuator Joint. In Proceedings of the International Simposium on Robotics and Automation, Querétaro, México, 25–27 August 2004; pp. 1–7. [CrossRef]

23. Feldman, A.G.; Levin, M.F. The EquilibriumPoint Hypothesis—Past, Present and Future. In *Progress in Motor Control. Advances in Experimental Medicine and Biology*; Sternad, D., Ed.; Springer: Boston, MA, USA, 2009; Volume 629, pp. 699–726. [CrossRef]
24. Spiers, A.; Khan, S.G.; Herrmann, G. *Biologically Inspired Control of Humanoid Robot Arms*, 1st ed.; Springer International Publishing: Cham, Switzerland, 2016; p. 276.
25. Hill, A.V. The possible effects of the aggregation of the molecules of haemoglobin on its dissociation curves. *J. Physiol.* **1910**, *40*, 1–7.
26. Hall, S. *Basic Biomechanics*, 4th ed.; McGrawHill: New York, NY, USA, 2003; p. 556.
27. Zajac, F. Muscle and tendon: Properties, models, scaling and application to biomechanics and motor control. *Crit. Rev. Biomed.* **1989**, *17*, 359–411.
28. Zakaryan, N.B.; Ghazaryan, S.D.; Harutyunyan, M.G.; Sargsyan, Y.L. Dynamic modeling of a new overactuated compliant joint mechanism for human limb rehabilitation. In Proceedings of the 15th IFToMM 2019 World Congress, Krakow, Poland, 1–4 July 2019; pp. 1781–1788. [CrossRef]
29. Bin Hj Shukor, A.Z.; bin Ali Ibrahim, F.; bin Miskon, M.F.; bin Md Saat, M.S. Musculoskeletal Robotics Modeling and Simulation. In *Proceedings of the 8th International Conference on Robotic, Vision, Signal Processing & Power Applications, Singapore, 27 February 2014*; Mat Sakim, H., Mustaffa, M., Eds.; Springer: Cham, Switzerland; pp. 15–21. [CrossRef]
30. Hirai, H.; Miyazaki, F.; Naritomi, H. On the Origin of Muscle synergies: Invariant Balance in the coactivation of agonist and antagonist Muscle Pairs. *Front. Bioeng. Biotechnol.* **2015**, *3*, 1–16. [CrossRef] [PubMed]
31. Tahara, K.; Arimoto, S.; Sekimoto, M.; Luo, Z.W. On Control of Reaching Movements for MusculoSkeletal Redundant Arm Model. *Appl. Bionics Biomech.* **2009**, *6*, 11–26. [CrossRef]
32. Storn, R.; Price, K. Differential Evolution—A Simple and Efficient Heuristic for global Optimization over Continuous Spaces. *J. Glob. Optim.* **1997**, *11*, 341–359. [CrossRef]
33. Fan, Z.; You, Y.; Zheng, H.; Zhu, G.; Li, W.; Chen, S.; Deb, K.; Goodman, E. Modeling and MultiObjective Optimization of a Kind of Teaching Manipulator. *arXiv* **2018**, arXiv:abs/1801.10599.
34. Rodriguez, E.; Saha, B.N.; RomeroHdz, J.; Ortega, D. A Multiobjective Differential Evolution Algorithm for Robot Inverse Kinematics. *SSRG Int. J. Comput. Sci. Eng.* **2016**, *3*, 61–69. [CrossRef]
35. Mohamed, A.W. Solving largescale global optimization problems using enhanced adaptive differential evolution algorithm. *Complex Intell. Syst.* **2017**, *3*, 205–231. [CrossRef]
36. AbdulAdheem, W.; Ibraheem, I. Decoupled control scheme for output tracking of a general industrial6 nonlinear MIMO system using improved active disturbance rejection scheme. *Alex. Eng. J.* **2019**, *58*, 1145–1156. [CrossRef]
37. Hill, A. The Heat of Shortening and the Dynamic Constant of Muscle. *Proc. R. Soc. Lond. Ser. B. Biol. Sci.* **1938**, *126*, 136–195. [CrossRef]
38. Jäntsch, M. Non−Linear Control Strategies for Musculoskeletal Robots. Ph.D. Thesis, Technical University of Munich, Munich, Germany, 2013.
39. Togashi, J.; Mitobe, K.; Capi, G. Control of LowCost Customizable Robot Arm Actuated by Elastic Tendons. *J. Robot. Mechatron.* **2016**, *28*, 509–522. [CrossRef]

robotics

MDPI

Article

Multi-Domain Dynamic Modelling of a Low-Cost Upper Limb Rehabilitation Robot

Adam G. Metcalf *, Justin F. Gallagher, Andrew E. Jackson and Martin C. Levesley

School of Mechanical Engineering, University of Leeds, Woodhouse, Leeds LS2 9JT, UK; J.F.Gallagher@leeds.ac.uk (J.F.G.); A.E.Jackson@leeds.ac.uk (A.E.J.); M.C.Levesley@leeds.ac.uk (M.C.L.)
* Correspondence: mn14agm@leeds.ac.uk

Abstract: Tracking patient progress through a course of robotic tele-rehabilitation requires constant position data logging and comparison, alongside periodic testing with no powered assistance. The test data must be compared with previous test attempts and an ideal baseline, for which a good understanding of the dynamics of the robot is required. The traditional dynamic modelling techniques for serial chain robotics, which involve forming and solving equations of motion, do not adequately describe the multi-domain phenomena that affect the movement of the rehabilitation robot. In this study, a multi-domain dynamic model for an upper limb rehabilitation robot is described. The model, built using a combination of MATLAB, SimScape, and SimScape Multibody, comprises the mechanical electro-mechanical and control domains. The performance of the model was validated against the performance of the robot when unloaded and when loaded with a human arm proxy. It is shown that this combination of software is appropriate for building a dynamic model of the robot and provides advantages over the traditional modelling approach. It is demonstrated that the responses of the model match the responses of the robot with acceptable accuracy, though the inability to model backlash was a limitation.

Keywords: dynamic modelling; rehabilitation robotics; computational modelling; simulation; MATLAB; Simulink; SimScape; SimScape Multibody

Citation: Metcalf, A.G.; Gallagher, J.F.; Jackson, A.E.; Levesley, M.C. Multi-Domain Dynamic Modelling of a Low-Cost Upper Limb Rehabilitation Robot. *Robotics* **2021**, *10*, 134. https://doi.org/10.3390/robotics10040134

Academic Editor: Raffaele Di Gregorio

Received: 15 November 2021
Accepted: 10 December 2021
Published: 15 December 2021

1. Introduction

There is increasing interest in the use of robots to assist the rehabilitation of stroke patients, partly due to the repetitive nature of post-stroke rehabilitation and the integration of stimulating video games [1] to increase patient motivation and partly due to a projected increase in strokes due to an ageing population, causing a strain on rehabilitation services [2]. Stroke can have severe consequences for the patient's quality of life, with reports estimating that up to 20% of patients require intensive rehabilitation programs [3].

Rehabilitation robots may be broadly classified into two categories: Class 1 or Class 2 [4,5]. Class 1 robots are high cost and are designed for clinical use with a therapist. Class 2 robots are low cost and intended for home use. Rehabilitation robots that are commercially available are Class 1 robots, and therefore high cost. Examples include the InMOTION (previously known as MIT Manus [6,7]) and the Armeo Power (previously known as the ARMin [8,9]). However, a large study has recently shown that the use of expensive robotics in a clinical setting provides little additional benefit to traditional rehabilitation alone and is not cost effective [10]. Essentially, if there is a choice between traditional physiotherapist-led post-stroke rehabilitation or robot-based rehabilitation, there is little benefit to selecting robot-based rehabilitation. Both options require travel to a clinical setting and face-to-face contact with a physiotherapist.

As a response to the COVID-19 pandemic, many countries imposed national lockdowns, with non-essential travel banned and social mixing criminalised. Health services were put under considerable strain dealing with the effects of COVID-19, with many services limited or cancelled. There was a noted decrease in stroke patient admissions, particularly patients presenting with minor symptoms and transient ischaemic attack (TIA) in Italy, France, Germany [11], Canada [12], and Norway [13]. In the U.K., there was a 12% reduction in the number of admissions for stroke patients between 1 October 2019 and 30 April 2020 compared to the same period in the three previous years [14]. It was noted that quality of care did not decrease for admitted stroke patients, but there are no data on access to rehabilitation or long-term outcomes. It is postulated that hospital avoidance is the likely cause of the decrease in stroke patients rather than a decrease in the number of strokes.

There is little doubt that social distancing has had an effect on post-stroke rehabilitation. Indeed, in Canada, it was noted that access to rehabilitation care has been significantly reduced [12]. According to the Stroke Association, around half of stroke survivors in the U.K. have had therapy appointments cancelled or postponed [15]. These cancellations may have occurred due to an increase in protective measures, which require more turnaround time between patients, a reduced capacity in rehabilitation centres due to social distancing requirements, and staff absence due to self-isolating or infection. Further to this, 56% of stroke patients have not felt safe to attend scheduled appointments. This is likely due to fear of becoming infected with COVID-19, especially considering that stroke patients are at higher risk [16].

The COVID-19 pandemic has created a need for social distancing and a new paradigm of hospital avoidance due to the fear of infection. It was noted by [12] that telerehabilitation is an effective and well-accepted method for providing access to therapy, though it was considered that this requires family members and care givers to be given additional training. Virtual reality (VR) technologies and existing computer game systems such as the Nintendo Wii could be used to supplement rehabilitation [17], though it was noted that most home-based exercises require oversight from a physiotherapist. It is clear from the literature that there is a need for Class 2 rehabilitation robotics in the current climate of social distancing.

1.1. MyPAM Rehabilitation Robot

MyPAM, shown by Figure 1, is a low-cost planar robot designed for upper limb rehabilitation of stroke patients in the home. The robot workspace exists in the x-y plane, with gravity acting in the z-plane. The robot is a 2 DoF powered arm, which assists the patient to reach targets presented to them on a computer screen. It is important to monitor patient progress during the course of rehabilitation to ensure that the recovery program is appropriate. Tracking patient progress through a course of tele-rehabilitation requires continuous logging of the robot position data during exercise and periodic testing with no robotic assistance. These test data must be compared with previous test attempts and an ideal baseline, which requires a good understanding of the dynamics of the robot when connected to a human arm.

Joint 0 is driven using a MAXON RE40 DC motor and has gear reduction of 40:1, composed of a planetary gear set attached to the motor with a gear reduction of 15:1 driving a spur gear pair with a gear reduction of 8:3. Joint 1 is driven using a MAXON DCX32L 24V DC motor and has a gear reduction of 70:1, composed of a planetary gear set attached to the motor with a gear reduction of 35:1 driving a bevel gear pair with a gear reduction of 2:1. The motors are powered using bespoke motor control boards, which handle current protection and disconnect the motors from power when the motor demand is 0V. Disconnecting the motors in this way prevents the braking effect caused by back EMF and allows the back-drivability of the robot, which is necessary for rehabilitation robotics. Joint position feedback is provided by quadrature encoders. MyPAM is controlled using a National Instruments MyRIO programmed using LabVIEW.

Whilst most upper limb rehabilitation robots use an admittance control or an impedance control approach, the high cost of force sensors required makes admittance control or impedance control inappropriate for a low-cost Class 2 robot such as MyPAM. As such, MyPAM uses position control as the control paradigm. A previous iteration of MyPAM (known as HCARR [5]) using a similar position control approach was tested in a clinical trial. In this trial, seventeen participants used the HCAAR for eight weeks, and statistically significant improvements were observed in kinematic and clinical outcomes.

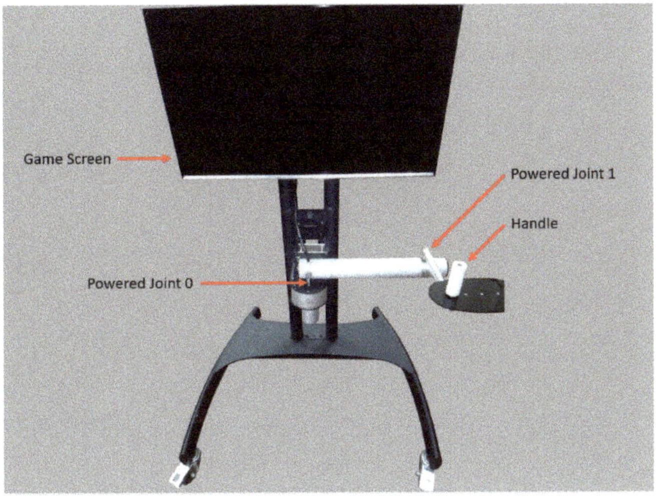

Figure 1. MyPAM upper limb rehabilitation robot.

1.2. Dynamic Modelling

The traditional approach to modelling involves forming dynamic equations (equations of motion), typically using Newtonian or Lagrangian formulations [18]. This becomes cumbersome and error prone as more components and more degrees of freedom are modelled. There has been a growing trend towards simulation-based dynamic modelling due to the increased capability of physics-based simulation software and the increased availability of computing power. Furthermore, efficient recursive algorithms have been developed, increasing the performance of simulation software [19].

The model presented in this paper incorporates three domains (mechanical, electro-mechanical, and control) and is constructed using a combination of software packages from MATLAB. The mechanical components were modelled using SimScape Multibody, which allows three-dimensional dynamic modelling and simulation of mechanical systems. The motors in each of the powered joints of MyPAM were modelled using SimScape, which allows modelling and simulation of electrical and electro-mechanical systems. Simulink and MATLAB were used to apply control to the model. SimScape and SimScape Multibody utilise a physical network approach, whereby block connections communicate information about power [20] and are presented within the Simulink environment. The SimScape Multibody MultiPhysics library was used to provide an interface between SimScape and SimScape Multibody, such that the motor models may be used to drive the joints.

1.3. Human Arm Proxy

A human arm proxy capable of replicating the seven degrees of freedom of a human arm was used for this work. The human arm proxy was previously developed for the validation of a different upper limb rehabilitation robot [21]. The human arm proxy emulates posterior/anterior translation and superior/inferior translation at the shoulder using a flexible steel rod secured at a distance away from the shoulder joint. Extension/flexion of the upper arm is achieved using a rotational joint at the shoulder. Abduction/adduction is achieved using hinge at the shoulder. Extension/flexion of the lower arm is achieved by a hinge at the elbow, with the addition of a spring contributing to the stiffness of the elbow joint. External/internal rotation and pronation/supination are achieved with rotational joints, with stiffness of these rotations facilitated by the addition of friction clutches.

The human arm proxy is shown by Figure 2, with four of the seven degrees of freedom illustrated in the front view and the remaining three degrees of freedom shown in the lateral view.

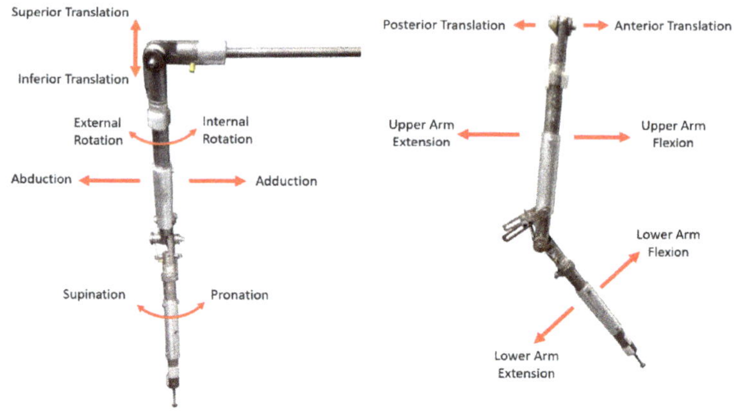

Figure 2. The 7 degrees of freedom of the human arm proxy.

1.4. Aim

The aim of this study is to create a multi-domain dynamic model capable of characterising movement to better understand patient progress.

2. Computational Modelling

2.1. Computational Modelling and Simulation—SimScape Multibody and the Mechanical Domain

The mechanical components of MyPAM were identified and replicated in SimScape Multibody to create a three-dimensional dynamic model in the mechanical domain. Joint 0 was constrained to rotate about the global origin. Similarly, a 3D dynamic model was created for the human arm proxy. The shoulder of the human arm proxy was constrained to rotate about an appropriate point in the global workspace, and the hand was connected to the MyPAM model with a 6-DoF rotational joint. The modelling procedure for each model followed the workflow:

1. Define the global reference point, coordinate system, and simulation settings by placing the world frame block in parallel with the solver configuration block and the mechanism configuration block;
2. In turn, place and configure a solid body block for a component or place and configure a joint;
3. Connect the block input/output ports, ensuring that rigid body transforms are used where appropriate to translate or rotate frames.

All solid bodies were configured with dimensions and assigned a material density, from which the inertia and mass were automatically calculated. Coordinate frames were assigned at suitable locations for each solid body, which created the input/output connection ports required for components and joints to be connected. Joints 0 and 1 of the MyPAM model were not assigned internal mechanics (friction) because these were instead applied to the motor model for each joint. The two joints were configured to receive a torque input, which was provided by the motor model. All joints in the human arm proxy model were configured with internal mechanics. Finally, Joint 1 and the end-effector of MyPAM were configured to the output position in the global workspace, which was externally logged to MATLAB as the output of each test. The SimScape Multibody model for the unloaded MyPAM and the SimScape Multibody model for MyPAM connected to the human arm proxy are shown by Figure 3.

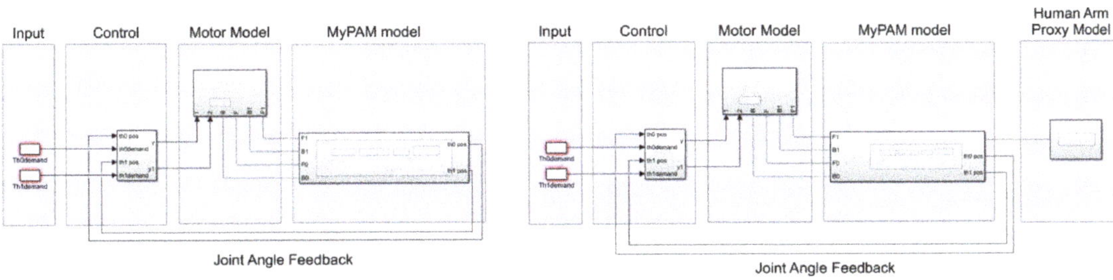

(a) (b)

Figure 3. (**a**) SimScape Multibody model for the unloaded MyPAM. (**b**) SimScape Multibody model for MyPAM connected to the human arm proxy.

2.2. Computational Modelling and Simulation—SimScape and the Electro-Mechanical Domain

A multi-domain motor model for each powered joint of MyPAM was created in SimScape using a combination of electrical and mechanical blocks. For each powered joint, the motor demand from the control model was provided to a controllable voltage source block. The output of the controllable voltage source block was provided to a DC motor block.

The DC motor block in each motor model was configured with the parameters defined in the datasheet for each joint, which are shown by Table 1.

Table 1. The DC motor block parameters.

Parameter	DC Motor Block (Joint 0)	DC Motor Block (Joint 1)
Armature Resistance (Ohm)	0.299	0.331
Armature Inductance (mH)	0.082	0.103
Torque Constant (mNm/A)	30.2	27.3
No-load Current (A)	0.137	0.164
Nominal Voltage (V)	24	24
Rotor Inertia (gcm^2)	142	72.8

The output of the motor was passed through a gear ratio block configured with the correct gear ratio. A friction block was placed in parallel across the motor and gear blocks to model the friction across the joint. The friction block in each motor model was configured as the friction parameters found in a previous experiment, which are shown by Table 2. Note that the breakaway friction velocity was set to the default value, which is close to zero. Finally, a conversion block from the SimScape Multibody MultiPhysics library was placed so that the torque output of the motor modelled in SimScape may be provided to the relevant joint of MyPAM modelled in SimScape Multibody. The motor model is shown by Figure 4.

Table 2. Friction parameters.

Parameter	Joint 0	Joint 1
Breakaway Friction Torque (Nm)	0.25	0.29
Breakaway Friction Velocity (rad^{-1})	0.1	0.1
Coulomb Friction Torque (Nm)	0.18	0.23
Viscous Friction Torque (Nm/rad^{-1})	0.181	0.372

2.3. Computational Modelling and Simulation—MATLAB, Simulink, and the Control Domain

The control scheme was implemented in MATLAB and Simulink and followed the block diagram shown by Figure 5.

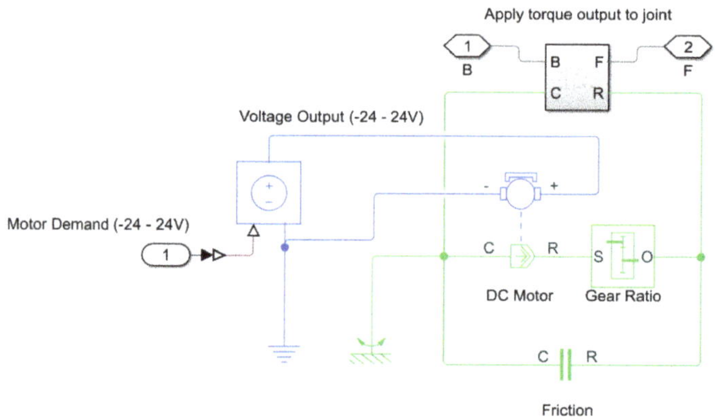

Figure 4. The SimScape motor model.

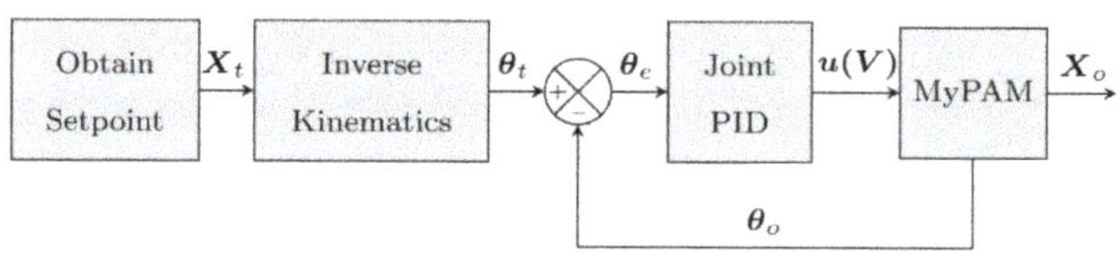

Figure 5. Control scheme block diagram.

A discretised minimum jerk trajectory between the start and end point of the reaching movement was initially generated using MATLAB to produce a set of Cartesian position vectors for the end-effector, which were used as an input to the whole Simulink model. A MATLAB function block was used in Simulink to perform the inverse kinematics, converting the Cartesian position vectors into joint position demand vectors used for control. Joint position feedback from the MyPAM SimScape Multibody models was used to evaluate the joint position error using standard Simulink mathematics blocks. Simulink PID blocks were used to generate a motor control signal u (in Volts) for each joint using the respective joint position errors. MyPAM uses PI position control only, as stated in Section 1.1. In the unloaded case for both the robot and the simulation, the gains were P = 1, I = 0.01. In the loaded case for both the robot and the simulation, the gains were P = 1.5, I = 0.01.

3. Testing and Validation

Methodology

A series of tests was performed to validate the performance of the model against the performance of MyPAM. The first pair of tests compared the performance of the MyPAM model against the performance of the MyPAM in the unloaded condition, with no external loading applied. In the first of these tests, a desired trajectory was applied only in the x-direction, and in the second test, a desired trajectory was applied only in the y-direction. The second pair of tests compared the performance of the MyPAM model against the performance of the MyPAM in the loaded condition, with the human arm proxy model connected. The same desired trajectories were applied only in the x-direction and only in the y-direction for each test, respectively. A summary of the testing is provided by Table 3.

Figure 6 shows MyPAM with the human arm proxy and a simulation of the MyPAM and human arm proxy models.

Table 3. Model validation test summary.

Test	MyPAM Loading Condition	Trajectory Direction
1	Unloaded	X-Direction
2	Unloaded	Y-Direction
3	Loaded	X-Direction
4	Loaded	Y-Direction

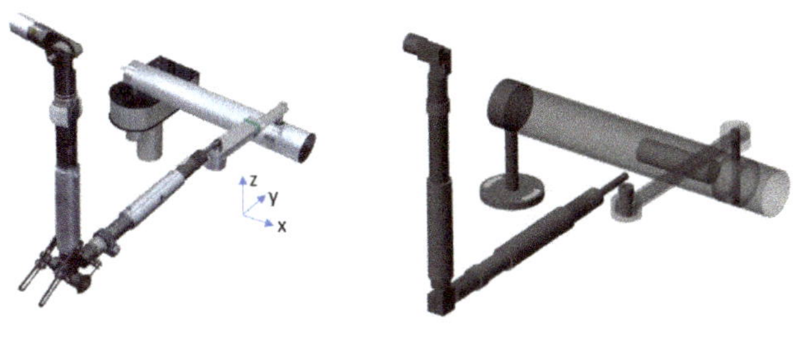

(a) (b)

Figure 6. (**a**) MyPAM with the human arm proxy. (**b**) Simulated MyPAM with the human arm proxy.

4. Results

4.1. Unloaded MyPAM: Tests 1 and 2

The graph shown by Figure 7 shows the simulated response and the mean real response of MyPAM when subjected to a minimum jerk trajectory in the x-direction, with the mean x- and y-positions against time shown by Figure 8. It may be observed that both responses show a characteristic curve, caused by the small demand of the motor at Joint 0 and the large demand of the motor at Joint 1. The graph shown by Figure 9 shows the simulated response and the real responses of MyPAM when subjected to a minimum jerk trajectory in the x-direction, with the x- and y-positions across all repeats against time shown by Figure 10.

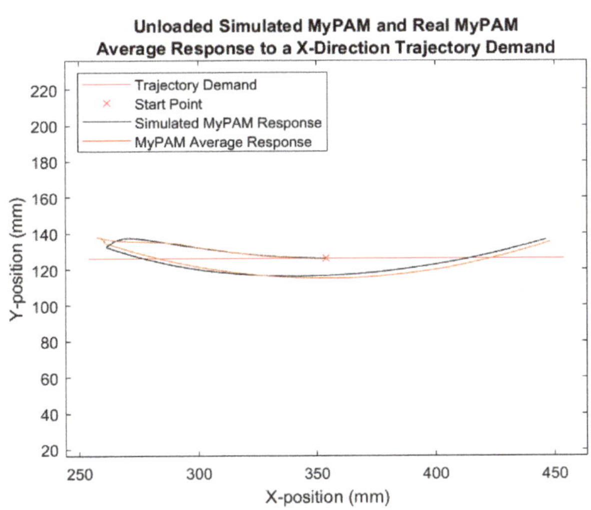

Figure 7. Unloaded simulated MyPAM and average MyPAM x-direction trajectory response.

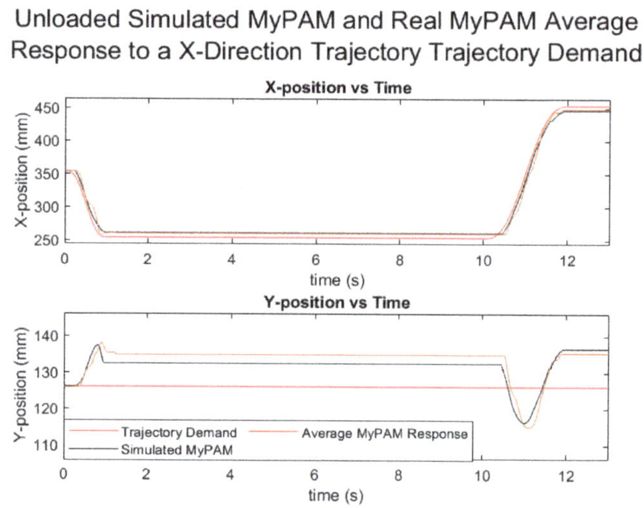

Figure 8. Unloaded simulated MyPAM and average MyPAM x-direction trajectory response against time.

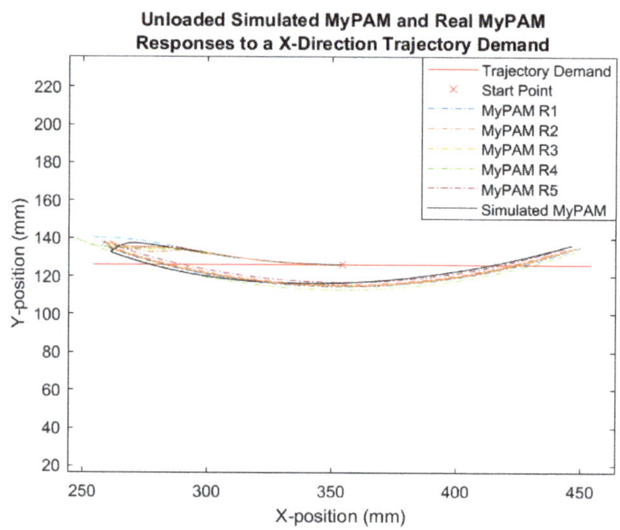

Figure 9. Unloaded simulated MyPAM and MyPAM x-direction trajectory response.

The graph given by Figure 11 shows the simulated response and the mean real response of MyPAM when subjected to a minimum jerk trajectory in the y-direction, with the mean x- and y-positions against time shown by Figure 12. The graph given by Figure 13 shows the simulated response and the real responses of MyPAM when subjected to a minimum jerk trajectory in the y-direction, with the x- and y-positions across all repeats against time shown by Figure 14.

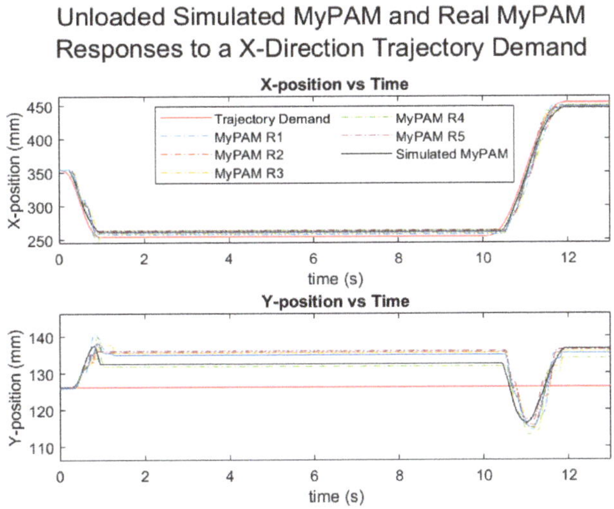

Figure 10. Unloaded simulated MyPAM and MyPAM x-direction trajectory response against time.

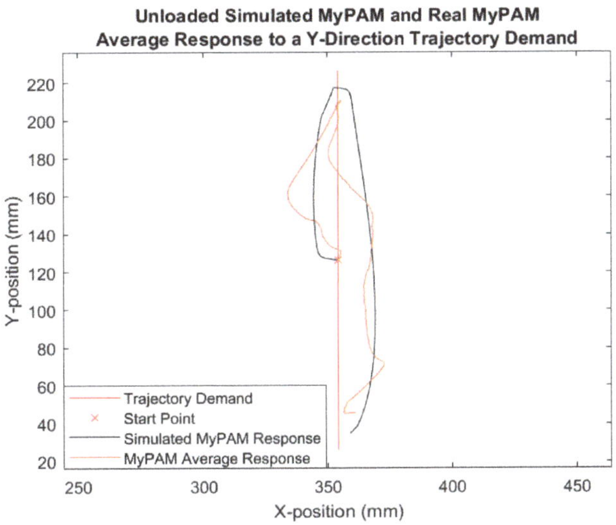

Figure 11. Unloaded simulated MyPAM and average MyPAM y-direction trajectory response.

4.2. Loaded MyPAM: Tests 3 and 4

The graph given by Figure 15 shows the simulated response and the mean real response of MyPAM when subjected to an x-direction minimum jerk trajectory, with the mean x- and y-positions against time shown by Figure 16. It may be observed that the curved response in Figure 7 is present. The graph given by Figure 17 shows the simulated response and the real responses of MyPAM when subjected to an x-direction minimum jerk trajectory, with the x- and y-positions across all repeats against time shown by Figure 18.

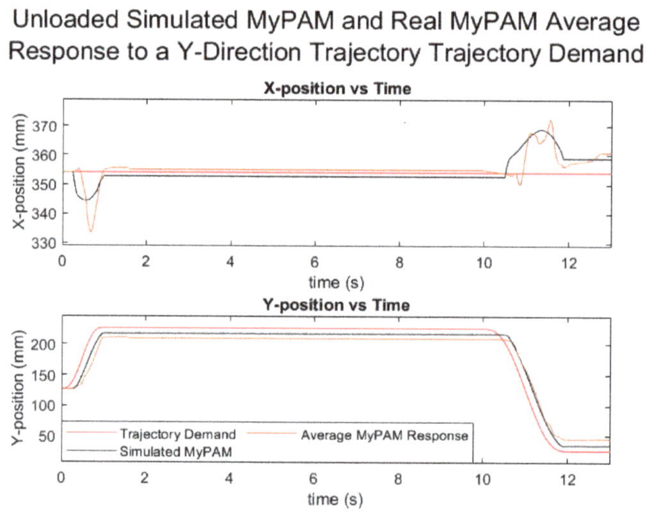

Figure 12. Unloaded simulated MyPAM and average MyPAM y-direction trajectory response against time.

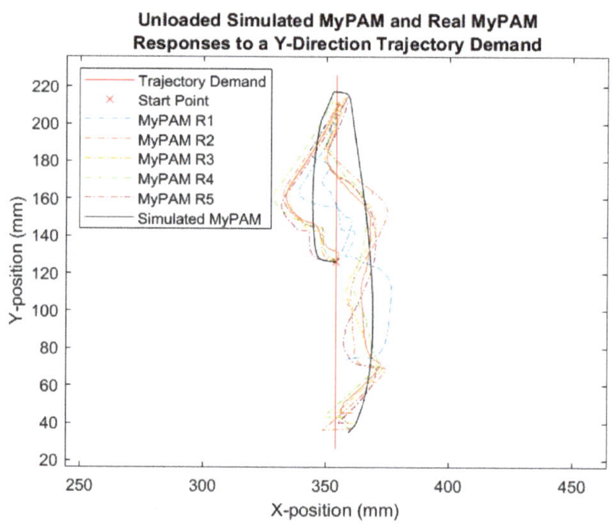

Figure 13. Unloaded simulated MyPAM and MyPAM y-direction trajectory response.

The graph given by Figure 19 shows the simulated response and the mean real response of MyPAM when subjected to a y-direction minimum jerk trajectory, with the mean x- and y-positions against time shown by Figure 20. The graph given by Figure 21 shows the simulated response and the real responses of MyPAM when subjected to a y-direction minimum jerk trajectory, with the x- and y-positions across all repeats against time shown by Figure 22.

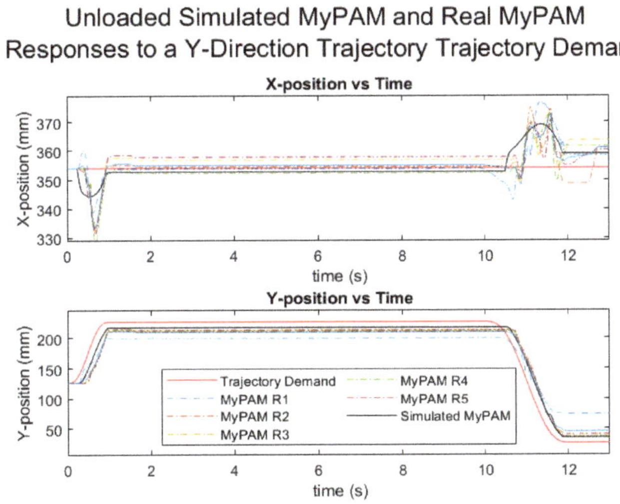

Figure 14. Unloaded simulated MyPAM and MyPAM y-direction trajectory response against time.

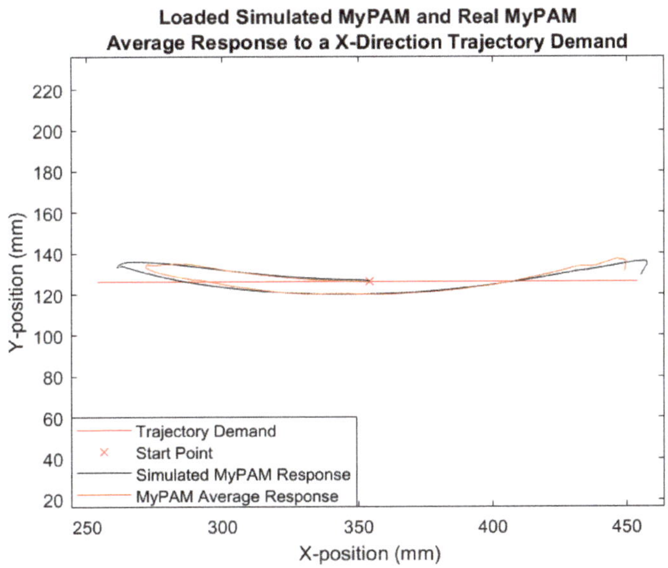

Figure 15. Loaded simulated MyPAM and average MyPAM x-direction trajectory response.

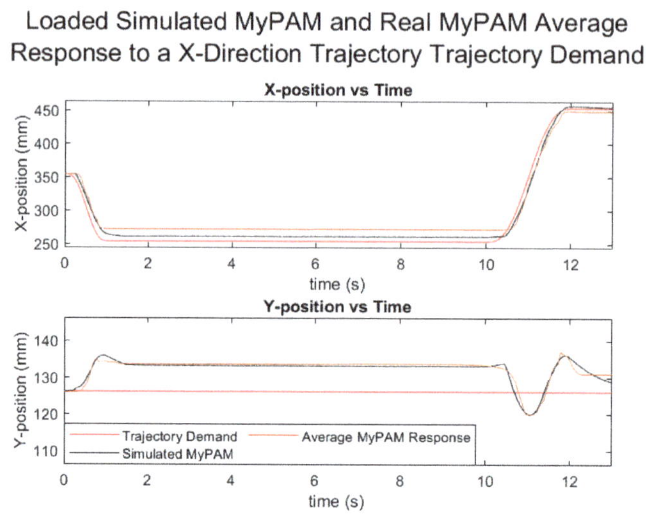

Figure 16. Loaded simulated MyPAM and average MyPAM x-direction trajectory response against time.

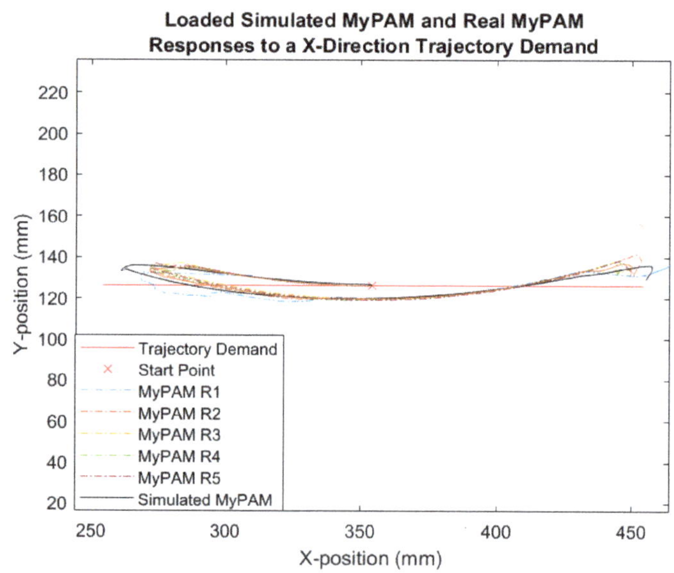

Figure 17. Loaded simulated MyPAM and MyPAM x-direction trajectory response.

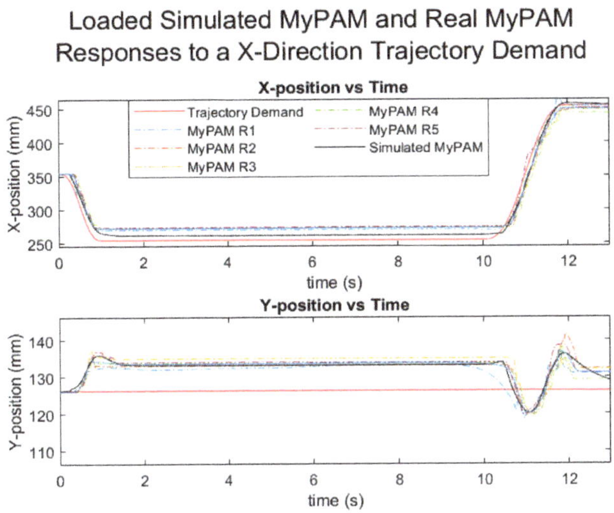

Figure 18. Loaded simulated MyPAM and MyPAM x-direction trajectory response against time.

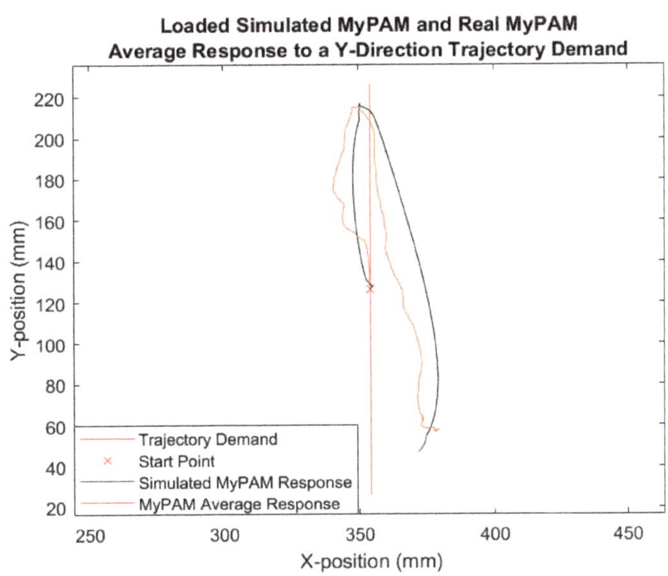

Figure 19. Loaded simulated MyPAM and average MyPAM y-direction trajectory response.

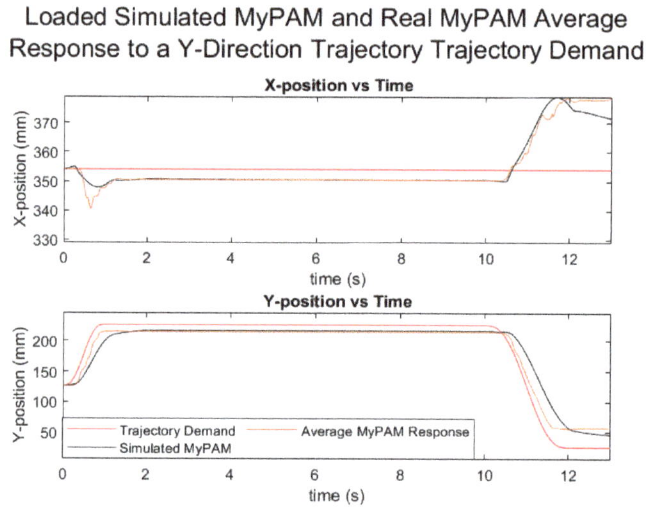

Figure 20. Loaded simulated MyPAM and average MyPAM y-direction trajectory response against time.

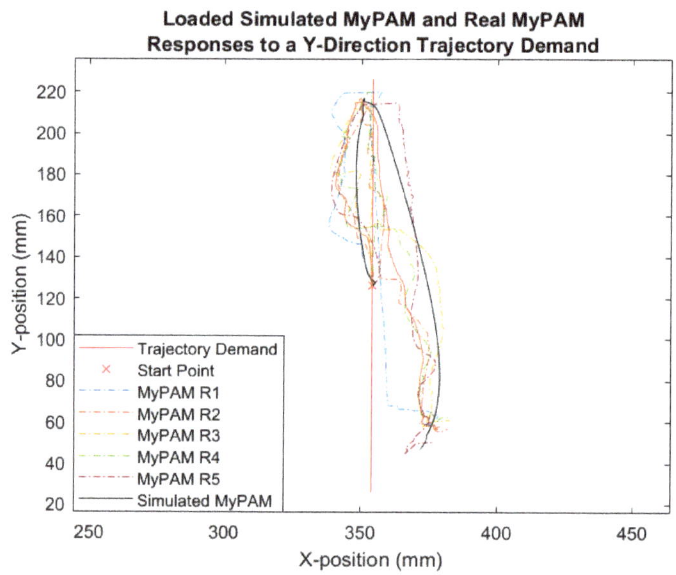

Figure 21. Loaded simulated MyPAM and MyPAM y-direction trajectory response.

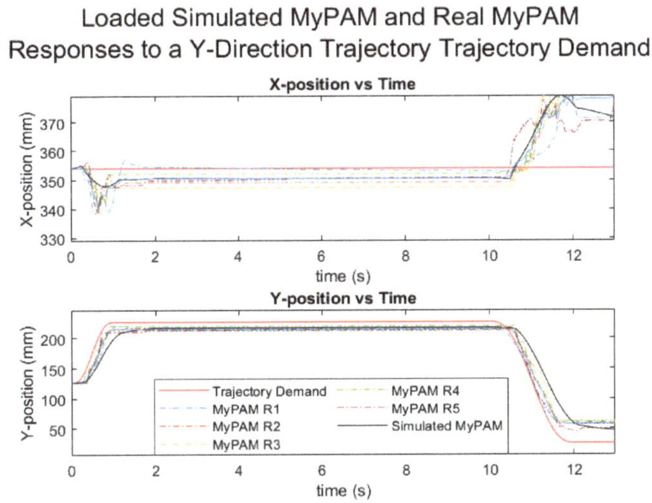

Figure 22. Loaded simulated MyPAM and MyPAM y-direction trajectory response against time.

5. Discussion

In both the unloaded and loaded cases, the x-direction trajectory tracking shown by Figures 7 and 15 (Tests 1 and 3) is much smoother than the y-direction trajectory tracking shown by Figures 11 and 19 (Tests 2 and 4) for both the simulated response and the robot response. This is because in this area of the robot workspace, x-direction movement is achieved mainly by rotation of Joint 1, with very little movement required by Joint 0. As expected, the effects of inertia are less pronounced when the majority of the movement occurs at Joint 1 because the majority of the mass of the robot is located between Joint 0 and Joint 1. Similarly, the y-direction trajectory tracking was poorer because most of the movement is achieved moving Joint 0, where the effects of inertia are more pronounced.

In the unloaded case, the x-direction trajectory tracking (Test 1), seen in Figure 8, shows good x-position agreement between the simulated response and the robot response, with both closely following the trajectory demand. It should be noted, however, that the robot response shows a small amount of overshoot that the simulated response does not show. There is a similar good x-position agreement between the simulated response and the robot response in the loaded case (Test 3) seen in Figure 16, though there is a greater difference between the robot response and the simulated response than is present in the unloaded case. This is likely caused by imperfect modelling of the human arm proxy. The human arm proxy has a greater number of joints than MyPAM, making it considerably more difficult to correctly account for the effects of friction.

The y-position agreement between the robot response and the simulated response for the unloaded x-direction trajectory tracking (Test 1) shown by Figure 8 is satisfactory since both lie within the 10 mm dead-zone allowed between the position of the end-effector and the target, though it is noted that the robot response shows greater overshoot and steady-state error than the simulated response. This may be accounted for by the absence of backlash modelling in the MyPAM model. There is a greater y-position agreement between the robot response and the simulated response for the loaded x-direction trajectory tracking (Test 3) shown by Figure 16, though the simulated response shows a small overshoot at around 1 s that the robot does not. This is again likely caused by imperfect modelling of friction in the human arm proxy, which added less damping to the simulated response than the human arm proxy added to the robot response.

In the unloaded case, the y-direction trajectory tracking (Test 2), seen in Figure 11, shows poorer x-position agreement between the simulated response and the robot response than may be observed in the response to the x-direction trajectory (Test 1). This is likely due to the lack of backlash modelling, which is relatively large at Joint 1. The effect of backlash is pronounced in the robot response because tracking a y-direction trajectory in this area of the robot workspace requires small movements of Joint 1 where the backlash to the x-position demand ratio is high. Despite this, it was observed that the steady-state response in the x-position, seen in Figure 12, is closely matched between the robot response and the simulated response. In the loaded case, the x-position agreement between the simulated response and the robot response while tracking the y-direction trajectory (Test 4), seen in Figure 20, is poorer than was observed in the unloaded case. This is again likely caused by imperfect modelling of friction in the joints of the human arm proxy model.

In the unloaded case, the y-direction trajectory tracking (Test 2), seen in Figure 11, shows reasonable y-position agreement between the simulated response and the robot response, though it was observed that the model tracks the trajectory demand better than the robot. This likely occurred due to the backlash at Joint 1, which caused greater error in the x-position with corresponding inertial effects on the robot. The loaded case (Test 4) shows a better y-position agreement between the simulated response and the robot response, seen in Figure 19, due to the damping effects of the friction in the human arm proxy.

It has been noted that SimScape Multibody was poor at modelling instability. This became apparent when attempting to tune the PID gains for the motor controllers using the Ziegler–Nichols method on the unloaded MyPAM model, where it was not possible to introduce instability even with an extremely large derivative gain. Indeed, the model response with large PID gains followed the trajectory perfectly. This means that the PID tuning strategy required a balance of rough tuning with the models to identify sensible gains, followed by finer tuning and validation on the robot. The set of software cannot be relied upon for tuning the control system.

SimScape Multibody does not provide a function for modelling backlash. This was a problem because the backlash in Joint 1 of MyPAM is relatively large. It would be preferred to model the backlash because it has a significant effect on the dynamics, particularly in circumstances where Joint 0 must make frequent changes in direction to track the trajectory. The effects of the backlash on MyPAM can be clearly seen by the jagged motion of the real MyPAM in Figure 10. In the same figure, the jagged motion, which would be caused by the effects of the backlash, is absent from the response of the simulated MyPAM, which instead shows a smoother response.

The agreement between the robot response and the model response is acceptable in the primary direction of travel in all cases (i.e., for motion primarily in the x-direction, there is good agreement in the x-position against time). The agreement between the robot response and the model response perpendicular to the primary direction of travel is less good, though it is considered acceptable because this particular region of the workspace was selected for testing as it is where the robot has the most difficulty. It is apparent that the performance of the model more closely matches the performance of the robot when loaded than when unloaded, which is a useful outcome from a rehabilitation perspective since the main purpose of the model is to provide a baseline against which patient data may be compared.

6. Conclusions

The combination of MATLAB, SimScape, and SimScape Multibody provides an appropriate tool for multi-domain modelling of MyPAM. While the response of the simulation is closer to the response of the robot in the loaded case than in the unloaded case, the model has sufficient accuracy to allow the dynamics of the robot to be accounted for when analysing patient movement data. Importantly, it can be seen that the shape and direction of the response curves produced by the models adequately match the shape and direction of the response curves produced by MyPAM. Modelling the robot using this combination of software allowed the creation of a model with a greater deal of both fidelity and complexity than traditional mathematical modelling alone would allow, though it was noted that the absence of backlash from the model is a limitation. The failure of the models to produce instability in response to a large control input means that this set of software is not appropriate for tuning the control system.

Author Contributions: Conceptualization, A.G.M., M.C.L., J.F.G. and A.E.J.; methodology, A.G.M.; software, A.G.M.; validation, A.G.M.; formal analysis, A.G.M., M.C.L., J.F.G. and A.E.J.; data curation, A.G.M.; writing—original draft preparation, A.G.M.; writing—review and editing, M.C.L., J.F.G., A.E.J. and A.G.M.; supervision, M.C.L., J.F.G. and A.E.J. All authors have read and agreed to the published version of the manuscript.

Funding: This research is funded by EPSRC (the Engineering and Physical Sciences Research Council).

Institutional Review Board Statement: Not applicable.

Informed Consent Statement: Not applicable.

Data Availability Statement: Data are available on GitHub (DOI:10.5281/zenodo.5648266) or available by correspondence with the authors.

Conflicts of Interest: The authors declare no conflict of interest.

References

1. Flores, E.; Tobon, G.; Cavallaro, E.; Cavallaro, F.I.; Perry, J.C.; Keller, T. Improving patient motivation in game development for motor deficit rehabilitation. In Proceedings of the 2008 International Conference on Advances in Computer Entertainment Technology, Yokohama, Japan, 3–5 December 2008; pp. 381–384.
2. Wafa, H.A.; Wolfe, C.D.; Emmett, E.; Roth, G.A.; Johnson, C.O.; Wang, Y. Burden of stroke in Europe: Thirty-year projections of incidence, prevalence, deaths, and disability-adjusted life years. *Stroke* **2020**, *51*, 2418–2427. [CrossRef] [PubMed]
3. Alrabghi, L.; Alnemari, R.; Aloteebi, R.; Alshammari, H.; Ayyad, M.; Al Ibrahim, M.; Alotayfi, M.; Bugshan, T.; Alfaifi, A.; Aljuwayd, H. Stroke types and management. *Int. J. Community Med. Public Health* **2018**, *5*, 3715–3720. [CrossRef]
4. Sulzer, J.S.; Peshkin, M.A.; Patton, J.L. Design of a mobile, inexpensive device for upper extremity rehabilitation at home. In Proceedings of the 10th International Conference on Rehabilitation Robotics, Noordwijk, The Netherlands, 12–15 June 2007; pp. 933–937.
5. Sivan, M.; Gallagher, J.; Makower, S.; Keeling, D.; Bhakta, B.; O'Connor, R.J.; Levesley, M. Home-based Computer Assisted Arm Rehabilitation (hCAAR) robotic device for upper limb exercise after stroke: Results of a feasibility study in home setting. *J. Neuroeng. Rehabilit.* **2014**, *11*, 163. [CrossRef] [PubMed]
6. Krebs, H.I.; Hogan, N.; Aisen, M.L.; Volpe, B.T. Robot-aided neurorehabilitation. *IEEE Trans. Rehabilit. Eng.* **1988**, *6*, 75–87. [CrossRef] [PubMed]
7. Krebs, H.I.; Ferraro, M.; Buerger, S.P.; Newbery, M.J.; Makiyama, A.; Sandmann, M.; Lynch, D.; Volpe, B.T.; Hogan, N. Rehabilitation robotics: Pilot trial of a spatial extension for MIT-Manus. *J. Neuroeng. Rehabilit.* **2004**, *1*, 5. [CrossRef] [PubMed]
8. Nef, T.; Guidali, M.; Riener, R. ARMin III–arm therapy exoskeleton with an ergonomic shoulder actuation. *Appl. Bionics Biomech.* **2009**, *6*, 127–142. [CrossRef]
9. Klamroth-Marganska, V.; Blanco, J.; Campen, K.; Curt, A.; Dietz, V.; Ettlin, T.; Felder, M.; Fellinghauer, B.; Guidali, M.; Kollmar, A.; et al. Three-dimensional, task-specific robot therapy of the arm after stroke: A multicentre, parallel-group randomised trial. *Lancet Neurol.* **2014**, *13*, 159–166. [CrossRef]
10. Rodgers, H.; Bosomworth, H.; Krebs, H.I.; van Wijck, F.; Howel, D.; Wilson, N.; Aird, L.; Alvarado, N.; Andole, S.; Cohen, D.L.; et al. Robot assisted training for the upper limb after stroke (RATULS): A multicentre randomised controlled trial. *Lancet* **2019**, *394*, 51–62. [CrossRef]
11. Bersano, A.; Kraemer, M.; Touzé, E.; Weber, R.; Alamowitch, S.; Sibon, I.; Pantoni, L. Stroke care during the COVID-19 pandemic: Experience from three large European countries. *Eur. J. Neurol.* **2020**, *27*, 1794–1800. [CrossRef] [PubMed]
12. Smith, E.E.; Mountain, A.; Hill, M.D.; Wein, T.H.; Blacquiere, D.; Casaubon, L.K.; Linkewich, E.; Foley, N.; Gubitz, G.; Simard, A.; et al. Canadian stroke best practice guidance during the COVID-19 pandemic. *Can. J. Neurol. Sci.* **2020**, *47*, 474–478. [CrossRef] [PubMed]
13. Kristoffersen, E.S.; Jahr, S.H.; Faiz, K.W.; Storstein, A.M.; Winsvold, B.S.; Sandset, E.C. Acute stroke care during the first phase of COVID-19 pandemic in Norway. *Acta Neurol. Scand.* **2021**, *143*, 349–354. [CrossRef] [PubMed]
14. Douiri, A.; Muruet, W.; Bhalla, A.; James, M.; Paley, L.; Stanley, K.; Rudd, A.G.; Wolfe, C.D.; Bray, B.D.; SSNAP Collaboration. Stroke care in the United Kingdom during the COVID-19 pandemic. *Stroke* **2021**, *52*, 2125–2133. [CrossRef] [PubMed]
15. Stroke Recoveries at Risk Report. Available online: https://www.stroke.org.uk/stroke-recoveries-at-risk-report (accessed on 5 December 2021).
16. y Galán, J.T.G. Stroke as a complication and prognostic factor of COVID-19. *Neurología (Engl. Ed.)* **2020**, *35*, 318–322. [CrossRef]

17. Wang, C.C.; Chao, J.K.; Wang, M.L.; Yang, Y.P.; Chien, C.S.; Lai, W.Y.; Yang, Y.C.; Chang, Y.H.; Chou, C.L.; Kao, C.L. Care for patients with stroke during the COVID-19 pandemic: Physical therapy and rehabilitation suggestions for preventing secondary stroke. *J. Stroke Cerebrovasc. Dis.* **2020**, *29*, 105182. [CrossRef] [PubMed]
18. Lynch, K.M.; Park, F.C. *Modern Robotics*, 1st ed.; Cambridge University Press: Cambridge, UK, 2017; p. 231.
19. Erez, T.; Tassa, Y.; Todorov, E. Simulation tools for model-based robotics: Comparison of bullet, havok, mujoco, ode and physx. In Proceedings of the 2015 IEEE International Conference on Robotics and Automation (ICRA), Seattle, DC, USA, 26–30 May 2015; pp. 4397–4401.
20. Das, S. *Modeling and Simulation of Mechatronic Systems Using Simscape*, 1st ed.; Morgan and Claypool Publishers: San Rafael, CA, USA, 2020; pp. 11–23.
21. Jackson, A.E.; Levesley, M.C.; Culmer, P. Development of a mechanical arm model of the human arm for use with an exercise robotic system being developed for people with stroke. In Proceedings of the 2nd Cambridge Workshop on Universal Access and Assistive Technologies, Cambridge, UK, 22–24 March 2004; pp. 91–98.

robotics

MDPI

Article

Development and Usability Testing of a Finger Grip Enhancer for the Elderly

Dominic Wen How Tan [1], Poh Kiat Ng [1,*], Ervina Efzan Mhd Noor [1], Adi Saptari [2], Chee Chen Hue [3] and Yu Jin Ng [4]

1 Faculty of Engineering and Technology, Multimedia University, Jalan Ayer Keroh Lama, Bukit Beruang, Melaka 75450, Malaysia; dominictan80@gmail.com (D.W.H.T.); ervina.noor@mmu.edu.my (E.E.M.N.)
2 Department of Industrial Engineering, President University, Jababeka Education Park, Jl. Ki Hajar Dewantara, RT.2/RW.4, Mekarmukti, North Cikarang, Bekasi 17530, Indonesia; adi.saptari@president.ac.id
3 Department of Orthopaedic Surgery, University of Malaya Medical Centre (UMMC), Jln Profesor Diraja Ungku Aziz, Kuala Lumpur 59100, Malaysia; huecc@ummc.edu.my
4 Department of Social Science and Humanities, Putrajaya Campus, College of Energy Economics and Social Sciences, Universiti Tenaga Nasional (UNITEN), Jalan IKRAM-UNITEN, Kajang 43000, Malaysia; yujin@uniten.edu.my
* Correspondence: pkng@mmu.edu.my

Citation: Tan, D.W.H.; Ng, P.K.; Noor, E.E.M.; Saptari, A.; Hue, C.C.; Ng, Y.J. Development and Usability Testing of a Finger Grip Enhancer for the Elderly. *Robotics* **2022**, *11*, 5. https://doi.org/10.3390/robotics11010005

Academic Editors: Raffaele Di Gregorio and Saïd Zeghloul

Received: 21 October 2021
Accepted: 28 December 2021
Published: 30 December 2021

Abstract: As people age, their finger function deteriorates due to muscle, nerve, and brain degeneration. While exercises might delay this deterioration, an invention that enhances elderly people's pinching abilities is essential. This study aims to design and develop a finger grip enhancer that facilitates the day-to-day pinching activities of elderly people. This research is an extension of a previous study that conceptualised a finger grip enhancer. The device facilitates finger flexion on the thumb and index finger, and weighs 520 g, allowing for improved portability and sufficient force exertion (13.9 N) for day-to-day pinching. To test for usability, eleven subjects aged 65 years and above performed a pinch-lift-hold test on various household objects. The pinch force before and after utilising the device was measured. Using Minitab 18, the statistical significance of using this device was analysed with a paired-samples *t*-test. With this device, the elderly people's pinching abilities significantly improved in both pinch force and pinch force steadiness ($p < 0.05$). The proposed device has the potential to enhance elderly people's quality of life by supporting a firm pinch in the handling of everyday objects. This research has applicational value in developing exoskeleton devices for patients who require rehabilitation.

Keywords: finger grip; elderly; ergonomics; pinch assistant; pinch force; usability

1. Introduction

The elderly's struggle in maintaining an active and healthy lifestyle in a fast-growing population of elderly people is becoming a serious concern in the world [1,2]. Degradation of fine motor skills and hand functions occur among elderly people due to nerve, brain and muscle degeneration [3,4]. In studies concerning elderly people's involvement in basic activities of daily living (ADL), poor hand strength causes even the screwing or unscrewing of medicine bottle caps to be daunting tasks [5,6]. Similarly, elderly people can struggle with simple pinch grip activities.

Researchers found that the handgrip strength and ADL scale values of elderly adults were greatly improved by the combined use of finger-movement exercises and correct finger weight-lifting training, and these rehabilitation activities can be utilised to assist the elderly in maintaining their ability to care for themselves [7]. Through finger exercise training regimens, researchers found that trained older adults showed significant gains compared to untrained senior citizens in the ability to control submaximal pinch force, maintain a steady hand posture, and transfer small objects with finger grips following expert finger movement training [8]. Similar benefits from hand exercise programmes have even been observed for people with hand-related rheumatoid arthritis [9,10], dementia [11], and chronic upper limb conditions [12].

While exercises can aid the survival of motor units (MU), reinnervation and neuromuscular junction morphology, the ageing MU has diminished plasticity in response to physical activity, particularly in an advanced age where exercise training might over-burden the remodelled surviving MUs and exacerbate their demise [13]. Hence, frail elderly people who experience severe weakness in their hands might need urgent medical attention. One possible solution to address this weakness is by researching exoskeleton devices. Hand exoskeletons can be broadly divided into two categories, namely rehabilitation and assistive exoskeletons. Rehabilitation exoskeletons are often used to help stroke patients [14–17], patients with limited hand motor functions [16,18], and patients who suffer from hand injuries [18,19]. Due to the similarity in construction, assistive components are sometimes incorporated into rehabilitation devices [20,21] or constructed to treat a specific condition such as paralysis of the fingers [22,23].

In a recent National Health and Morbidity Survey conducted by Mahmud, et al. [24], elderly peoples' functional limitations in performing ADL were investigated. Most of the activities required good hand and finger mobility such as grooming, toilet use, feeding and dressing. The findings revealed that 17% of the elderly people had functional limitations of which females (21.2%) had a higher prevalence compared to males (12.7%). These statistical findings highlighted the neglected quality of life among the elderly due to poor finger abilities.

Numerous pieces of research have been done on finger gripping through population-based data [13–15], determinants that affect pinch strength [16–18], and pinch strength effects on ADL [18,19]. Many of these studies are statistical by nature and look into providing fundamental inferences of pinch health along with their effects on people and society. However, exoskeleton devices specifically designed to assist in finger pinching are few and uncommon. In previous research, the theory of inventive problem solving (TRIZ) was used to systematically synthesise and select a finger grip enhancer concept powered by a miniature linear actuator that actuates the fingers through tendon cables triggered by a push start button [25]. Therefore, the aim of the present study is to extend from this previous research by designing and developing a finger grip enhancer that facilitates the day-to-day pinching activities of elderly people.

2. Literature Review

2.1. Conceptual Design

The concept of the present study was conceptualised and selected from a previous study [25]. When designing a finger exoskeleton, the anatomy of the hand is a priority. The anatomical measurements include length, width and circumference [26,27]. The range of motion (ROM) of fingers and wrist (flexion, extension, abduction, adduction) is also a common functional measurement. With the objective of assisting elderly people in daily pinch activities, the ROM that the device must cover is that of pinch prehension.

The palmar pinch was found to be the most commonly used prehension. It was found to be used 50% of the time for picking-up situations and 88% of the time for hold-for-use situations when considering palmar, tip and lateral prehensions [28]. During palmar prehension, the hand system is found to have the following characteristics:

a. Wrist is in slight extension.
b. Metacarpophalangeal flexion with slight phalangeal flexion.
c. Flexion of the thumb's interphalangeal (IP) joint and index finger's distal interphalangeal (DIP) joint [29].

The constraints of the human hands were technically presented by Lin, et al. [30]. The three types of constraints are listed below. The mechanical limitation of the hand was considered during the design phase by studying the max ROM of each finger joint.

a. Each finger has a limited motion range due to the mechanical limitation of hand anatomy.
b. In each finger, the DIP joint and proximal interphalangeal (PIP) joint always move together.
c. People have a habit of making standard gestures, where they usually bend all the fingers at the same time to make a fist, instead of one by one.

Generally, a hand exoskeleton can have one of three main objectives, namely rehabilitation, assistive use or haptic use [31]. An assistive exoskeleton is for patients with hand disabilities to assist them during activities of daily living (ADL) which can include holding a key, plugging a power adapter, buttoning and unbuttoning a shirt, or picking up small fruit.

A direct matching mechanism includes a series of linkages connected to mirror the movement of fingers. This concept is one of the simplest underactuated mechanisms attached to the side and back of the fingers. The current study has adopted the tendon-driven mechanism that imitates the flexor digitorum tendons (profundus and superficialis), which is the tendon responsible for finger flexion [32].

The underactuated mechanism in this study's application is not required for finger extension as this action is done naturally by the fingers. Actuation of the fingers is accomplished through the tensioning of artificial tendons or string cables on the palmar side of the hand. The device needs to assist in finger flexion, and includes wires underneath the fingers and pivot points at the finger joints as seen in Figure 1.

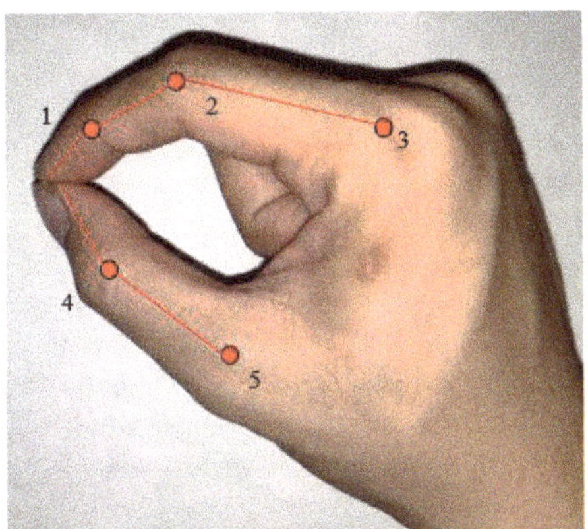

Figure 1. Linkage and joint mechanism of a human finger.

The human hand has 27 degrees of freedom (DOF): 5 DOF in the thumb, 4 (3 for extension/flexion and 1 for abduction/adduction) for each of the other 4 fingers and 6 more in the wrist for translation and rotation [33]. However, for the purpose of designing an exoskeleton or prosthetic device for the hand, many researchers proposed simplified models with less DOF. Lisini, et al. [34] used a simplified kinematic structure with 23 degrees of freedom while Lince [35] designed the ReHand exoskeleton device to accommodate 24 DOF of the hand.

With the device actuated by flexible cables, it gives users freedom in both finger and hand movements. This concept enables the researchers to ignore some DOFs in the prototype while still allowing users to freely control all the ROM of the fingers. The ideal wrist position during any pinching action is a neutral position as deviation from that position was found to reduce pinch force [36]. Thus, the wrist joint is not included in the design of the device. This strategy was also applied by Worsnopp, et al. [37] as most pinching movements involve fingers with neutral abduction or adduction.

Furthermore, a two-finger pinch relies mostly on the flexion of the finger joints which allowed the researchers to ignore the abduction/adduction of joints. Thus, to achieve the goal of assisting the elderly in pinching, the device assisted in 5 DOF, of which two are in the thumb and three are in the index finger.

2.2. Design Subassembly

Each subassembly of the pinch enhancer is explained in this section for a more detailed understanding of how different parts with different functions come together. Figure 2 is the subassembly tree diagram of the pinch enhancer which is broken down into three main components, namely the control system, actuation assembly and palm assembly.

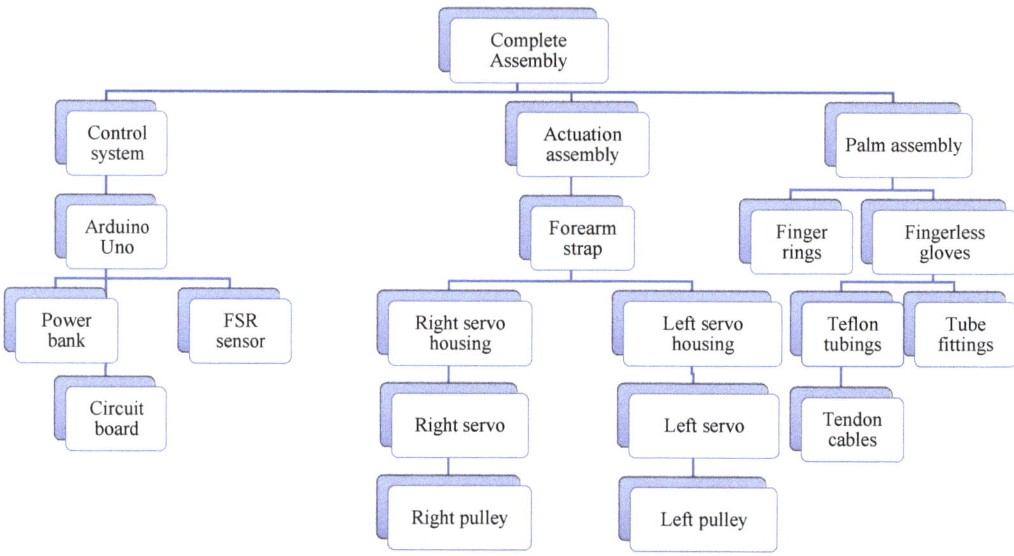

Figure 2. Pinch enhancer tree diagram.

Figure 3 shows the device divided into its subassemblies. With the user's right hand equipped with the device, the first pinch is initiated. Once their fingers have pinched the objects, the force-sensitive resistor is pressed using their left hand. This operation will indicate that a pinch action is taking place and trigger the device to rotate. The rotation of the servo is based on the amount of force applied from the left hand on the force-sensitive resistor. This method of using both hands to control the action was also studied by Leonardis, et al. [38] and Rahman and Al-Jumaily [15]. For different digit sizes, different finger rings are used to ensure a good fit between the fingers and the device. In a hand exoskeleton, Sarac, Solazzi and Frisoli [31] categorised the design aspect into four different categories: mobility, mechanical design, actuation and operational strategies. A similar categorisation is also discussed in the following sections.

2.3. Actuation

Figure 4 illustrates the transmission of a phalange in a finger flexion to achieve a pulp pinch. The device does not control the rotation of the thumb's carpometacarpal (CMC) joint which is only required for hard linkage designs. The reason is due to the flexibility of the current system which is similar to the flexibility in soft robotics. The CMC joint has its uninterrupted natural ROM during pinching tasks. This condition also applies to the abduction and adduction of both fingers.

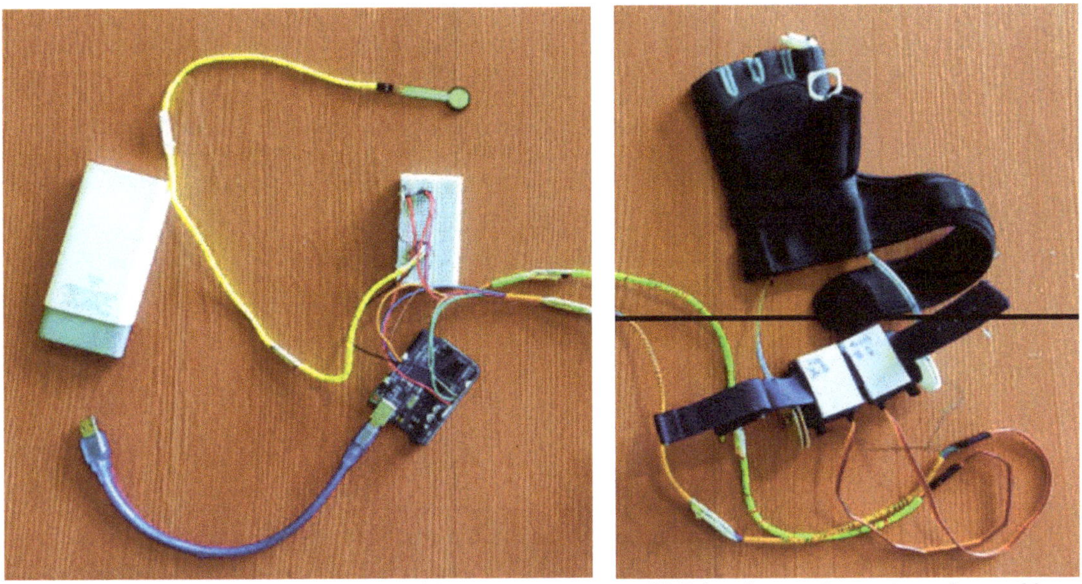

(**a**) Control system subassembly (**b**) Palm glove and actuation subassembly

Figure 3. Finger grip enhancer divided into subassemblies (control system, palm glove, and actuation).

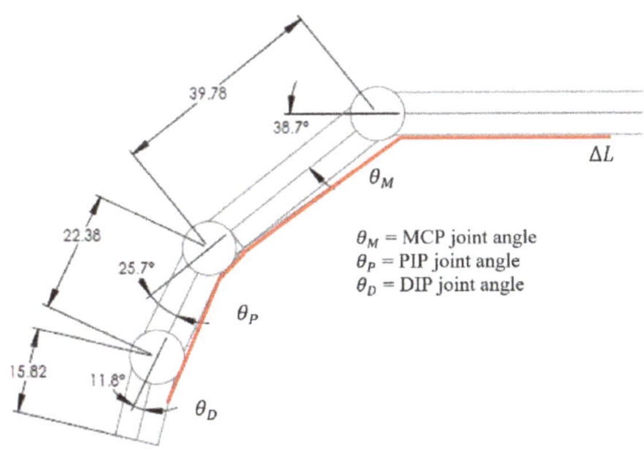

Figure 4. Mechanical design of the phalange transmission. Note: MCP—metacarpophalangeal; PIP—proximal interphalangeal; DIP—distal interphalangeal.

Alexander and Kotiuk [39] investigated the length of hand segments of 66 adult patients aged 19 to 78 years. After the age of 18 years, changes in the proportions of the hand segments are normally considered to be insignificant [40]. Table 1 shows the extraction of segment lengths for the thumb and index finger [39].

Table 1. Lengths of phalanges, metacarpals and distal phalanges.

Finger	Distal Phalanx (mm)	Medial Phalanx (mm)	Proximal Phalanx (mm)
Thumb	21.67 ± 1.60	-	31.57 ± 3.13
Index Finger	15.82 ± 2.26	22.38 ± 2.51	39.78 ± 4.94

The underactuated design depends on the change in length, L. The change in length, ΔL, will translate to the rotational displacement of the phalanges about their respective joints. This movement is estimated to follow a perfect arc trajectory thus allowing for the use of circular equations. In order to measure the ROM of the fingers, Hume, et al. [41] used both a standard goniometric and electro goniometric method to obtain maximum joint measurements of the fingers. Jahn, et al. [42] used a video motion capture technology to measure the ROM of joints while sustaining a pulp pinch. Table 2 shows the extraction of their experimental results [41,42].

Table 2. Maximum and pinch range of motion for the thumb (T) and index finger (I) (degrees).

Fingers	DIP/IP	PIP	MCP
Max ROM (T)	5–73	-	0–56
Max ROM (I)	0–85	0–105	0–100
Pinch ROM (T)	23	-	11.1
Pinch ROM (I)	11.8	25.7	38.7

The ΔL is a linear displacement that is achieved by the rotation of a DC servo motor. When ΔL increases, the tendon cables are pulled, creating deflection joint angles θ_M, θ_P and θ_D. Therefore, the rotation translation of the motor must be sufficient to transform the linear displacement of the fingers such that the pinch joint ROM can be achieved. For safety purposes, the ΔL must not cause the tendon cables to excessively pull the finger until the joints exceed their maximum ROM.

The average ΔL required to rotate all three joints for a pulp pinch can be calculated through Equations (1) and (2). The servo motor requires a rotational displacement which corresponds to the 40.16 mm linear displacement.

$$\Delta L = s_M + s_P + s_D \quad (1)$$

$$\Delta L = r_M\theta_M + r_P\theta_P + r_D\theta_D \quad (2)$$

$$\Delta L = 40.16 \text{ mm}$$

This requirement is determined by calculating the arc of the pulley (see Equation (3)). The potentiometer servo motor has a 180-degree rotation. Therefore, the motor pulley requires a minimum diameter of 25.57 mm but is finally designed with a 45 mm diameter. A larger diameter pulley produces a larger moment and also greater adjustment flexibility.

$$s = r\theta \quad (3)$$

$$2r = d = 25.57 \text{ mm}$$

2.4. Servo Unit

The servo unit is tasked to pull the cables connected to the fingers which mimic how the hand muscles function. The selected servo requires a stall torque that is greater than the extension force of the fingers. This requirement is important to prevent the fingers from losing their grip which may result in the pinched object falling from the grasp. Table 3 presents the mean extension force of the index finger for both men and women [43].

Table 3. Extension force in the index fingers.

Extension Force of Index Fingers Mean ± SD, N	
Men	Women
40.6 ± 2.9	17.2 ± 1.6

The servo pulley was previously designed to have a diameter of 45 mm. Using the mean extension force of men at 40.6 N, the servo's required stall torque is calculated with Equation (4).

$$\tau_{\text{stall}} > Fr \quad (4)$$

$$\tau_{\text{stall}} > 913.5 \text{ Nmm}$$

$$\tau_{\text{stall}} > 9.31 \text{ kg cm}$$

The stall torque of the servo should be at least 9.31 kg cm. Therefore, a Feetech metal gears analogue servo with a 15 kg cm stall torque is selected. The maximum holding force that the motor can generate is 65.4 N.

A servo housing unit is designed to hold the servos in place and is strapped onto the forearm to minimise mass added to the fingers. The force is transmitted through the cables. The cables follow the cable guide from the fingers through the glove and all the way to each servo motor (see Figure 5). The unit is 3D printed with PLA plastic which is durable and lightweight. The housing unit does not fully enclose the servos which enable easy access for technical adjustments.

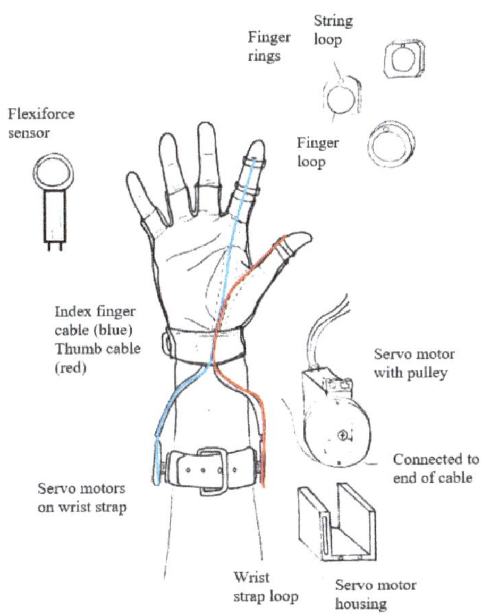

Figure 5. Sketched design of device prototype with cable travel.

2.5. Control System

Figure 6 shows a diagram for determining the most suitable control device to be used in this project. From this control flow, the number of I/O (input/out) pins required are listed in Table 4. The two-finger control device is simply a single input single output control system that receives input data in the form of finger pressure force, processes it before rotating the servo motors based on the required angle (output). The control units will then receive the analogue input from the fingers through the force-sensitive resistors, indicating that a pinch action has begun (Figure 6b). From Figure 6c, the "if" statement is used as a logic decision that constantly checks for the input value. This analogue value will then be processed by the control unit which maps that value to the corresponding servo output (Figure 6d). The servo rotates to that specific position which provides tension to the fingers (Figure 6e). The control units will then continue to check the input signal until the analogue value becomes zero which indicates that the pinch action has been completed. Correspondingly, it will signal the servo to return to the zero position which relaxes the fingers.

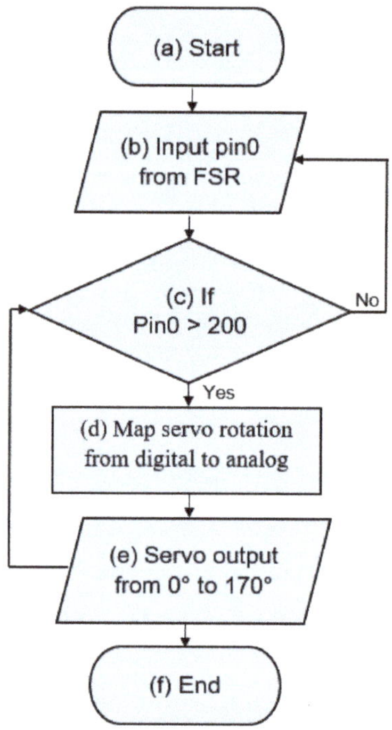

Figure 6. Control flow of the device.

Table 4. I/O pins required for pinch assistant device.

Input Pins	Output Pins
Thumb sensor (analogue)	Thumb servo motor (analogue)
Index finger sensor (analogue)	Index finger servo motor (analogue)

To further elaborate on Figure 6c,d which are the devices main programming logic, the map function used to determine the servo rotation has the following syntax:

$$\text{map}(\text{value, from Low, from High, to Low, to High})$$

With the map function, the Arduino is programmed as follows:

$$\text{Servoposition0} = \text{map}(\text{fsrReading } 0, 200, 500, 0, 170)$$

This map function is used to re-map the input value from the force-sensitive resistor to the servo output rotation. The sensor is connected to the 10-bit analogue pin0 which has a digital value between 0 to 1023. The digital value ranging from 200 to 500 is obtained through calibrations. The lower limit of 200 is used to prevent the device from mistaking any slight mistouches as actual pinches. The sensor does not produce values above 500 even after excessive pressing. In other words, the force-sensitive resistor (FSR) converts the analogue signals received by the sensors into digital signals. The microcontroller receives digital signals between 200 and 500 which are the digital representation of the fingers' applied pressure. Finally, mathematical logic is used to scale the 200 to 500 reading into servo rotation which in this case is between 0 and 170 degrees. This control is made possible due to the inclusion of the servo library while using the servo.write(angle) function. In this case, the "map to" values between 0 to 170 translate to a 0 to 170 degree rotation of the servo motor.

Figure 7 shows the electrical design of the device. The system can be divided into the Arduino unit, servo motors, force-sensitive resistors, a force-to-voltage circuit and a negative voltage generator circuit. The Arduino is the central unit of this design receiving analogue input from the Flexiforce sensor (Tekscan, Inc., South Bosto, MA, USA) and converting it into digital output for the servo motors. Besides using a force-sensitive resistor (FSR), other measuring devices include torque sensors and EMG sensors. The FSR was selected in this application because of its low price and its ease of use compared to EMG sensors which can be highly sensitive to changes in skin conductivity. Besides these sensors, Shojaei Barjuei, et al. [44] used a dynamic bond graph modeling method to analyse the effects of variables on an industrial back exoskeleton.

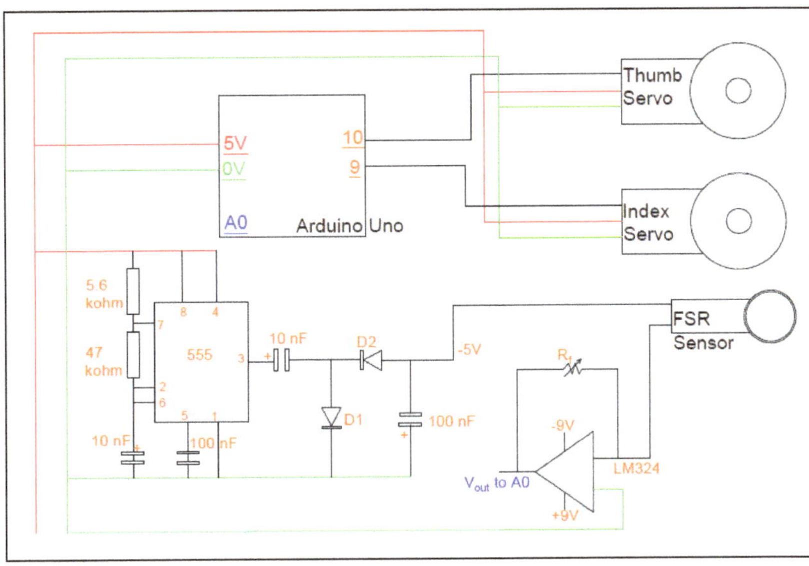

Figure 7. Electrical circuit design of the device prototype.

Based on the sensor manufacturer, an inverting operational amplifier (LM324) is needed to produce an analogue output as a means of calibration. This circuit also requires a −5 V DC excitation voltage which is the purpose of the negative voltage generator circuit. The 555-oscillator is used to supply a constant negative voltage to the capacitors.

3. Materials and Methods

This section clarifies the design analyses, test plans and proof of concept for the proposed invention. The design analyses and test plans which are under the materials and methods section account for the material selection of the finger rings, finite element analysis, and usability experiment. The proof of concept which is under Section 4 accounts for experimental results, *t*-test results comparisons, prototype design discussion, device comparison, and novelty and industrial applicability. Figure 8 presents the framework of the entire process.

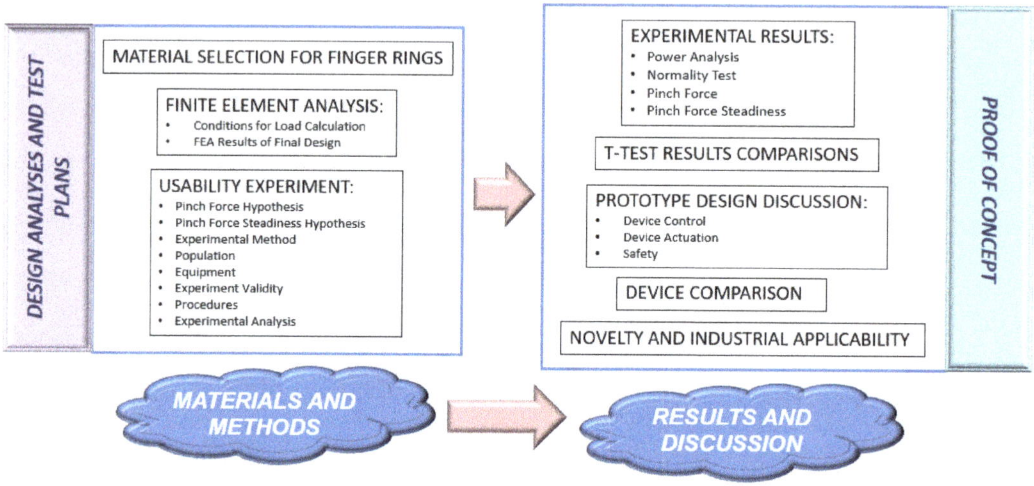

Figure 8. Framework of design analyses, test plans and proof of concept for the finger grip enhancer.

The following section discusses the material selection process for the finger ring component. After establishing the design requirements and objectives, the Ashby chart was used for initial material screening. Prospective materials and their properties were tabulated. The pairwise comparison table was used to determine the weightage of each criterion as not all criteria are equally important in achieving the design objective. Finally, the materials' properties were scaled accordingly and a cost-adjusted weighted property index became the final selection score. The methodology applied here was referenced from Lampman and Dieter [45].

3.1. Material Selection Finger Rings

The function, constraints and objective in Table 5 guided the material selection process in choosing the lightest and strongest possible material.

Table 5. Component function, constraints and objective for material selection.

Criteria	Details
Function	Act as Pivot Points to Guide the Finger Movements
Constraints	Must not fail under stress Water-resistant Length
Objective	As small and light as possible

The versatility of FDM rapid prototyping allows for a variety of building materials. Polymers are the most common category due to the high strength-to-weight ratio, resistance to chemical corrosions and low density [46]. Polymers can be further divided into three different categories which are thermoplastic polymers, thermosetting polymers and elastomers. Although elastomers are flexible and elastic (thus comfortable to be used for day-to-day activities), they have a low Young's modulus which reduces its rigidity [47]. Thermosetting polymers are very strong and can withstand high temperatures [48]. However, they degrade when heated, hence are unsuited to be used in FDM. Therefore, the thermoplastic group of polymers was brought forward to the next stage of the selection process.

The cost per unit property method was applied in this study as it is a reasonable method to measure how much material will cost per required function. In this case, the most important property is flexural strength (bending strength) as the component is more likely to experience bending stress compared to tensile stress. The mass of the material can be calculated with Equation (5).

$$m = A\,L\,\rho \tag{5}$$

From the equation, A is the cross-sectional area, L is the length and ρ. is its density. Both the length and pressure were applied as constraints and specified by design requirements. The cross-sectional area was, however, a free variable (see Equation (6)).

$$\sigma = \frac{3PL}{2bd^2} \tag{6}$$

From the equation, σ. is the flexural strength for a rectangular sample under a load in a three-point bending condition, P is the axial load, b is the width and d is the depth. Eliminating the cross-sectional area in both Equations (5) and (6) results in the following expression:

$$m = (P)\left(L^2/d\right)\left(\rho/\sigma^{2/3}\right)$$

The term m is the performance of the finger ring described by three factors. The first term is the functional requirement followed by the geometric parameter. The third factor in the bracket is the material performance index. The performance index M when the design criteria are at minimum cost instead of minimum weight are given by Equation (7).

$$M = \left(\sigma^{2/3}/C\rho\right) \tag{7}$$

From the equation, C. is the cost per unit mass. Five different polymers were shortlisted for further selection after the initial strength-verses-density Ashby chart screening process was done. This screening was achieved using the $\sigma^{2/3}/\rho$. reference line together with the minimum required strength by the material which deemed the polymers and elastomers bubble most appropriate for this study. A list of criteria (see Table 6) that should be considered for this particular design was listed for each material.

Table 6. Properties of candidate materials for the finger ring.

Criteria	ABS	Polylactic Acid (PLA)	Polycarbonate (PC)	Polyamides (Nylon)	Polypropylene (PP)
Tensile strength (MPa)	22	41	68	53	21.4
Flexural modulus (MPa)	1834	2800	2234	1300	698
Flexural strength (MPa)	41	103	104	69	55
IZOD impact (J/m)	107	139	53	150	75
Density (g/cm^3)	1.05	1.25	1.2	1.00	0.84
Heat deflection temperature @ 0.45 MPa	90	80.3	138	97	121
Elongation at break (%)	6	4.7	4.8	9.5	529
Sources	[49]	[50–52]	[49]	[53]	[49,54]

Table 7 is a comparison table that assigns different weightage to each criterion according to its importance. The flexural strength of the material was the most important criterion as it should not break during the application to protect the user. With this calculation, the weighted property index for each material is given in Table 8. Polycarbonate (PC) was found to have the highest index, which was slightly edging above polylactic acid (PLA).

Table 7. Pairwise comparison table of finger ring (FDM) material.

Criteria	Decision Number																					Positive Decisions	Weighting Factors
	1	2	3	4	5	6	7	8	9	10	11	12	13	14	15	16	17	18	19	20	21		
Tensile strength	2	1	2	2	2	2																11	0.17
Flexural modulus	1						1	1	1	2	2											8	0.13
Flexural strength		2					2					2	2	2	2							12	0.19
IZOD impact			1					2				1				1	2	2				9	0.14
Density				1					2				1			2			2	2		10	0.16
Heat deflection temp.					1					1				1			1		1		2	7	0.11
Elongation at break						1					1				1			1		1	1	6	0.10
Total Number of Positive Decisions																						63	1.00

Table 8. Scaled values of properties and calculated weighted property index.

Materials	Tensile Strength	Flexural Modulus	Flexural Strength	IZOD Impact	Density	Heat Deflection Temperature	% Elongation	Weighted Property Index
	0.17 *	0.13 *	0.19 *	0.14 *	0.16 *	0.11 *	0.1 *	
ABS	32.35	65.5	39.42	71.33	80	65.22	1.13	51.58
PLA	60.29	100	99.04	92.67	67.20	58.19	0.88	72.28
PC	100	79.79	100	35.33	70	100	0.91	73.61
Nylon	77.94	46.43	66.35	100	84	70.29	1.80	67.24
PP	31.47	24.93	52.88	50	100	87.68	100	61.28

Note: * weighting factors.

With the relative cost of each material, the figure of merit which is a weighted property index after considering the cost per unit flexural strength was tabulated in Table 9. PLA was ranked the highest and selected as the material for the finger rings.

Table 9. Relative cost, the figure of merit and ranking of materials.

Material	Relative Cost	Cost of Unit Strength × 100	Weighted Property Index	Figure of Merit	Rank
ABS	1	8.83	51.58	5.84	2
PLA	1	5.69	72.28	12.70	1
PC	2.80	15.19	73.61	4.84	3
Nylon	2.80	16.64	67.24	4.04	4
PP	3.25	18.88	61.28	3.24	5

The same material selection process was used to compare Nylon 6, Polyester (PET), Kevlar 49 and Polypropylene for the cable material. While only considering the physical criteria, Kevlar 49 had the highest weighted property index due to its superior property in tenacity (strength) and initial moduli (stiffness). However, after factoring in the cost, Polypropylene (PP) was considered instead.

3.2. Finite Element Analysis

3.2.1. Conditions for Load Calculation

The pinch enhancer used a servo and cable combination to apply tension force onto the fingers that assist in enabling a stronger and more stable pinch. The user is required to put on a three-piece set of PLA-made rings that attaches the cables onto their fingers. The final design was first presented in the previous section and is the product of a Finite Element Analysis (FEA) method used to optimise and ensure the safety and suitability of its shape and size. Figure 9 exhibits the initial and final design of the enhancer rings.

The static analysis was carried out in three different loading conditions to simulate the real-world applications. The first condition was the external load caused by the compressive force of the fingers (see Figure 10). The purpose of this analysis was to test if the design fails under extreme loads during component handling. This test intended to simulate compression caused by finger pinching actions as shown below. This analysis was carried out under different conditions due to the various orientations about the centre axis.

Pinch force was obtained through various literature studies. Mathiowetz, et al. [55] and Mohammadian, et al. [56] found that men between the ages of 60 to 64 years have an average palmar pinch force of 98 N and 96 N respectively. On the other hand, Nilsen, et al. [57] found that the force was lower (63 N) for men between the ages of 60 to 69 years. Lastly, on the other extreme, men who regularly deal with heavy manual work above the age of 60 years were found to possess an average palmar pinch force of 120 N [58].

Therefore, for this experiment, a pinch force of 120 N was used. The boundary conditions used are:

1. Axial load of 120 N acting across a single surface of the ring.
2. Fixed support at the outer surface opposite of the axial load.

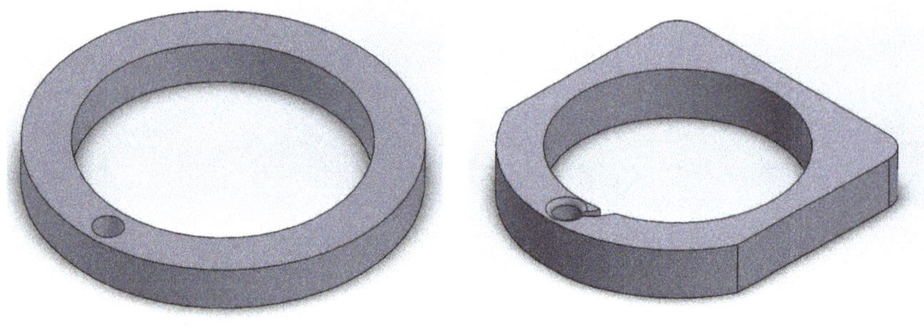

(**a**) initial design (**b**) final design

Figure 9. Initial and final design after Finite Element Analysis (FEA).

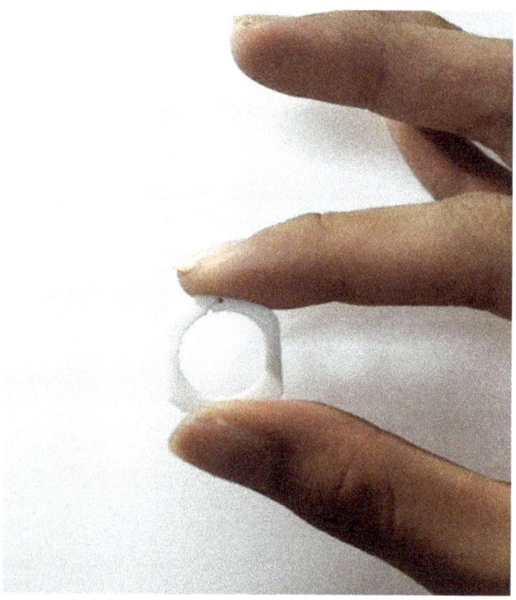

Figure 10. External compressive pinch force applied on component.

The second condition would be the analysis of force around the cable loop (see Figure 11). The servo motors will produce torque to rotate and hold the fingers into position through the force transfer of the cables. This section was considered critical due to the constant tension force experienced by the PLA rings. The magnitude of the tension force applied on the cable loop was estimated to be 70 N, calculated from the maximum stall torque of the servo motor. The boundary conditions used are:

1. Axial load of 70 N acting around the edge of the string loop.
2. Fixed cylindrical support at the inner surface of the finger loop.

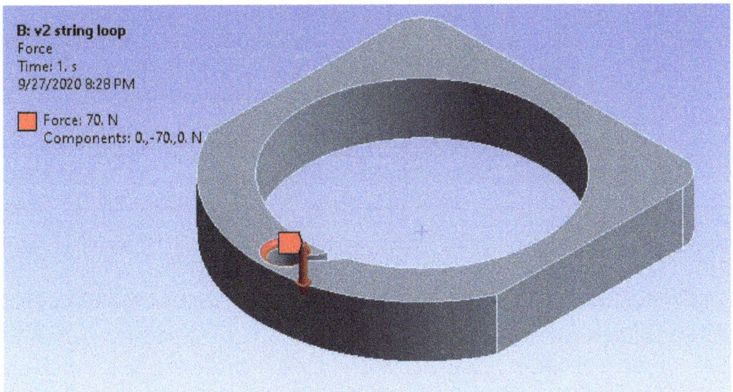

Figure 11. Force applied during FEA analysis for the second condition.

The final FEA condition was to simulate the actual pinching action while using the device. This condition was a combination of 120 N pinch force exerted by the fingers and 70 N tension force from the servo cables. The boundary conditions used were:

1. Axial load of 70 N acting around the edge of the string loop.
2. Axial load of 120 N acting across a single surface of the ring.
3. Fixed cylindrical support at the inner surface of the finger loop.

In the FEA for the initial design, it was found that the component structure was incapable of withstanding the applied stress thus causing it to fail. The complete analysis is revealed in the data availability sheet.

The overall shape of the finger ring was modified from a circular ring into a squarish shape (see Figure 12). Increasing the volume of an object would increase its sturdiness and resistance to stresses and strain. This motive was achieved by increasing the height and changing the shape from a simple circle into a more complex shape.

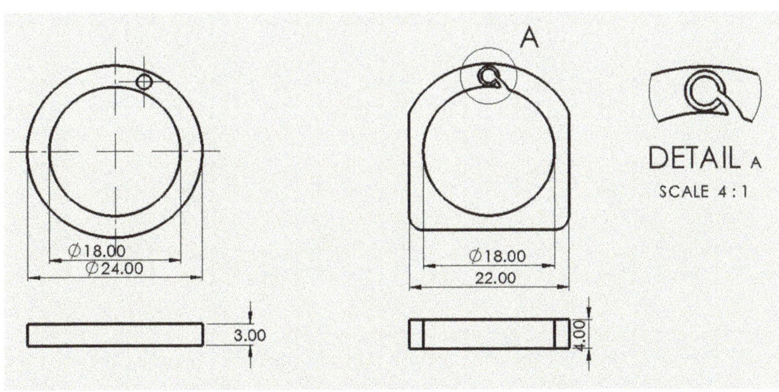

Figure 12. Drawing comparison of finger ring component before and after FEA.

The final design allowed for a narrower profile, reducing the outer length from 24 mm to 22 mm, which made the ring less encumbering. Furthermore, additional material was added to the four corners of the item, increasing the maximum wall thickness up to 5.73 mm. As shown in detail (A), a 0.4 mm fillet was added to the cable loop which increased stress distribution. Additionally, a curved slot was designed to allow for easy attachment and detachment of the finger rings depending on the user's finger size. The curved design prevented the tendon cables from escaping the loop during a normal application.

Table 10 compares the major differences before and after modifications. The design changes showed improved characteristics on all areas except in weight. The finger ring increased in weight by 42% though an object weighing one gram was still acceptable on the fingers as a typical gold ring weighs about 100 g.

Table 10. Physical design changes between initial and final finger ring components.

Parameters	Initial Design	Final Design
Mass	0.731 g	1.035 g
Volume	584.514 mm^3	828.12 mm^3
Top surface area	194.838 mm^2	204.28 mm^2
Wall thickness	3 mm	2–5.73 mm
Height	3 mm	4 mm

3.2.2. FEA Results of Final Design

The final design iteration used for the prototype was also tested using the FEA method. The boundary conditions and test results are tabulated in Table 11. For tests 1a and 1b, the boundary condition of 120 N force was applied on the surface with a fixed support on the opposite surface. The displacement contour and stress profile under the condition of test 1a yielded a maximum deformation of 0.225 mm and maximum stress at 28.649 MPa while for 1b, the values were 0.303 mm and 36.113 MPa respectively. Both tests 1a and 1b did not exceed the ultimate tensile stress limit of 41 MPa under these loadings.

Table 11. Deformation and stress profile for final finger ring component.

Test	Boundary Conditions	Total Deformation Profile	Von Mises Stress Profile
1a	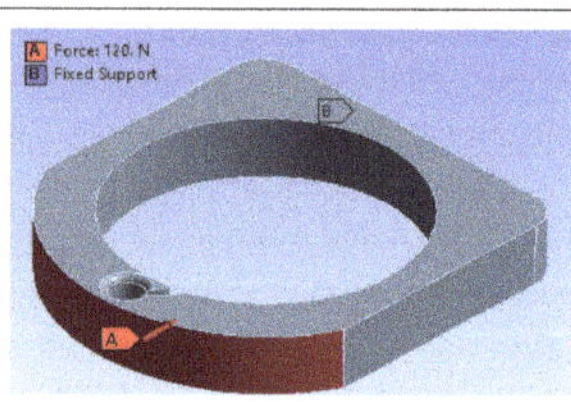	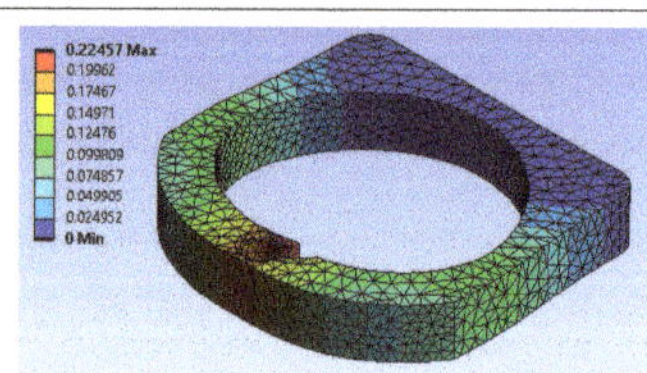	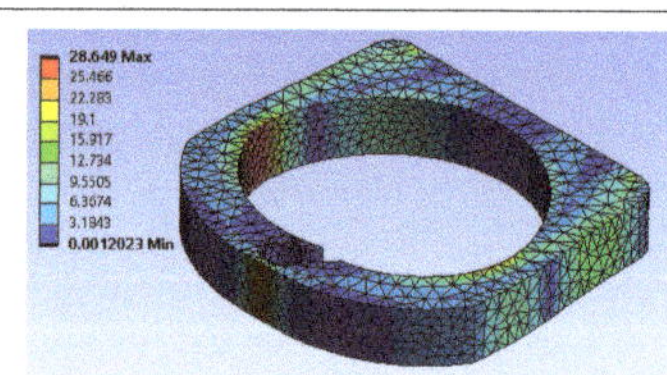
1b	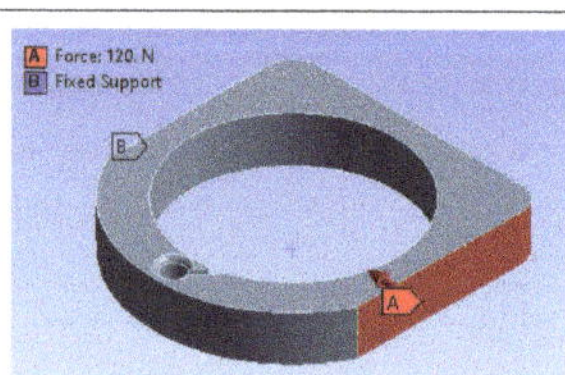	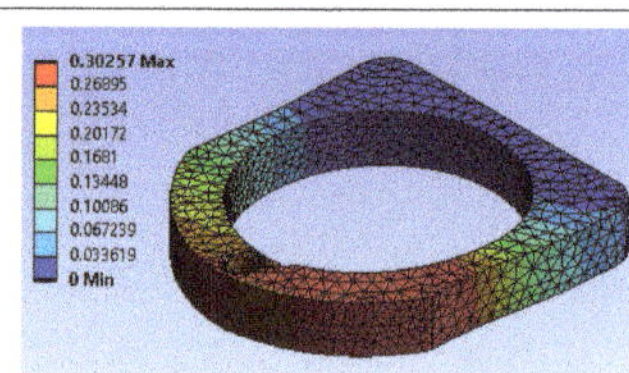	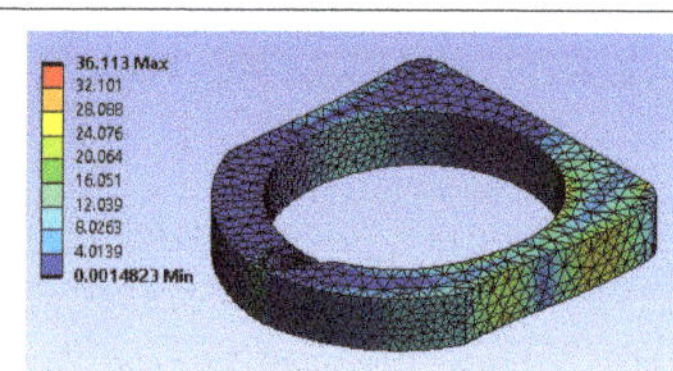
2	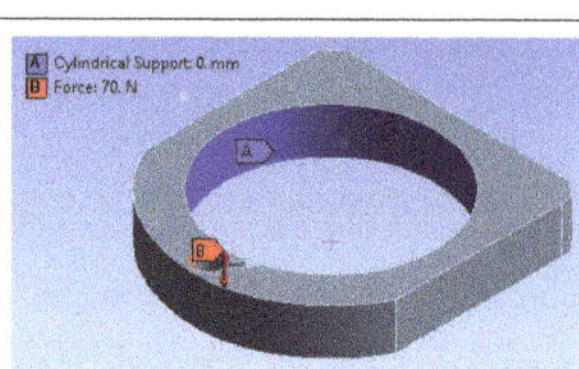	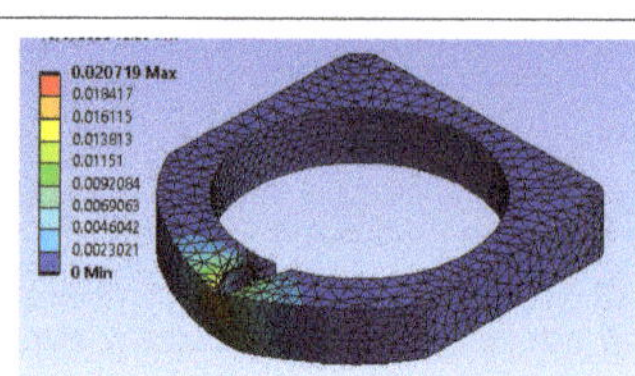	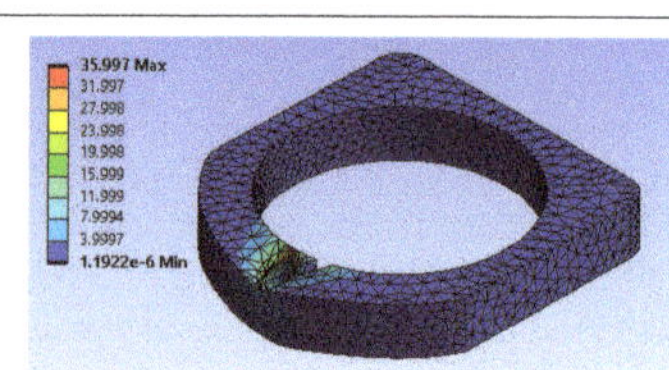

Table 11. *Cont.*

Test	Boundary Conditions	Total Deformation Profile	Von Mises Stress Profile
3	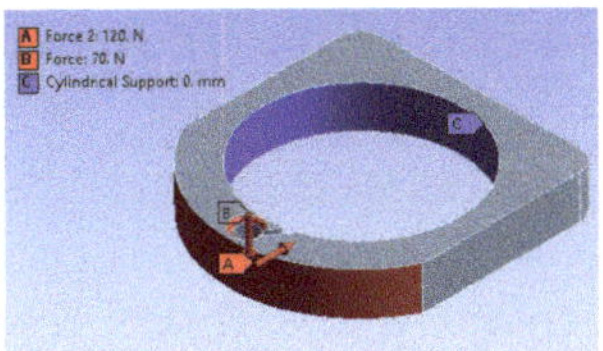	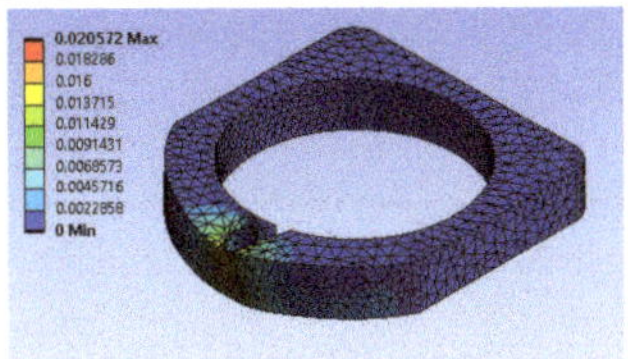	

Test 2 was designed to analyse the tension force of the tendon cable that reacts on the area of the string loop. The maximum von Mises stress that it experienced was 35.997 MPa which is less than 41 MPa (ultimate tensile strength). The maximum deformation was 0.027 mm. The design modification of adding a fillet around the loop increased the ability of the component to withstand stress from the tendon cables.

Test 3 on the other hand was a combination of test 1a and test 2. The displacement contours and stress profiles of the final finger ring design during a simulated pinch action (test 3) are also presented. The maximum displacement was 0.0206 mm while the maximum von Mises stress was 36.031 MPa.

The final simulation result is tabulated in Table 12. The improved design fulfilled the static analysis requirement of experiencing forces below the tensile limit.

Table 12. FEA results of the final finger ring design.

Simulation Conditions	Maximum Deformation (mm)	Maximum von Mises Stress (MPa)	Result
Compression force 1	0.2246	28.649	Pass
Compression force 2	0.3026	36.113	Pass
Cable reaction force	0.0207	35.997	Pass
Pinch action simulation	0.0206	36.031	Pass

3.3. *Usability Experiment*

Usability in this study was measured by the device's ability to produce a statistically positive outcome when elderly people pinch everyday objects while using it. The experiment design is discussed in this section.

3.3.1. Pinch Force Hypotheses

Null Hypothesis H_0. *There is no significant difference in the mean pinch force of elderly people with and without the use of the pinch enhancer.*

Alternative Hypothesis H_1. *There is a significant difference in the mean pinch force of elderly people with and without the use of the pinch enhancer.*

3.3.2. Pinch Force Steadiness Hypotheses

Null Hypothesis H_0. *There is no significant difference in the pinch force steadiness of elderly people with and without the use of the pinch enhancer.*

Alternative Hypothesis H_1. *There is a significant difference in the pinch force steadiness of elderly people with and without the use of the pinch enhancer.*

3.3.3. Experimental Method

A quasi-experiment aims to evaluate causation between an intervention and an outcome without randomisation [59,60]. This design method is prominent in medical informatics literature [61] and in studies of infectious diseases [59]. A counterbalanced design is a higher-order quasi-experiment design whereby every participant receives all the same interventions but in a random order [62]. This counterbalanced quasi-experiment method was chosen in this study as it is able to control participant variables while obtaining statistical significance at a smaller sample size [63].

In the present study, participants were required to pinch and hold six different household objects using the pulp pinch. In their effort to study challenges faced by elderly people in their homes, Fausset, et al. [64] developed seven sub-categories of tasks which included cleaning, outdoor, home upkeep, repair, indoor update remodelling and movement within the home. Among those selections, cleaning, outdoor and home upkeep were found to be the most challenging activities for the elderly people.

Based on these three categories, two everyday items commonly used by the elderly people are selected for each category. The six items included a detergent cup and clothes peg representing cleaning activities, a golf ball and insect repellent for the outdoor category, and a power plug and remote control for the home upkeep (see Figure 13). Smaby, et al. [65] conducted a similar experiment testing lateral pinch on six items including using a key, inserting and removing a plug from an outlet, using a zipper, pushing a television remote button, stabbing food with a fork and inserting an automated teller machine card.

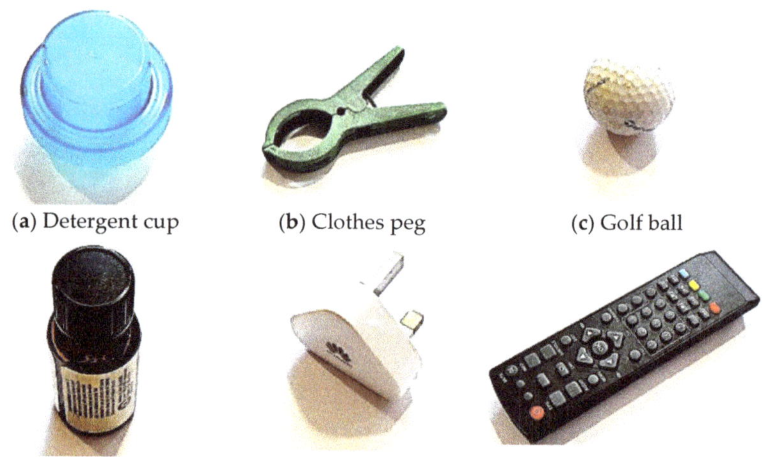

(**a**) Detergent cup (**b**) Clothes peg (**c**) Golf ball

(**d**) Insect repellent bottle (**e**) Power plug (**f**) Remote control

Figure 13. Test items for pinch force measurements (detergent cup, clothes peg, golf ball, insect repellent bottle, power plug, and remote control).

3.3.4. Population

The procedures of this study were submitted to the Technology Transfer Office of the Multimedia University, and a research ethics approval was granted by the research ethics committee on 13 July 2020 with the approval reference number EA0052020. Upon approval, the study was conducted at various homes of elderly individuals in Johor Bahru, Johor, Malaysia.

The research recruited 11 healthy elderly people above the age of 60 years (65.27 $\pm$ 5.76) consisting of five males and seven females. The test was not administrated to subjects who did not fulfil the general health criteria and who had prior injuries on the hands or fingers. These exclusion criteria were referenced from several research studies with a similar experimental aim of studying finger pinching abilities [55,58,66].

A non-probability sampling method was chosen for this qualitative study as the goal was not to describe a broad population [67]. The criterion-i strategy was used to identify and select all cases that fulfil a predetermined set of criteria [68]. The exclusion criteria included (a) experiencing acute pain in the right hand and fingers or having any form of hand-related injuries, (b) having medical conditions such as high blood pressure, post-stroke conditions or have been hospitalised during the last six months, and (c) unable to live an active lifestyle. All participating subjects voluntarily declared themselves to be healthy and able to live an active lifestyle independently. Participants were asked to report their age and gender through a consent form.

3.3.5. Equipment

The Tekscan® Flexiforce A201 sensor (Tekscan, Inc., Massachusetts, USA) is a force sensing resistor that reduces resistance when force is applied. It is a flexible, ultra-thin printer circuit that is only 0.208 mm in thickness. These advantages allow the flexiforce sensors to be integrated with objects of different shapes and sizes for pinch force measurements in this study which would otherwise be impossible with a traditional Jamar dynamometer. According to Werle, et al. [69], elderly people were found to have a higher standard deviation of pinch force when compared to younger people. They suggested that this difference was partially influenced by the characteristics of the pinch gauge which had a step scale of 0.5 kg. As the strength value of elderly people decreases, the force difference expressed in standard deviation in percent increases due to the low sensitivity of the measuring device. The Flexiforce sensor has the advantage of measuring forces up to 0.01 g [70]. Figure 14 shows the calibration of the pressure sensor using fixed calibrated weights of 50 g, 100 g, and 200 g.

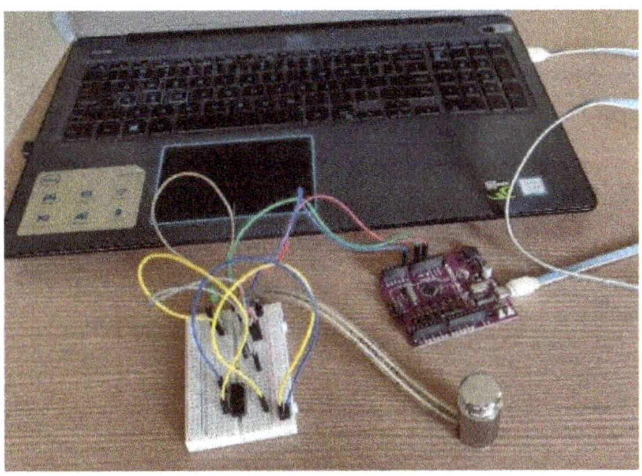

Figure 14. Force calibration of Tekscan® Flexiforce sensor.

3.3.6. Experiment Validity

For each pinch test, subjects were seated with their shoulder adducted and neutrally rotated, elbow flexed to 90° and their forearm and wrist in a neutral position (see Figure 15) [71]. A neutral wrist position was found to exert higher pinch force compared to extension or flexion positions [72]. These guidelines comply with the recommendations of the American Society of Hand Therapists (ASHT) for upper extremity and body position during grip strength measurements.

Each participant was given one warm-up trial to familiarise themselves with all six objects. Participants were then to proceed with pinching and lifting the six items for ten seconds, alternating between with and without using the device. A 30s recovery period was allocated between the sets. According to previous research [58,69,73,74], a 15 s rest was found to be sufficient. Trossman and Li [75] found no significant difference between the grip strength of participants who had 15, 30 or 60 s of recovery periods. However, the subjects in this particular study were between 19 to 41 years of age. Since the subjects of the present study were elderly people, it was reasonable to allocate a longer rest period of 30 s.

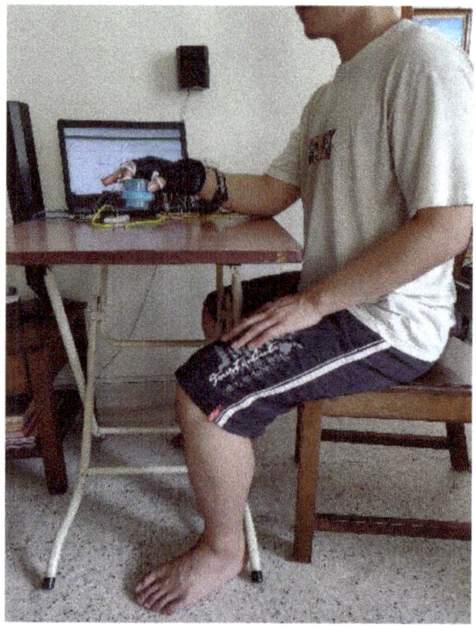

Figure 15. Sitting posture of participants during force measurements.

3.3.7. Procedures

Before the start of each action, participants would position their fingers with the Flexiforce sensor between their thumb and the object (see Figure 16). Once they were ready, the researcher would say "start" and participants would pinch the object firmly and lift it about 10 cm off the table, with their wrist leaving the table. The researcher would then say "hold" for the cue to sustain the pinch force and position for 10 s. During this 10 s, subjects were required to pinch the objects naturally as they would during everyday applications. After 10 s, the "stop" command would signal subjects to place the object back on the table, thus completing one measurement. Each measurement was then saved in an Excel spreadsheet for further analysis. After completing all six objects both without and with the device twice, participants were thanked for their contribution.

3.3.8. Experimental Analysis

The collected data was processed using inferential statistics from the Minitab 18 software. The parametric paired t-test was used to analyse the data for each pinch object experiment with a p-value of 0.05 to determine statistical significance. Besides that, support analysis was also conducted on the collected data. These support analyses included a priori power analysis to ensure that there is sufficient power in the data apart from sample size estimation, and normality test to check if the data was normally distributed before proceeding to the t-test.

Power in statistics refers to the number of tests that participants needed to avoid a type II error which is a false negative result. In designing a study, the a priori power analysis was used to ensure that the number of recruited test participants had adequate power, though over-testing was discouraged as it is a waste of time and resources [76]. The researchers decided to use the power analysis as a priori test to decide on the required sample size.

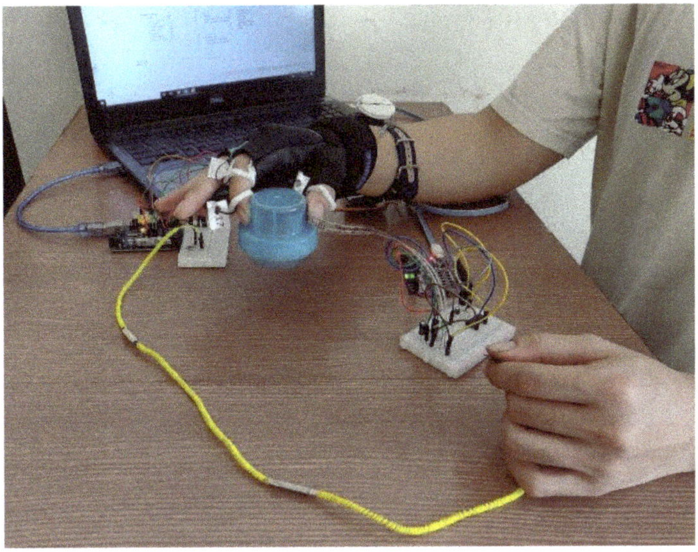

Figure 16. Starting position of participants during pinch force test.

In the power analysis calculation of the pilot study, the power was normally set at 0.80 (80% statistical power) which is often the recommended value [77]. The Cohen's d for the paired *t*-test was calculated with Equation (8) [78].

$$\text{Cohen's } d_z = \frac{M_{diff}}{S_p} \tag{8}$$

From the equation, M_{diff} is the mean difference between the two groups and S_p is the pooled standard deviation.

4. Results and Discussion

Figure 17 shows the final device assembly with its different subassemblies. The device's main parts are the glove component, actuator system and control system. The whole assistant device was powered by a 10,000 mAh power bank (Zhongshan Pineng Battery Co., Ltd., Guangdong, China) which enables portability. A description of how each component contributes to the overall device is shown in Table 13.

Table 13. Device components and functions.

Device Component	Function
Servo motor with pulley (Feetech RC Model Co., Ltd., Shenzhen, China) 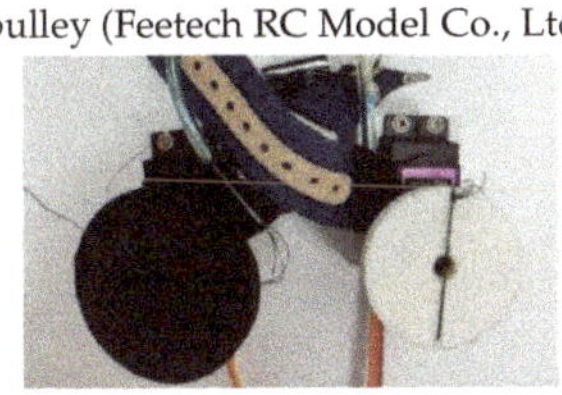	Main driver unit that provides the required torque to support the thumb and index finger. Servo motor: Fitec FS5115M

Table 13. *Cont.*

Device Component	Function
PLA finger rings with tendon cables 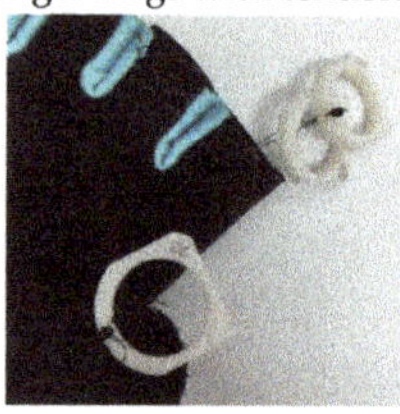	Used to transfer tension force from the servo motors to the fingers.
Flexiforce sensor (Tekscan, Inc., South Boston, MA, USA) 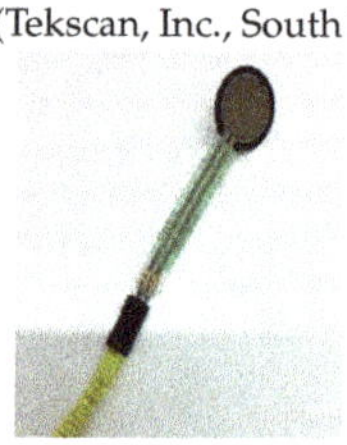	Measures the amount of pressure from the fingers during pinching activity. FSR: Tekscan FlexiForce A201 Sensor
Power bank (Zhongshan Pineng Battery Co., Ltd., Guangdong, China) 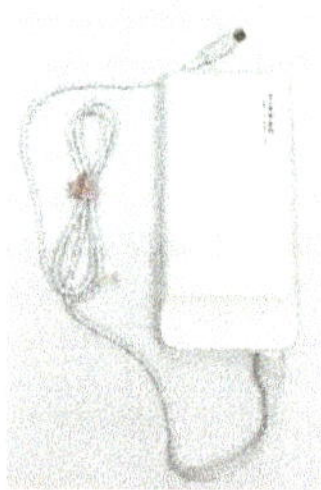	Supplies 5 VDC, 2.1 A to power the Arduino, sensor, and servo motor.
Arduino Uno and electronic circuit (Arduino SA, Chiasso, Switzerland) 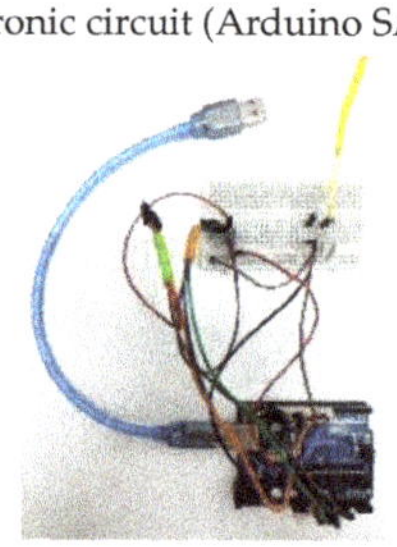	The control unit receives an analog signal from the sensor and instructs the servo motor to rotate accordingly. Control board: Arduino Uno Microcontroller: ATmega328P (part of the Arduino Uno board)
Glove 	Central piece of the device that keeps the tendon cables in place.

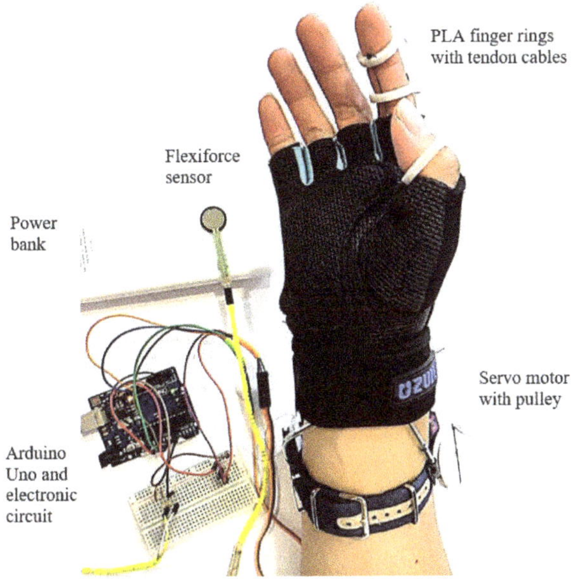

Figure 17. Pinch assistant prototype.

Table 14 highlights some performance details of the pinch assistant device. The total weight of the device prototype (excluding the power bank) was 320 g. This number represents the total amount of added weight on the users' hand while operating the device. This device was designed to facilitate three flexion DOFs (DIP, PIP, MCP) for the index finger and two flexion DOFs (IP, MCP) for the thumb. Both bending angles of the thumb and index finger were tuned to the fingers' range of motion but could be adjusted both mechanically and in the control programme.

Table 14. Pinch assistant performance details.

Characteristic	Device Performance
Weight of glove and actuators	320 g
Weight of device	520 g
DOFs for index finger	3
DOFs for thumb	2
Bending angle of index finger	~134°
Bending angle of thumb	~100.76°
Force range	8.5 to 13.9 N
Electrical input	5 V DC 2 A
Operating hours	~3.5 h continuously
Cost	~RM 296.5

4.1. *Experimental Results*

4.1.1. Power Analysis

Table 15 shows the a priori power analysis test results for all six test objects. The predicted sample sizes in the table are recommended for the statistical results to have a Type II error of only 20% probability or lower. The analysis also verifies that there is only a 5% change or less for a Type I error to occur. The power analysis of the pilot study suggests that at least seven samples should be collected to ensure sufficient statistical power (actual power = 0.91).

Table 15. Results of the a priori power analysis.

Test	Effect Size	Actual Power	Predicted Sample Size
T1 Detergent cup	1.409	0.81	4
T2 Clothes peg	1.364	0.98	4
T3 Golf ball	1.038	0.91	7
T4 Insect repellent	1.065	0.82	5
T5 Power plug	0.843	0.95	5
T6 Remote control	1.162	0.88	5

A total of 11 subjects between the ages of 60 to 70 years participated in the study, comprising of five males and seven female elderlies. This participation is a 50% increase from the seven participants estimated in the pilot study. The test was not conducted on subjects who did not fulfil the general health criteria and who had prior injuries on their hands or fingers. All subjects voluntarily declared themselves to be healthy and able to live an active lifestyle independently.

4.1.2. Normality Test

A *t*-test is a parametric test that assumes that the difference between the pairs is normally distributed. A normality test is done for each pair of datasets. The Ryan-Joiner test of normality (similar to the Shapiro-Wilk test) was selected because it was found to be more powerful with smaller sample sizes when compared with other tests such as the Anderson-Darling and Kolmogorov-Smirnov test [79,80]. The null hypothesis is that the collected data does not significantly differ from a normal distribution. In order to accept the null hypothesis, the *p*-value must be larger or equal to 0.05 as shown in Table 16. All other assumptions for using the paired *t*-test to determine the difference between the pinch force measured with and without using the pinch assistant were also met.

Table 16. The *p*-value of the normality test.

Test	*p*-Values	
	With Device	Without Device
T1 Detergent cup	>0.1	>0.1
T2 Clothes peg	>0.1	>0.1
T3 Golf ball	0.056	>0.1
T4 Insect repellent	>0.1	>0.1
T5 Power plug	0.05	>0.1
T6 Remote control	>0.1	>0.1

4.1.3. Pinch Force

Table 17 shows the paired mean and standard deviations of the pinch force measured during the *t*-test. The mean force exerted using the pinch device is recorded to be larger across all six tests. The *t*-test is conducted using a *p*-value of 0.05 for all statistical tests.

The pinch force generated together with the pinch assistive device is found to be significantly higher than the one generated without the device. The significant differences are found in Test 1 [$t(10)$ = t-statistic, $p < 0.001$], Test 2 [$t(10)$ = t-statistic, $p < 0.001$], Test 3 [$t(10)$ = t-statistic, $p = 0.001$], Test 4 [$t(10)$ = t-statistic, $p = 0.001$], Test 5 [$t(10)$ = t-statistic, $p = 0.004$], and Test 6 [$t(10)$ = t-statistic, $p < 0.001$]. The paired difference result for the pinch force is recorded in Table 18. The null hypothesis is rejected while the alternative hypothesis is accepted. Hence, there is a significant difference in the mean pinch force of elderly people with and without the use of the pinch enhancer.

Table 17. The paired sample pinch force test result in kg.

Test	Control		Device	
	Mean	**SD**	**Mean**	**SD**
T1 Detergent cup	0.35	0.29	1.20	0.59
T2 Clothes peg	0.31	0.16	0.68	0.33
T3 Golf ball	0.81	0.38	1.39	0.75
T4 Insect repellent	0.48	0.31	1.01	0.57
T5 Power plug	0.57	0.33	1.02	0.60
T6 Remote control	0.31	0.13	0.85	0.31

Table 18. The paired difference t-test for pinch force in kg.

Paired Sample	Paired Difference		
	Mean	**Std. Deviation**	***T*-Value**
T1 Detergent cup	0.85	0.42	6.71 *
T2 Clothes peg	0.36	0.23	5.26 *
T3 Golf ball	0.58	0.43	4.52 **
T4 Insect repellent	0.53	0.35	4.96 **
T5 Power plug	0.46	0.40	3.80 **
T6 Remote control	0.54	0.35	5.10 *

* $p < 0.001$; ** $p < 0.005$.

Figure 18 compares the force measured while the pinch test is conducted on all six test objects with and without the assistance of the pinch device. The average pinch force graph profile of the six test objects has a similar trend between the control set and the results obtained while using the pinch device. The golf ball pinch test is found to be the highest in both sets of tests while the clothes peg and remote-control tests have the two lowest average pinch forces. On the other hand, the detergent cup is found to produce the highest percentage of increase in average pinch force (more than 200%) with the help of the pinch enhancer.

The combination of p-values and effect sizes concludes that the use of this finger pinch enhancer has a statistically significant outcome on the user's pinch force and also indicates that the magnitude of the difference between the control and experimental pinch force is large. Results from the pinch force test indicate a significant increase in pinch force after using the pinch assistive device to pick up all six day-to-day objects. The device is hence considered able to assist the elderly people in applying a higher amount of pressure on those common everyday items.

This assisted increase in force during pinching activities can act as a reinforcement for the naturally exerted finger force. Given that elderly people often naturally exert a higher grip force [81] and pinch force [82,83] compared to younger people due to the slippery and insecure grip caused by changes in skin properties, this device would be useful in increasing the safety margin of their pinch force. With this device, elderly people would enjoy the benefits of a stronger pinch without overexerting themselves.

Although such an excess force might be considered as wasted energy, it might still be necessary in order to help elderly people pinch more effectively and safely. There are several changes in the elderly body that could explain the need for this excessive force during pinching. Firstly, when compared to younger subjects, elderly people were less capable of intentionally resetting neuro-synergistic reflexes while having excessive and random firing motor neurons to reach the same force output [84]. With the Thenar muscles being the most active muscles in grasping activities, reduced strength and range of thumb abduction could reduce the finger dexterity of elderly people [85]. Moreover, old people often have a diminished anticipatory build-up in pinch force even when perturbations (changes in load) are expected [86]. These factors could be possible reasons why elderly people typically use a probing strategy that lacks a smooth ramp in force during pinches. The inability to quickly control their finger force in response to changes would suggest the need for a device that actively provides pressure, thus creating a stronger and steadier pinch.

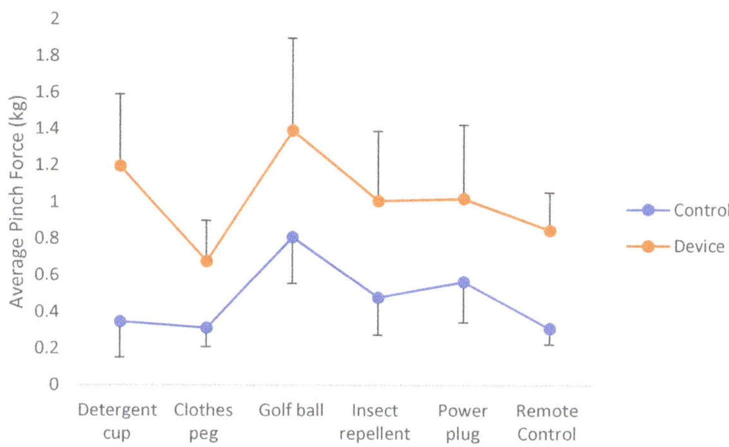

Figure 18. Comparison of force during pinch action for each test object.

The sudden spike in pinch force after using the pinch assistive device during the detergent cup test and the inconsistent graph profile between both sets of tests suggests some irregularity caused by the unfamiliarity with the device. This slight aberration is commonly preventable once the users have enough practice with the assistive device as suggested by past studies [87,88].

4.1.4. Pinch Force Steadiness

Figure 19 illustrates the pinch force performance of one of the participants during the clothes peg pinch test before and after using the pinch enhancer. From this comparison, after using the device, the subject produced better pinch abilities in both force (maximum) and steadiness (standard deviation).

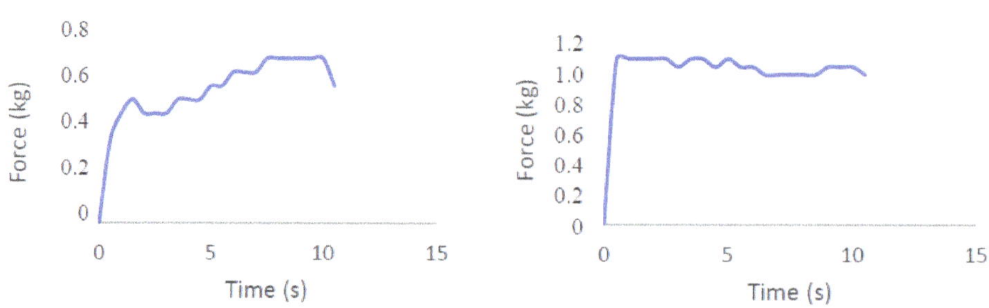

(**a**) Before using the pinch enhancer (**b**) After using the pinch enhancer

Figure 19. Pinch force performance of an elderly person during the clothes peg pinch test before and after using the prototype.

With the utilisation of the pinch enhancer, results of pinch force steadiness show significant improvements. Table 19 compares the mean and standard deviation of both the control results and results while using the device. The steadiness of pinch force is quantified by calculating the standard deviation of the force profile while sustaining the pinch for 10 s.

Table 19. The paired sample pinch test result for pinch force fluctuation (SD).

Test	Control		Device	
	Mean *	SD	Mean *	SD
T1 Detergent cup	46.01	19.46	33.09	17.71
T2 Clothes peg	50.14	28.99	32.96	16.47
T3 Golf ball	69.40	34.00	51.90	23.1
T4 Insect repellent	57.07	22.18	44.06	16.73
T5 Power plug	83.40	46.20	51.40	24.10
T6 Remote control	62.86	31.15	45.86	17.72

* Mean of standard deviation.

From Table 20, results of pinch force steadiness while receiving assistance from the device decreased significantly across all six test objects. The significant differences were found in Test 1 [$t(10)$ = t-statistic, $p = 0.004$], Test 2 [t(10) = t-statistic, $p = 0.004$], Test 3 [$t(10)$ = t-statistic, $p = 0.005$], Test 4 [t(10) = t-statistic, $p = 0.018$], Test 5 [$t(10)$ = t-statistic, $p = 0.007$], and Test 6 [$t(10)$ = t-statistic, $p = 0.019$]. The null hypothesis is rejected while the alternative hypothesis is accepted. Hence, there is a significant difference in pinch force steadiness of elderly people with and without the use of the pinch enhancer.

Table 20. The paired difference results for pinch force fluctuation (SD).

Paired Sample	Paired Difference		
	Mean	Std. Deviation	*T*-Value
T1 Detergent cup	12.92	11.65	3.68 *
T2 Clothes peg	17.17	15.00	3.80 *
T3 Golf ball	17.49	16.25	3.57 **
T4 Insect repellent	13.01	15.30	2.82 ***
T5 Power plug	31.96	31.42	3.37 **
T6 Remote control	17.00	20.17	2.80 ***

* $p < 0.005$; ** $p < 0.01$. *** $p < 0.05$.

Figure 20 illustrates the pinch force steadiness before and after using the device. Similar to the pinch force experiment, the graph profile between each device (before and after assistance) has a similar pattern. Across all six objects, the utilisation of the device can reduce the standard deviation of the pinch force, which infers that a steadier pinch grip is produced after using the device.

It was found that after using the device, the elderly people exerted a pinch force with a lower standard deviation, indicating a more stable pinch grip. The force steadiness graph for all six objects had a similar profile with and without the use of the device, indicating a uniform pattern that is not object-dependent. Cole and Rotella [89], De Serres and Fang [82], and Kinoshita and Francis [81] all found that elderly people have a large fluctuation in their precision grip force rate curve which also indicates a lower force control capacity. With the help of a pinch assistant such as the one in this study, they will have the ability to reduce the force fluctuation which would enable them to pinch with a steadier grip.

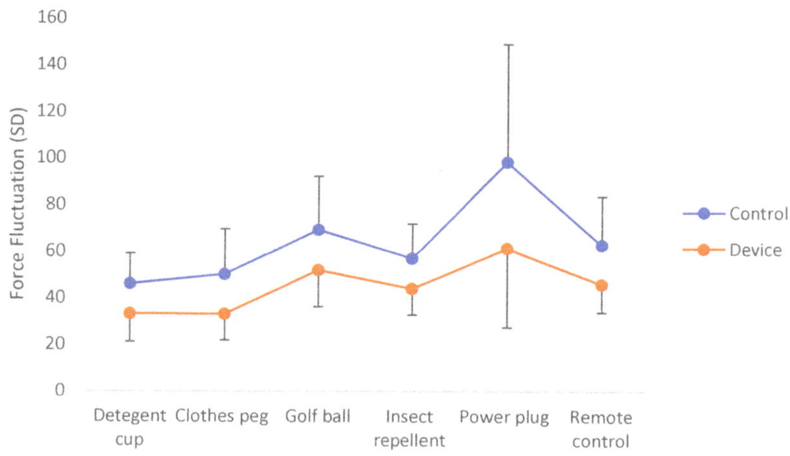

Figure 20. Pinch force fluctuation for each object test with and without the assistance device.

4.2. T-Test Results Comparison

Tables 21 and 22 compare the improvements measured in the current study with findings from other researchers who have applied other intervention methods mainly through exercises. Due to the disparities in key research methods, a direct comparison might not be possible. However, an indirect comparison could provide some level of justification between this novel pinch assistive device and other more traditional methods of improving pinch abilities.

Table 21. Comparison of improvements in pinch abilities with ball rolling training.

Output Variables	Ref. [8]	Current Study
Pinch force	$p > 0.05$	$p < 0.05$
Pinch force steadiness	$p > 0.05$	$p < 0.05$

Note: p refers to the probability value.

Table 22. Comparison of improvements in pinch abilities with muscle strength training.

Output Variables	Ref. [90]	Current Study
Pinch force	$p > 0.05$	$p < 0.05$
Pinch force steadiness	$p > 0.05$	$p < 0.05$

Note: p refers to the probability value.

The referenced studies [8] recorded the improvements in pinch abilities after subjects underwent a training task of holding two metal balls in the palm and rotating them clockwise and counterclockwise. The training frequency included two 10-min sessions each day, 6 days a week for 8 weeks. It was tested on elderly people between 65–79 years of age. In contrast to the current study's findings, these exercises have led to significant improvements in pinch force but not in pinch force steadiness.

In another study, elderly people between the ages of 59 and 74 years participated in a 12-week strength training programme with three training sessions per week as the intervention method [90]. To test for pinch force and pinch steadiness, a portable device that restrained the hand while permitting load variance in accordance with the needs of the subjects was used. The present study showed a significant increase in pinch force and pinch force steadiness after training. Therefore, it can be somewhat affirmed that the improvements in pinching abilities seen in the usage of the current study's device are akin to results obtained by someone who has undergone a strength training regime.

4.3. Prototype Design Discussion

The pinch assistant is compared with several research devices in terms of its control, actuation, and safety. The unique simplicity of the device actuation can be seen in its single tendon cable used to control finger flexion. With just one cable running along each finger at the front of the palm, this simplifies the structure of the device, as well as allows more user flexibility with a half glove design. The PLA finger rings attached at each of the finger joints act as pivot points of the cables and allow the device to assist the fingers while not obstructing finger mobility.

4.3.1. Device Control

The current study's device included a straightforward single-path control trigger where the actuators activate only when a certain force threshold is exceeded. The pros and cons of this method can be seen when compared with other devices such as the one developed by Hasegawa, et al. [91] which had a bioelectric potential-based switching control that switches between two algorithms for finger-following control and grasping force control. The device used its finger-following control mode during the pinching of small objects that do not require exoskeleton support and switched to grasping force control when the user activated a power grasp. While a two-stage control allowed for better movement control, it is actually a necessary function for the device to differentiate between a power grasp and a finger pinch to avoid restraining the user's finger movements. Such a function was not required in the current study's device due to its flexible structure which allowed the hands to move freely.

However, one area of improvement in the prototype control would be the integration of more sensitive sensors. Currently, the user must activate the device using their left hand. With force sensors at the fingertips, for instance, Nilsson, et al. [92] developed an assistive device that was able to recognise the moment a user begins to grasp an object and calculate the amount of force required for the activity.

4.3.2. Device Actuation

A successful design of a finger exoskeleton largely depends on its ability to kinematically control the movement of fingers through the control of their joints. The device by Hasegawa, Mikami, Watanabe and Sankai [91] is a device with eight degrees of freedom (DOFs) that requires eight DC motors. Using one motor for each joint, allows for better control of finger flexion and extension.

In the current design, one motor was used to control the thumb (two DOFs) and another was used for the index finger (three DOFs). This method is analogous to how a finger pinch works as it was found that most of the actuation force during a pulp pinch, tip pinch and grasp motion is transmitted through a single flexor digitorum profundus tendon [93]. Furthermore, each motor can be tuned to suit flexion of various degrees which defers with the pinch prehension type. This condition was still sufficient for control because the objective of the device was only to assist in pinching action.

With the fingers being such complex systems with many ROMs and DOFs, an exoskeleton must compliment the finger movements as naturally as possible. With this understanding in mind, soft robotics and tendon cables that actuate fingers have the advantage over mechanical linkage designs. The hand exoskeleton (HX) is a two-finger exoskeleton device with a similar concept but with four and three active DOFs for the thumb and index finger respectively [16]. The additional two DOFs as compared to the current design were required to facilitate finger abduction/adduction. This comparison between the HX design and the current device illustrates the limitation of using linkages although it does allow for better position control.

4.3.3. Safety

Elderly people are prone to injuries which makes user safety a priority in the device design. The current study's device has a ROM similar to that of a human's finger joint.

The servo motors are placed away from the finger and attached to the forearm. These conditions are safety benefits similar to that of soft robotic assistive devices like the SEM Glove. These devices excel in being lightweight, easy to use and safe for the fingers but at the cost of lower finger strength support [92].

Another benefit to using the tendon actuator mechanism includes the ability for customisation and easy adjustments for joint ROM. The UoA hand exoskeleton is an important device that uses pneumatic and electric actuators to control the fingers. It is a device for the whole hand with 19 DOFs and has the ability to control 81% of the overall ROM of the hand [94]. However, this advantage is still restricted by limitations in the mechanical joints. In contrast, a tendon cable actuator system, such as the one used in the current study, can be adjusted to achieve a larger ROM by using the appropriate gear transmission.

When compared with the earlier mentioned devices that use mechanical serial link manipulation to move the fingers, the current study's design has the advantage of improved ROM, lightweight attributes and a customisable design. The soft robotic glove by Polygerinos, Wang, Galloway, Wood and Walsh [21] has more control over finger movements compared with the current design though this control was achieved through a custom-made composite tubular construction reinforced by fibre walls which require a 3.3 kg belt pack around the waist to provide hydraulic pressure. The current study's device on the other hand weighs only about 520 g.

Safety systems were set in place to protect the individual's hand. With reference to the ROM of each finger joint as stated in the previous chapters, software limits were used to limit the servo position which prevented excessive joint rotation as also seen in an exoskeleton design by Worsnopp, Peshkin, Colgate and Kamper [37]. Besides that, the servos were mechanically limited to a maximum of 180° which presented the second level of safety for the user.

4.4. Device Comparison

Smaby, et al. [95] conducted a study to measure the average pinch force needed for ADL which included opening and closing zippers, inserting and removing a key, using a remote control button and others. Most of these actions required less than 10.5 N of force except for inserting and removing a plug and closing a large zipper which required between 15 to 30 N. The pinch force of the current study's device during the pinch action was measured at 13.9 N which is sufficient for most daily pinching activities.

Table 23 compares the current study's device with two other research devices and two commercial exoskeleton devices with regard to weight, number of DOFs, number of actuators and pinch force exerted. It is crucial to note that there are some disparities in the experiments done on the other devices. Hence, it may only be possible to use their findings as an indirect reference.

Table 23. Comparison of exoskeleton devices performance.

Devices		Weight (g)	No. of DOFs	No. of Actuators	Pinch Force (N)
Current study's device		520	5	2	13.9
Research	SoftHand Pro	520	2	2	20
	X-Limb	253	13	5	10.2
Commercial	Bebionic	539	6	5	12.47
	I-limb	504	6	5	6.54

Before using the device, the average pinch force of the elderly people was measured to be between 3.1 to 8.1 N. This lower pinch force could be partly due to the weak strength of the elderly test participants or due to the different shapes and sizes of the objects. The pinch assistant was able to increase that force to a region that is closer to the average pinch force generated from the other exoskeleton devices (see Figure 21). The force-to-weight ratio was used to compare the device performance as it fairly considered the output force with respect to the device weight. The current study's device has a ratio of 0.0248 N/g which outperformed both commercial devices, namely the Bebionic (0.0231 N/g) and I-limb (0.13 N/g) [96]. Although the current study's device has a higher force output as compared with the X-Limb, the X-Limb has a higher force-to-weight ratio at 0.0403 N/g, as found in the comparison with the SoftHand (0.0384 N/g) [97].

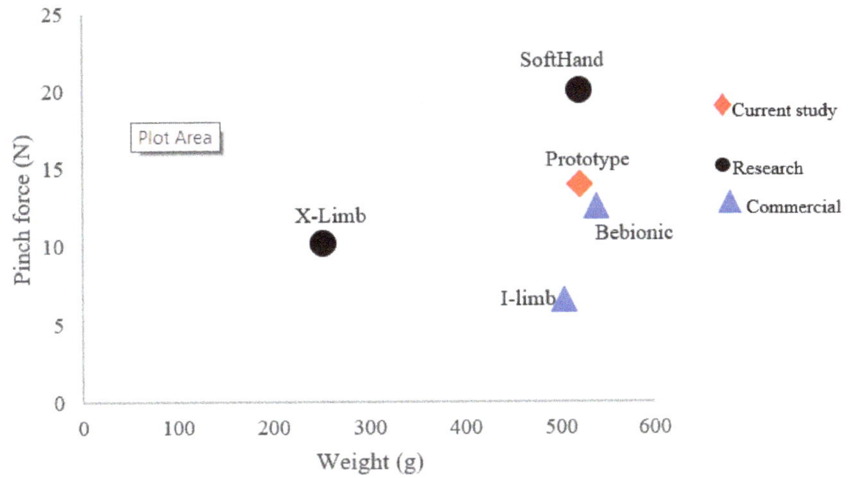

Figure 21. Pinch force vs. weight graph of different hand exoskeleton.

Takahashi, et al. [98] developed a soft exoskeleton glove with the ability to generatedexterous finger joint movement with little constraints. By using the penumatic artificial muscles, postural control of the joints (up to 20 DOFs) was enabled with a fingertip force of 8 N. The soft exoskeleton design has the benefit of flexibility but a generally lower output force of 8 N compared to the proposed invention of this study (13.9 N). The pneumatic system used by the authors also requires an air compressor unit which reduces user mobility, unlike mechanical designs which do not.

Finger exoskeletal devices are frequently measured by their force output and deflection angles [18,19,97]. The conventional method of obtaining device force would be with the use of a stationary load cell which measures the pressure exerted by the device. By using several day-to-day objects during the usability test, this study presents another dimension to the obtained results. The shape and size of an object can significantly affect how the elderly person pinches it [76]. During the golf ball experiment, it was found that elderly people pinched with a greater force (both with and without using the device) when compared to the force produced on the other objects. That being said, the device was still able to significantly improve the users' pinch both in strength and stability. Thus, the usability test coupled with statistical results indicated that the device significantly improves the strength and stability of elderly users while pinching common daily household items.

4.5. Novelty and Industrial Applicability

The novelty of the device is in its design which stems from the specific goal of assisting elderly people to pinch better. This goal allowed for a streamlined design process to solve the problem of pinch weakness first before extending it to other areas of hand functions such as three- or four-finger pinch and power grip. Together with the underactuated mechanism, additive manufacturing technology and off-the-shelf electronic components, a working prototype was created for the testing and proof of concept.

As observed from the *t*-test results, the device has applicational value for elderly people in their day-to-day pinching activities. A significant improvement in pinch force and pinch force steadiness would allow them to have a better overall pinch ability. Furthermore, the inexpensive production cost of each device would widen its accessibility to the public. Independent living elderly people would also be able to use the device due to its easy-to-use design that does not require extra assistance.

5. Conclusions

The main purpose of this study was to extend the development of a concept from previous research by designing and developing a finger grip enhancer that facilitates the day-to-day pinching activities of elderly people. In order to achieve this aim, a finger exoskeleton assistive design from a previous study was analysed using an FEA simulation to validate the device's safety and usability aspects. A working prototype capable of producing a max force of 13.9 N and a force-to-weight ratio of 0.0248 N/g was produced. The usability validation was done through an experiment that measured the exerted force and the force steadiness of elderly people when pinching everyday household objects. Overall, the *t*-test indicated that there were significant improvements ($p < 0.05$) in the elderly people's pinch abilities after using the proposed invention.

5.1. Contribution of Study

With elderly people experiencing functional limitations in performing activities of daily living, this device is an attempt to assist elderly people in pinching better. The approach taken was a streamlined design process to solve the problem of pinch weakness first before extending it to other areas of hand functions such as three- or four-finger pinch and power grip. Together with the underactuated mechanism, additive manufacturing technology and off-the-shelf electronic components, a working prototype was created for the testing and proof of concept.

As observed from the *t*-test results, the device has applicational value for elderly people in their day-to-day pinching activities. A significant improvement in pinch force and pinch force steadiness would allow them to have a better overall pinch ability. Furthermore, the inexpensive production cost of each device would widen its accessibility to the public. Independent living elderly people would also be able to use the device due to its easy-to-use design that does not require extra assistance.

5.2. Limitations of Research

There were several limitations identified in this study. Firstly, the usability of the device was only measured in pinch force and pinch force steadiness. In order to account for dynamic pinches in daily tasks more effectively, it would have been better to measure both force magnitude and directional control (dexterity). Therefore, finger dexterity could be another measurable which determines the effectiveness of the device.

Although usability tests were performed, there was a lack of formal feedback from the elderly users. Furthermore, while this study accounted for various mechanical design analyses, it lacked the analyses on anatomical aspects such as finger joint biomechanics and fingertip trajectory, which could have been done to verify the risk of injuries to the users.

Besides that, another limitation was found in the test procedures. When participants pinched and lifted objects using submaximal force, some inconsistencies in the results could have occurred when the assistive device was used. These inconsistencies were mainly due to the participants' unfamiliarity with using the device, which might have caused participants to pinch harder than required. In order to overcome this limitation, the participants would require more practice with the device over a longer time period.

The validity experiment uses a purposive sampling method which is not meant to represent an entire population but rather selectively engage participants that would benefit from this device. Elderly participants who fulfilled the predefined criteria could participant in the experiment. With the statistical support of the power analysis, the number of samples was found to be sufficient. However, the increase in sample size would generally improve the results by providing a more accurate mean value, smaller margin of error, and identifying outliers. Unfortunately, these benefits come at the cost of time and money which are limited resources.

5.3. Research Challenges

Throughout the testing process, the participants generally did not experience much discomfort. The palm gloves and finger rings were designed to fit most hand sizes and did not produce additional stress on the hand. That being said, the device assembly on the user's wrist is heavy and might lead to long-term complications. The two servo motors should be reduced before the device can truly be used for day-to-day activities. The reason for this challenge is due to the researcher's decision to use off-the-shelf components such as servo motors, a power supply, and a control board. This convenience came at the cost of heavier and bulkier components.

Furthermore, the device uses tension in the tendon cables to assist with pinch activities. Unlike some innovations that use pneumatic flexible soft robotics, the tendon cables, though strong, can strain the fingers if not properly controlled. Although the mechanism is only activated when the user initiates a pinch, pinching multiple items over an extended period of time might stress the finger joints. The tension of the cables, although closely mimicking finger tendons, presents additional challenges. With the cables running on the palm, the user's ease of movement is limited when picking up an item. If the control mechanism were to be fitted behind the hand, the device could then wrap over the fingers for a smoother control during flexion and extension of the finger joints.

5.4. Recommendation for Future Research

Based on the research outcomes and limitations of this study, some directions for future research are suggested. With regard to the device, more movement-perceptive sensors such as Electromyography (EMG) sensors could be used for motor activation. This approach would allow the system to detect the electrical signal produced by the hand muscles. EMG sensors were also used in exoskeleton devices in other areas such as a hip-assisted exoskeleton for semi-squat lifting [99]. EMG sensors were used to measure muscle activation so that the output torque can change in an adaptive manner according to the angular velocity of the wearer's joint. Furthermore, it would also be of interest to account for finger extension in the pinch assistive device although the extension of fingers can be accomplished by the users themselves. With the ability to manipulate both extension and flexion, the device control would be able to adjust the fingers to external perturbation, allowing the finger pinch to adapt naturally. The device is also limited to elderlies who have a certain degree of finger mobility. In the future, modifications in the device to accommodate hand dysfunctions from post-stroke patients would be beneficial.

The Purdue Pegboard Test and the Jebsen-Taylor Hand Function Test are recognised methods for testing manual dexterity and motor function. These tests were initially excluded from this study as they were considered to be secondary hand-eye or reach-to-grasp tests that are prone to adaptive strategies. However, with the understanding that the body is a complete system, tests involving the entire upper extremity would allow for a more holistic view of the improvements made in pinch abilities through the device.

The test analysis could be extended to a broader category of people, which can shed insights on how different people react to an assistive device such as the one in this study. Since conditions such as sarcopenia and carpal tunnel syndrome are fairly common among elderly people, further testing among these patients is encouraged. Testing the device with a younger population would also allow for more comparisons to be made with the elderly. This suggestion would serve as a benchmark or controlled study for improvements made by the device. Tests among the younger population would also open possibilities of developing an assistive device for manual workers such as factory packers that heavily rely on repeated precision pinching.

Incorporating the present invention with therapies based on virtual reality could be another way to improve elderly people's motor functions in the upper limb, though the most optimal treatment approaches, ideal duration, and intensity of the intervention still remains unclear [100]. Lastly, the invention can be integrated into therapies based on neuroplasticity and motor (re)learning through game-based rehabilitation, which involves intense, repeated, and task-based training for upper limb function improvement [101].

Author Contributions: Conceptualization, D.W.H.T., P.K.N., C.C.H. and Y.J.N.; Data curation, D.W.H.T.; Formal analysis, D.W.H.T. and P.K.N.; Funding acquisition, P.K.N.; Investigation, D.W.H.T., P.K.N., A.S., C.C.H. and Y.J.N.; Methodology, D.W.H.T., P.K.N. and A.S.; Project administration, P.K.N. and E.E.M.N.; Resources, P.K.N. and E.E.M.N.; Supervision, P.K.N. and E.E.M.N.; Validation, D.W.H.T. and C.C.H.; Visualization, A.S.; Writing—original draft, D.W.H.T. and P.K.N.; Writing—review & editing, P.K.N., E.E.M.N., A.S., C.C.H. and Y.J.N. All authors have read and agreed to the published version of the manuscript.

Funding: The article processing charge for this manuscript is supported by Multimedia University (Reference ID: MMU/RMC/PC/2021/53612).

Institutional Review Board Statement: The study was conducted according to the guidelines of the Declaration of Helsinki. All participants gave their written informed consent prior to the experiments. All procedures and protocols have been approved by the Research Ethics Committee (REC) from the Technology Transfer Office (TTO) of Multimedia University. The research ethics approval for the project has been granted with the approval number EA0052020, and the approval letter has been endorsed by the TTO Director cum REC Secretariat of the university.

Informed Consent Statement: Informed consent was obtained from all subjects involved in the study.

Data Availability Statement: This project contains the following underlying data: Data Availability Sheet.docx (facts, comparisons, preliminary results, data and discussions). The data can be found at Figshare: https://doi.org/10.6084/m9.figshare.15027975. Data are available under the terms of the Creative Commons Attribution 4.0 International license (CC-BY 4.0).

Acknowledgments: This research article is in partial fulfilment of the requirements for the Master of Engineering Science degree at the Faculty of Engineering and Technology, Multimedia University (MMU), Malaysia. The researchers sincerely thank the faculty and university for their support in allowing this research to be carried out. The researchers also thank the Technology Transfer Office of MMU for granting the public disclosure approval for this paper. Special thanks go to Chiew Fen Ng for her constructive criticism of the manuscript.

Conflicts of Interest: The authors declare no conflict of interest.

References

1. Gul, R.; Alam, I.; Ibrar, A.; Saleem, K.; Pervez, I.P. Aging Affects the Number of WBC and Its Subsets in A Pakistani Cohort of Young and Elderly Women. *ARPN J. Eng. Appl. Sci.* **2013**, *8*, 638–641.
2. Kurnianingsih, K.; Nugroho, L.; Widyawan, W.; Lazuardi, L. Reflective Sensing and Conditioning System in Ubiquitous Home Care for Elderly People. *ARPN J. Eng. Appl. Sci.* **2015**, *10*, 998–1007.
3. Seidler, R.; Bernard, J.; B Burutolu, T.; Fling, B.; Gordon, M.; Gwin, J.; Kwak, Y.; Lipps, D. Motor Control and Aging: Links to Age-Related Brain Structural, Functional, and Biochemical Effects. *Neurosci. Biobehav. Rev.* **2009**, *34*, 721–733. [CrossRef]
4. Young Hoogendam, Y.; van der Lijn, F.; Vernooij, M.W.; Hofman, A.; Niessen, W.J.; Lugt, A.; Arfan Ikram, M.; van der Geest, J.N. Older Age Relates to Worsening of Fine Motor Skills: A Population-Based Study of Middle-Aged and Elderly Persons. *Front. Aging Neurosci.* **2014**, *6*, 259. [CrossRef]
5. Rahman, N.; Thomas, J.J.; Rice, M.S. The Relationship between Hand Strength and the Forces Used to Access Containers by Well Elderly Persons. *Am. J. Occup. Ther.* **2002**, *56*, 78–85. [CrossRef]
6. Rice, M.S.; Leonard, C.; Carter, M. Grip strengths and required forces in accessing everyday containers in a normal population. *Am. J. Occup. Ther.* **1998**, *52*, 621–626. [CrossRef]
7. Chen, X.P.; Lu, Y.M.; Zhang, J. Intervention study of finger-movement exercises and finger weight-lift training for improvement of handgrip strength among the very elderly. *Int. J. Nurs. Sci.* **2014**, *1*, 165–170. [CrossRef]
8. Ranganathan, V.K.; Siemionow, V.; Sahgal, V.; Liu, J.Z.; Yue, G.H. Skilled Finger Movement Exercise Improves Hand Function. *J. Gerontol. Ser. A* **2001**, *56*, 518–522. [CrossRef]

9. Ellegaard, K.; von Bülow, C.; Røpke, A.; Bartholdy, C.; Hansen, I.S.; Rifbjerg-Madsen, S.; Henriksen, M.; Wæhrens, E.E. Hand exercise for women with rheumatoid arthritis and decreased hand function: An exploratory randomized controlled trial. *Arthritis Res. Ther.* **2019**, *21*, 158. [CrossRef]
10. Rodríguez-Sánchez-Laulhé, P.; Luque-Romero, L.G.; Blanquero, J.; Suero-Pineda, A.; Biscarri-Carbonero, Á.; Barrero-García, F.J.; Heredia-Rizo, A.M. A mobile app using therapeutic exercise and education for self-management in patients with hand rheumatoid arthritis: A randomized controlled trial protocol. *Trials* **2020**, *21*, 777. [CrossRef]
11. Liu, B.; Chen, X.; Li, Y.; Liu, H.; Guo, S.; Yu, P. Effect of passive finger exercises on grip strength and the ability to perform activities of daily living for older people with dementia: A 12-week randomized controlled trial. *Clin. Interv. Aging* **2018**, *13*, 2169–2177. [CrossRef]
12. Cole, T.; Robinson, L.; Romero, L.; O'Brien, L. Effectiveness of interventions to improve therapy adherence in people with upper limb conditions: A systematic review. *J. Hand Ther. Off. J. Am. Soc. Hand Ther.* **2019**, *32*, 175–183.e7. [CrossRef]
13. Hepple, R.T.; Rice, C.L. Innervation and neuromuscular control in ageing skeletal muscle. *J. Physiol.* **2016**, *594*, 1965–1978. [CrossRef] [PubMed]
14. Rahim, A.A.; Patar, M.N.A.A.; Amin, A.M.; Mahmud, J. The Development of Finger Rehabilitation Device for Stroke Patients. *Appl. Mech. Mater.* **2013**, *393*, 604–610. [CrossRef]
15. Rahman, M.A.; Al-Jumaily, A. Design and Development of a Hand Exoskeleton for Rehabilitation Following Stroke. *Procedia Eng.* **2012**, *41*, 1028–1034. [CrossRef]
16. Cempini, M.; De Rossi, S.M.M.; Lenzi, T.; Cortese, M.; Giovacchini, F.; Vitiello, N.; Carrozza, M.C. Kinematics and design of a portable and wearable exoskeleton for hand rehabilitation. In Proceedings of the 2013 IEEE 13th International Conference on Rehabilitation Robotics (ICORR), Seattle, WA, USA, 24–26 June 2013; pp. 1–6. [CrossRef]
17. Zaid, A.; Tee, C.C.; Sukor, J.A.; Hanafi, D. Development of hand exoskeleton for rehabilitation of post-stroke patient. *AIP Conf. Proc.* **2017**, *1891*, 020103. [CrossRef]
18. Rudd, G.; Daly, L.; Jovanovic, V.; Cuckov, F. A Low-Cost Soft Robotic Hand Exoskeleton for Use in Therapy of Limited Hand–Motor Function. *Appl. Sci.* **2019**, *9*, 3751. [CrossRef]
19. Wege, A.; Hommel, G. Development and Control of A Hand Exoskeleton for Rehabilitation of Hand Injuries. In Proceedings of the 2005 IEEE/RSJ International Conference on Intelligent Robots and Systems, Edmonton, AB, Canada, 2–6 August 2005; pp. 3046–3051.
20. In, H.; Cho, K.-J.; Kim, K.; Lee, B. Jointless structure and under-actuation mechanism for compact hand exoskeleton. In Proceedings of the 2011 IEEE International Conference on Rehabilitation Robotics, Zurich, Switzerland, 29 June–1 July 2011; Volume 2011, pp. 1–6.
21. Polygerinos, P.; Wang, Z.; Galloway, K.C.; Wood, R.J.; Walsh, C.J. Soft robotic glove for combined assistance and at-home rehabilitation. *Robot. Auton. Syst.* **2015**, *73*, 135–143. [CrossRef]
22. DiCicco, M.; Lucas, L.; Matsuoka, Y. Comparison of Control Strategies for an EMG Controlled Orthotic Exoskeleton for the Hand. In Proceedings of the IEEE International Conference on Robotics and Automation 2004, New Orleans, LA, USA, 26 April–1 May 2004; pp. 1622–1627. [CrossRef]
23. Hasegawa, Y.; Mikami, Y.; Watanabe, K.; Firouzimehr, Z.; Sankai, Y. Wearable Handling Support System for Paralyzed Patient. In Proceedings of the 2008 IEEE/RSJ International Conference on Intelligent Robots and Systems, Nice, France, 22–26 September 2008; pp. 741–746. [CrossRef]
24. Mahmud, N.A.; Yoep, N.; Paiwai, F.; Shahein, N.A.; Tan, M.P.; Yusof, M.; Muhamad, N.A. Functional Limitations and Falls. In *National Health and Morbidity Survey 2018: Elderly Health*; Chan, Y.Y., Tan, L., Ahmad, F.H., Yusoff, M.F.M., Lodz, N.A., Shawal, N., Sooryanarayana, R., Aris, T., Robert, T., Saminathan, T.A., Eds.; Institute for Public Health, National Institutes of Health (NIH), Ministry of Health: Selangor, Malaysia, 2018; Volume 2, pp. 8–13.
25. Tan, D.; Ng, P.; Noor, E. A TRIZ-driven conceptualisation of finger grip enhancer designs for the elderly [version 1; peer review: 1 approved, 1 approved with reservations]. *F1000Research* **2021**, *10*, 392. [CrossRef]
26. Garrett, J.W. The adult human hand: Some anthropometric and biomechanical considerations. *Hum. Factors* **1971**, *13*, 117–131. [CrossRef] [PubMed]
27. Greiner, T.M. *Hand Anthropometry of US Army Personnel*; Army Natick Reseach Development and Engineering Center: Natick, MA, USA, 1991.
28. Taylor, C.L.; Schwarz, R.J. The anatomy and mechanics of the human hand. *Artif. Limbs* **1955**, *2*, 22–35.
29. Duncan, S.F.M.; Saracevic, C.E.; Kakinoki, R. Biomechanics of the Hand. *Hand Clin.* **2013**, *29*, 483–492. [CrossRef]
30. Lin, J.; Wu, Y.; Huang, T.S. Modeling the constraints of human hand motion. *Proc. Workshop Hum. Motion* **2000**, 121–126. [CrossRef]
31. Sarac, M.; Solazzi, M.; Frisoli, A. Design Requirements of Generic Hand Exoskeletons and Survey of Hand Exoskeletons for Rehabilitation, Assistive or Haptic Use. *IEEE Trans. Haptics* **2019**, *12*, 400–413. [CrossRef]
32. Tanaka, T.; Amadio, P.C.; Zhao, C.; Zobitz, M.E.; An, K.-N. Flexor digitorum profundus tendon tension during finger manipulation. *J. Hand Ther.* **2005**, *18*, 330–338. [CrossRef]
33. ElKoura, G.; Singh, K. Handrix: Animating the human hand. In Proceedings of the 2003 ACM SIGGRAPH/Eurographics Symposium on Computer Animation, San Diego, CA, USA, 26–27 July 2003; pp. 110–119. [CrossRef]

34. Lisini, T.; Scheggi, S.; Meli, L.; Mohammadi, M.; Prattichizzo, D. GESTO: A Glove for Enhanced Sensing and Touching Based on Inertial and Magnetic Sensors for Hand Tracking and Cutaneous Feedback. *IEEE Trans. Hum.-Mach. Syst.* **2017**, *47*, 1066–1076. [CrossRef]
35. Lince, A.R. A Portable Assistive Rehabilitation Hand Exoskeleton. Ph.D. Thesis, Polytechnic University of Turin, Turin, Italy, 2016.
36. Imrhan, S.N. The influence of wrist position on different types of pinch strength. *Appl. Ergon.* **1991**, *22*, 379–384. [CrossRef]
37. Worsnopp, T.T.; Peshkin, M.; Colgate, J.; Kamper, D. An Actuated Finger Exoskeleton for Hand Rehabilitation Following Stroke. In Proceedings of the 2007 IEEE 10th International Conference on Rehabilitation Robotics, Noordwijk, The Netherlands, 13–15 June 2007; pp. 896–901. [CrossRef]
38. Leonardis, D.; Barsotti, M.; Loconsole, C.; Solazzi, M.; Troncossi, M.; Mazzotti, C.; Castelli, V.P.; Procopio, C.; Lamola, G.; Chisari, C.; et al. An EMG-Controlled Robotic Hand Exoskeleton for Bilateral Rehabilitation. *IEEE Trans Haptics* **2015**, *8*, 140–151. [CrossRef]
39. Alexander, B.; Kotiuk, V. Proportions of Hand Segments. *Int. J. Morphol.* **2010**, *28*, 755–758.
40. Richterman, I.E.; DuPree, J.; Thoder, J.; Kozin, S.H. The radiographic analysis of web height. *J. Hand Surg.* **1998**, *23*, 1071–1076. [CrossRef]
41. Hume, M.; Gellman, H.; McKellop, H.; Brumfield, R. Functional range of motion of the joints of the hand. *J. Hand Surg.* **1990**, *15*, 240–243. [CrossRef]
42. Jahn, J.; Janes, W.E.; Saheb-Al-Zamani, M.; Burbank, C.M.; Brown, J.M.; Engsberg, J.R. Identification of three movement phases of the hand during lateral and pulp pinches using video motion capture. *Hand* **2013**, *8*, 123–131. [CrossRef] [PubMed]
43. Brorsson, S. Biomechanical Studies of Finger Extension Function. Analysis with a New Force Measuring Device and Ultrasound Examination in Rheumatoid Arthritis and Healhty Muscles. Ph.D. Thesis, University of Gothenburg, Gothenburg, Sweden, 2008.
44. Shojaei Barjuei, E.; Caldwell, D.G.; Ortiz, J. Bond Graph Modeling and Kalman Filter Observer Design for an Industrial Back-Support Exoskeleton. *Designs* **2020**, *4*, 53. [CrossRef]
45. Lampman, S.; Dieter, G.E. *ASM Handbook: Materials Selection and Design*; Taylor & Francis: Abingdon, UK, 1997; Volume 20.
46. Novakova-Marcincinova, L.; Novak-Marcincin, J.; Barna, J.; Torok, J. Special Materials Used in FDM Rapid Prototyping Technology Application. In Proceedings of the 2012 IEEE 16th International Conference on Intelligent Engineering Systems (INES), Lisbon, Portugal, 13–15 June 2012; pp. 73–76. [CrossRef]
47. McKeen, L. Elastomers. In *The Effect of Sterilization Methods on Plastics and Elastomers*, 4th ed.; McKeen, L., Ed.; William Andrew Publishing: Norwich, NY, USA, 2018; pp. 305–351.
48. Kulkarni, G.S. Introduction to Polymer and Their Recycling Techniques. In *Recycling of Polyurethane Foams*; Thomas, S., Rane, A.V., Kanny, K., V.K, A., Thomas, M.G., Eds.; William Andrew Publishing: Norwich, NY, USA, 2018; pp. 1–16.
49. Materialise. 3D Printing Materials. Available online: https://bit.ly/3o4HM03 (accessed on 27 March 2020).
50. Camargo, J.; Machado, A.; Almeida, E.; Moura, E. Mechanical properties of PLA-graphene filament for FDM 3D printing. *Int. J. Adv. Manuf. Technol.* **2019**, *103*, 2423–2443. [CrossRef]
51. Rodriguez-Panes, A.; Claver, J.; Camacho, A.M. The Influence of Manufacturing Parameters on the Mechanical Behaviour of PLA and ABS Pieces Manufactured by FDM: A Comparative Analysis. *Materials* **2018**, *11*, 1333. [CrossRef]
52. MatWeb. Overview of Materials for Polylactic Acid (PLA) Biopolymer. Available online: https://bit.ly/2KTcbQy (accessed on 27 March 2020).
53. Stratasys. Nylon (Polyamide) Parts on Demand. Available online: https://www.stratasysdirect.com/materials/thermoplastics/nylon#nylon-12 (accessed on 27 March 2020).
54. Hyie, K.; Budin, S.; Martinus, N.; Salleh, Z.; Masdek, N. Tensile and flexural investigation on polypropylene recycling. *J. Phys. Conf. Ser.* **2019**, *1174*, 012005. [CrossRef]
55. Mathiowetz, V.; Kashman, N.; Volland, G.; Weber, K.; Dowe, M.; Rogers, S. Grip and Pinch Strength: Normative data for adults. *Arch. Phys. Med. Rehabil.* **1985**, *66*, 69–74.
56. Mohammadian, M.; Choobineh, A.; Haghdoost, A.; Hasheminejad, N. Normative data of grip and pinch strengths in healthy adults of Iranian population. *Iran J. Public Health* **2014**, *43*, 1113–1122.
57. Nilsen, T.; Hermann, M.; Eriksen, C.; Dagfinrud, H.; Mowinckel, P.; Kjeken, I. Grip force and pinch grip in an adult population: Reference values and factors associated with grip force. *Scand. J. Occup. Ther.* **2011**, *19*, 288–296. [CrossRef]
58. Harth, A.; Vetter, W.R. Grip and pinch strength among selected adult occupational groups. *Occup. Ther. Int.* **1994**, *1*, 13–28. [CrossRef]
59. Eliopoulos, G.M.; Harris, A.D.; Bradham, D.D.; Baumgarten, M.; Zuckerman, I.H.; Fink, J.C.; Perencevich, E.N. The Use and Interpretation of Quasi-Experimental Studies in Infectious Diseases. *Clin. Infect. Dis.* **2004**, *38*, 1586–1591. [CrossRef]
60. Maciejewski, M.L. Quasi-experimental design. *Biostat. Epidemiol.* **2020**, *4*, 38–47. [CrossRef]
61. Harris, A.D.; McGregor, J.C.; Perencevich, E.N.; Furuno, J.P.; Zhu, J.; Peterson, D.E.; Finkelstein, J. The use and interpretation of quasi-experimental studies in medical informatics. *J. Am. Med Inform. Assoc.* **2006**, *13*, 16–23. [CrossRef] [PubMed]
62. Campbell, D.T.; Stanley, J.C. *Experimental and Quasi-Experimental Designs for Research*; Ravenio Books: Cambridge, UK, 2015.
63. Walukonis, K.; Beasley, J.; Boerema, R.; Powers, J.; Anderson, K. The impact of finger position on pinch strength. *Hand Ther.* **2018**, *23*, 70–76. [CrossRef]

64. Fausset, C.; Kelly, A.; Rogers, W.; Fisk, A. Challenges to Aging in Place: Understanding Home Maintenance Difficulties. *J. Hous. Elder.* **2011**, *25*, 125–141. [CrossRef] [PubMed]
65. Smaby, N.; Baker, B.; Johanson, M.E.; Towles, J.D.; Murray, W.M. Determination of lateral pinch force requirements for six common activities of daily living. In Proceedings of the Third National Meeting for Rehabilitation Research and Development Conference, Arlington, VA, USA, 10–12 February 2002; p. 104.
66. Stegink Jansen, C.W.; Simper, V.K.; Stuart, H.G., Jr.; Pinkerton, H.M. Measurement of maximum voluntary pinch strength: Effects of forearm position and outcome score. *J. Hand Ther.* **2003**, *16*, 326–336. [CrossRef]
67. Ames, H.; Glenton, C.; Lewin, S. Purposive sampling in a qualitative evidence synthesis: A worked example from a synthesis on parental perceptions of vaccination communication. *BMC Med. Res. Methodol.* **2019**, *19*, 26. [CrossRef]
68. Palinkas, L.A.; Horwitz, S.M.; Green, C.A.; Wisdom, J.P.; Duan, N.; Hoagwood, K. Purposeful Sampling for Qualitative Data Collection and Analysis in Mixed Method Implementation Research. *Adm. Policy Ment. Health* **2015**, *42*, 533–544. [CrossRef]
69. Werle, S.; Goldhahn, J.; Drerup, S.; Simmen, B.R.; Sprott, H.; Herren, D.B. Age- and Gender-Specific Normative Data of Grip and Pinch Strength in a Healthy Adult Swiss Population. *J. Hand Surg.* **2009**, *34*, 76–84. [CrossRef]
70. Parmar, S.; Khodasevych, I.; Troynikov, O. Evaluation of Flexible Force Sensors for Pressure Monitoring in Treatment of Chronic Venous Disorders. *Sensors* **2017**, *17*, 1923. [CrossRef]
71. Mathiowetz, V.; Weber, K.; Volland, G.; Kashman, N. Reliability and validity of grip and pinch strength evaluations. *J. Hand Surg.* **1984**, *9*, 222–226. [CrossRef]
72. Shih, Y.-C.; Ou, Y.-C. Influences of span and wrist posture on peak chuck pinch strength and time needed to reach peak strength. *Int. J. Ind. Ergon.* **2005**, *35*, 527–536. [CrossRef]
73. Hanten, W.P.; Chen, W.-Y.; Austin, A.A.; Brooks, R.E.; Carter, H.C.; Law, C.A.; Morgan, M.K.; Sanders, D.J.; Swan, C.A.; Vanderslice, A.L. Maximum grip strength in normal subjects from 20 to 64 years of age. *J. Hand Ther.* **1999**, *12*, 193–200. [CrossRef]
74. Mathiowetz, V. Effects of three trials on grip and pinch strength measurements. *J. Hand Ther.* **1990**, *3*, 195–198. [CrossRef]
75. Trossman, P.B.; Li, P.-W. The Effect of the Duration of Intertrial Rest Periods on Isometric Grip Strength Performance in Young Adults. *Occup. Ther. J. Res.* **1989**, *9*, 362–378. [CrossRef]
76. Jones, S.R.; Carley, S.; Harrison, M. An introduction to power and sample size estimation. *Emerg. Med. J.* **2003**, *20*, 453–458. [CrossRef] [PubMed]
77. Brysbaert, M. How Many Participants Do We Have to Include in Properly Powered Experiments? A Tutorial of Power Analysis with Reference Tables. *J. Cogn.* **2019**, *2*, 16. [CrossRef] [PubMed]
78. Lakens, D. Calculating and reporting effect sizes to facilitate cumulative science: A practical primer for t-tests and ANOVAs. *Front. Psychol.* **2013**, *4*, 863. [CrossRef]
79. Ghasemi, A.; Zahediasl, S. Normality tests for statistical analysis: A guide for non-statisticians. *Int. J. Endocrinol. Metab.* **2012**, *10*, 486–489. [CrossRef]
80. Nosakhare, U.H.; Bright, A.F. Evaluation of Techniques for Univariate Normality Test Using Monte Carlo Simulation. *Am. J. Theor. Appl. Stat.* **2018**, *6*, 51–61.
81. Kinoshita, H.; Francis, P.R. A comparison of prehension force control in young and elderly individuals. *Eur. J. Appl. Physiol. Occup. Physiol.* **1996**, *74*, 450–460. [CrossRef]
82. De Serres, S.J.; Fang, N.Z. The accuracy of perception of a pinch grip force in older adults. *Can. J. Physiol. Pharmacol.* **2004**, *82*, 693–701. [CrossRef]
83. Spirduso, W.W.; Choi, J. Age and Practice Effects on Force Control of the Thumb and Index Fingers in Precision Pinching and Bilateral Coordination. In *Sensorimotor Impairment in the Elderly*; Stelmach, G.E., Hömberg, V., Eds.; Springer: Dordrecht, The Netherlands, 1993; pp. 393–412.
84. Stelmach, G.E.; Hömberg, V. *Sensorimotor Impairment in the Elderly*; Springer: Heidelberg, Germany, 2012.
85. Boatright, J.R.; Kiebzak, G.M.; O'Neil, D.M.; Peindl, R.D. Measurement of thumb abduction strength: Normative data and a comparison with grip and pinch strength. *J. Hand Surg.* **1997**, *22*, 843–848. [CrossRef]
86. Carmeli, E.; Patish, H.; Coleman, R. The Aging Hand. *J. Gerontol. Ser. A* **2003**, *58*, 146–152. [CrossRef]
87. Kornatz, K.W.; Christou, E.A.; Enoka, R.M. Practice reduces motor unit discharge variability in a hand muscle and improves manual dexterity in old adults. *J. Appl. Physiol.* **2005**, *98*, 2072–2080. [CrossRef]
88. Portnoy, S.; Halaby, O.; Dekel-Chen, D.; Dierick, F. Effect of an auditory feedback substitution, tactilo-kinesthetic, or visual feedback on kinematics of pouring water from kettle into cup. *Appl. Ergon.* **2015**, *51*, 44–49. [CrossRef] [PubMed]
89. Cole, K.; Rotella, D. Old age affects fingertip forces when restraining an unpredictably loaded object. *Exp. Brain Res.* **2001**, *136*, 535–542. [CrossRef] [PubMed]
90. Keen, D.A.; Yue, G.; Enoka, R. Training-related enhancement in the control of motor output in elderly humans. *J. Appl. Physiol.* **1995**, *77*, 2648–2658. [CrossRef] [PubMed]
91. Hasegawa, Y.; Mikami, Y.; Watanabe, K.; Sankai, Y. Five-fingered assistive hand with mechanical compliance of human finger. In Proceedings of the 2008 IEEE International Conference on Robotics and Automation, Pasadena, CA, USA, 19–23 May 2008; pp. 718–724. [CrossRef]
92. Nilsson, M.; Ingvast, J.; Wikander, J.; Holst, H. The Soft Extra Muscle system for improving the grasping capability in neurological rehabilitation. In Proceedings of the 2012 IEEE-EMBS Conference on Biomedical Engineering and Sciences, Langkawi, Malaysia, 17–19 December 2012; pp. 412–417. [CrossRef]

93. An, K.N.; Chao, E.Y.; Cooney, W.P.; Linscheid, R.L. Forces in the normal and abnormal hand. *J. Orthop. Res.* **1985**, *3*, 202–211. [CrossRef] [PubMed]
94. Tjahyono, A.; Aw, K.; Devaraj, H.; Surendra, W.; Hämmerle, E.; Travas-Sejdic, J. A five-fingered hand exoskeleton driven by pneumatic artificial muscles with novel polypyrrole sensors. *Ind. Robot* **2013**, *40*, 251–260. [CrossRef]
95. Smaby, N.; Johanson, M.E.; Baker, B.; Kenney, D.E.; Murray, W.M.; Hentz, V.R. Identification of key pinch forces required to complete functional tasks. *J. Rehabil. Res. Dev.* **2004**, *41*, 215–224. [CrossRef] [PubMed]
96. Belter, J.T.; Segil, J.L.; Dollar, A.M.; Weir, R.F. Mechanical design and performance specifications of anthropomorphic prosthetic hands: A review. *J. Rehabil. Res. Dev.* **2013**, *50*, 599–618. [CrossRef] [PubMed]
97. Mohammadi, A.; Lavranos, J.; Zhou, H.; Mutlu, R.; Alici, G.; Tan, Y.; Choong, P.; Oetomo, D. A practical 3D-printed soft robotic prosthetic hand with multi-articulating capabilities. *PLoS ONE* **2020**, *15*, e0232766. [CrossRef]
98. Takahashi, N.; Furuya, S.; Koike, H. Soft Exoskeleton Glove with Human Anatomical Architecture: Production of Dexterous Finger Movements and Skillful Piano Performance. *IEEE Trans Haptics* **2020**, *13*, 679–690. [CrossRef] [PubMed]
99. Wei, W.; Zha, S.; Xia, Y.; Gu, J.; Lin, X. A Hip Active Assisted Exoskeleton That Assists the Semi-Squat Lifting. *Appl. Sci.* **2020**, *10*, 2424. [CrossRef]
100. Webster, A.; Poyade, M.; Rooney, S.; Paul, L. Upper limb rehabilitation interventions using virtual reality for people with multiple sclerosis: A systematic review. *Mult. Scler. Relat. Disord.* **2021**, *47*, 102610. [CrossRef]
101. Sulfikar Ali, A.; Arumugam, A.; Kumaran, D.S. Effectiveness of an intensive, functional, gamified Rehabilitation program in improving upper limb motor function in people with stroke: A protocol of the EnteRtain randomized clinical trial. *Contemp. Clin. Trials* **2021**, *105*, 106381. [CrossRef] [PubMed]

robotics

MDPI

Article

Six-Bar Linkage Models of a Recumbent Tricycle Mechanism to Increase Power Throughput in FES Cycling

Nicholas A. Lanese [1], David H. Myszka [1], Anthony L. Bazler [2] and Andrew P. Murray [1,*]

1 Mechanical and Aerospace Engineering, University of Dayton, 300 College Park, Dayton, OH 45469, USA; lanesen1@udayton.edu (N.A.L.); dmyszka1@udayton.edu (D.H.M.)
2 University of Dayton Research Institute, 1700 S Patterson Blvd, Dayton, OH 45469, USA; Anthony.Bazler@udri.udayton.edu (A.L.B.)
* Correspondence: murray@udayton.edu

Abstract: This paper presents the kinematic and static analysis of two mechanisms to improve power throughput for persons with tetra- or paraplegia pedaling a performance tricycle via FES. FES, or functional electrical stimulation, activates muscles by passing small electrical currents through the muscle creating a contraction. The use of FES can build muscle in patients, relieve soreness, and promote cardiovascular health. Compared to an able-bodied rider, a cyclist stimulated via FES produces an order of magnitude less power creating some notable pedaling difficulties especially pertaining to inactive zones. An inactive zone occurs when the leg position is unable to produce enough power to propel the tricycle via muscle stimulation. An inactive zone is typically present when one leg is fully bent and the other leg is fully extended. Altering the motion of a cyclist's legs relative to the crank position can potentially reduce inactive zones and increase power throughput. Some recently marketed bicycles showcase pedal mechanisms utilizing alternate leg motions. This work considers performance tricycle designs based on the Stephenson III and Watt II six-bar mechanisms where the legs define two of the system's links. The architecture based on the Stephenson III is referred to throughout as the CDT due to the legs' push acting to coupler-drive the four-bar component of the system. The architecture based on the Watt II is referred to throughout as the CRT due to the legs' push acting to drive the rocker link of the four-bar component of the system. The unmodified or traditional recumbent tricycle (TRT) provides a benchmarks by which the designs proposed herein may be evaluated. Using knee and hip torques and angular velocities consistent with a previous study, this numerical study using a quasi-static power model of the CRT suggests a roughly 50% increase and the CDT suggests roughly a doubling in average crank power, respectively, for a typical FES cyclist.

Keywords: functional electrical stimulation; six-bar linkage; Watt II; Stephenson III; performance tricycle; mechanism optimization

Citation: Lanese, N.A.; Myszka, D.H.; Bazler, A.L.; Murray, A.P. Six-Bar Linkage Models of a Recumbent Tricycle Mechanism to Increase Power Throughput in FES Cycling. *Robotics* **2022**, *11*, 26. https://doi.org/10.3390/robotics11010026

Academic Editor: Raffaele Di Gregorio

Received: 15 November 2021
Accepted: 4 February 2022
Published: 11 February 2022

1. Introduction

A five-bar mechanism can be used to effectively model a cyclist's upper and lower leg actuating a conventional crank-driven bicycle [1]. When restricting the motion and torque at the cyclist's ankle, the kinematics and dynamics are described with a four-bar model. The ubiquitous four-bar mechanism has four pairs of coordinates defining the locations of its revolute (R) joints for a total of eight design variables. A potential way to improve the performance of a system is to replace its four-bar by a six-bar mechanism. A six-bar mechanism has seven R joints resulting in 14 design parameters. The six-bar performs in a manner that is similar to the four-bar, but has extra design variables that can be used to improve the performance of the mechanism, typically supplying amplification or subtler control over forces and/or displacements. Several recent examples of using the six-bar to provide such design refinement include a furniture hinge mechanism [2], an adjustable mechanical forming press [3], a biomimetic leg mechanism [4], a body guidance mechanism [5], a hand rehabilitation robot [6], and a compact bellow globe valve design [7]. When locking the motion of the ankle, a human pedaling a bicycle is seen as a four-bar loop, with the thigh as a rocking input, the cnemis, or lower leg, as the coupler, and the crank connecting to sprocket and chain as the fully rotating output. With improved performance in mind, the thigh and cnemis can be viewed as two links of a six-bar mechanism instead of two links in a four-bar where, instead of the foot driving the output crank directly, additional links are available between foot and output crank. Two such alternates are the Watt II and the Stephenson III.

Prior work investigated the Watt II design [8], referred to therein as the CRT, short for crank-rocker tricycle. That is, the thigh and calf drive a rocker arm that moves a coupler and rotates the output crank. Together, the motion of the legs and the crank-rocker define a pair of connected four-bars as in a Watt II mechanism. The significant change to this work is to consider the leg as pushing on the coupler of a four-bar rather than its output link, producing the CDT or coupler-driven tricycle. The overall architecture resembles that of the Stephenson III. As in the work of Bazler et al. [8], the focus here is on riders using functional electrical stimulation (FES). Although FES and its use in cycling are reviewed here, that work contains a more detailed examination of these topics.

FES uses low-energy electrical pulses to artificially generate muscle contractions. This stimulation can create movement for individuals who have been paralyzed due to spinal cord injuries (SCI) or other lower-body neurological impairments [9]. Outcomes include the ability to stand, grasp objects and exercise in people with physically disabling conditions [10]. FES cycling can be a successful means of exercise and rehabilitation for these patients. A variety of physical and mental health benefits have been observed [11–13]. The biomechanical challenges of FES cycling can be observed in the bike race event of the Cybathlon [14,15]. The challenges arise as FES cyclists are observed to produce an order of magnitude less power than able-bodied (AB) cyclists [16], limiting FES cycling to stationary bikes or horizontal surfaces. Biomechanical studies of cycling include Hull et al. [1] and Yamazaki et al. [17] demonstrating accurate dynamic simulations for AB cyclists, Fregly et al. examining steady state pedaling [18], and Gfohler et al. developing models in SCI cyclists that relate joint torques and angles [19,20].

The pedaling power model used in this paper is derived from the results of Szecsi et al. [21] developed on stationary tricycles wherein data was collected in two phases for 16 persons with SCI. The ankle was placed in a boot to constrain the foot to a 90° angle with the cnemis and restricting motion to the sagittal plane, typical of FES cycling. Their experiments involved a passive pedaling phase followed by an active phase in which stimulation of the subject's legs rotated the crank at a cadence of 60 rpms against a machine-controlled resistance of 30 W (15 W per leg). An inverse dynamic approach was used to determine the knee and hip moments from the pedal contact forces [22]. Their data is measured from top dead center (TDC), an absolute crank angle of 22° as used in this work. Based on their prior work showing that AB cyclists have four distinct power peaks (P1, P2, P3, and P4) during one rotation of the crank, FES cyclists were found to fall into two groups [23]. Of the 16 SCI subjects, 75% produced the power alignment with the P1 and P2 phases (the P1P2 group). The remaining 25% produced power in alignment with the P1 and P3 phases (the P1P3 group). The knee and hip joint moments, calculated from the inverse dynamics, as a function of crank angle are shown for each group in Figure 1.

The literature on riders using FES typically involves a tricycle featuring a traditional drivetrain, designed for the power stroke of an AB cyclist using voluntary movement of the lower limbs. Some innovative approaches have been attempted to address the challenges of FES cycling. An energy storage device proposed by Ibrahim et al. sought to store power in an elastic band to evenly redistribute positive power at the crank through the full duration of the pedaling sequence [24]. Shan [25] and Gfohler et al. [20] proposed ovate pedaling motion in an effort to increase power production in both AB and SCI cyclists. In line with these efforts, this work proposes that an alternative drivetrain could achieve greater power throughput than the traditional recumbent tricycle (TRT).

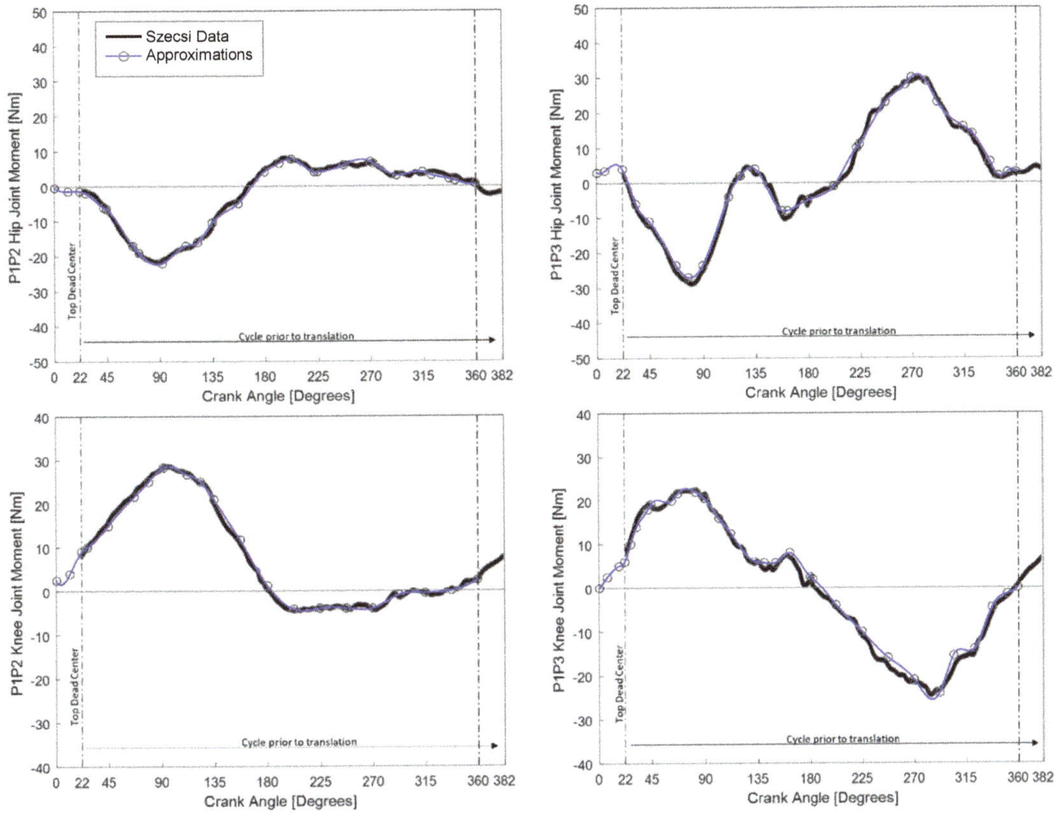

Figure 1. Knee and hip moment data reported in Szecsi P1P2 and P1P3 groups, adapted from Ref. [21], 2014 Clarivate Analytics Web of Science, shown as the black curve. The numerical approximations are shown in blue. The curves are plotted from an absolute crank angle of 22° to correspond with Szecsi's data, which is measured from top dead center.

The contribution of this paper is to present a design methodology and recumbent tricycle architectures with the capacity to increase the power throughput for FES cycling. Previous research [8] proposed a six-bar pedalling linkage, with a coupler-rocker configuration, where analysis shows over a 50% power increase when compared to a traditional tricycle design. This paper introduces another six-bar linkage, with a coupler-driver configuration, that promises to increase power throughput by over 100%. The paper is organized as follows. The design approach is outlined in detail in Section 2, following closely the work of Bazler et al. [8]. Kinematic and quasi-static models are described in Section 3 as derived from the traditional recumbent tricycle. Section 4 discusses transferring the torque into a form that can be used in alternate tricycle deigns. Section 5 presents the considerations in the six-bar linkage-based designs, and Section 6 details its optimization. Section 7 compares the results of the three tricycle architectures. Section 8 concludes the paper.

2. Design Approach

This work investigates alternative tricycle drivetrain configurations to improve the effectiveness of FES cycling as a means of exercise and rehabilitation for persons with SCI. The investigation encompasses the following high-level tasks:

1. *Model the traditional recumbent tricycle (TRT).* Kinematic and quasi-static models are constructed to simulate the experimental work published by Szecsi et al. The TRT includes a continuously rotating crank, which serves as the dependent variable for the published hip-joint and knee-joint moment data as shown in Figure 1. Along with the joint moment vs. crank angle data, the TRT model also accepts the tricycle dimensions and the lower limb dimensions of the cyclist.
2. *Power output of the P1P2 and P1P3 riders.* The performance of the TRT riders is analyzed, using the respective joint moment data, over a full revolution of the crank angle θ_2. The moment at the crank center M_A produced by one one leg is determined throughout $0 \leq \theta_2 < 360\,^{\circ}$. The average power produced by one leg of the cyclist is

$$\overline{P} = \frac{\dot{\theta}_2}{2\pi} \int_0^{2\pi} M_A d\theta_2, \tag{1}$$

   where $\dot{\theta_2}$ is the constant rotational velocity of the crank.
3. *Transform the muscle data.* The kinematics of alternative tricycle designs will differ from the TRT. The crank angle is an inappropriate dependent variable for the exploration presented in this paper. In order to apply the joint moment data to alternative designs, the data dependency is shifted from crank angle to the appropriate joint angle.
4. *Model the crank rocker tricycle (CRT) and the coupler driver tricycle (CDT).* Kinematic and quasi-static models are created for the CRT and CDT, both utilizing tricycle dimensions, rider dimensions and transformed muscle data.
5. *Optimize the CRT and CDT dimensions.* Optimizations are performed on the CRT and CDT models with the objective of improving upon the cycle-averaged power produced by each group (P1P2 and P1P3) of riders.

3. Traditional Recumbent Tricycle Model

The kinematics of the interaction between the cyclist and the TRT is modeled as a four-bar linkage. Figure 2b displays the vector diagram. The recumbent tricycle frame is designated as $\mathbf{R}_1$.

The thigh $\mathbf{R}_4$ serves as an oscillating link. The cnemis $\mathbf{R}_5$ and foot $\mathbf{R}_6$ are held at a constant relative angle by a rigid boot and are combined to form the coupler $\mathbf{R}_3$. The cycle pedal arm $\mathbf{R}_2$ is a fully-rotating crank. The frame dimensions are $\mathbf{R}_1 = [-73.6, -30.4]^{\top}$ cm and the crank length is $R_2 = 15.0$ cm.

The loop closure equation for the TRT model is

$$R_1\vec{\theta}_1 + R_2\vec{\theta}_2 - R_3\vec{\theta}_3 - R_4\vec{\theta}_4 = 0, \tag{2}$$

where R_i is the length of vector i, $\vec{\theta}_i = [\cos\theta_i \quad \sin\theta_i]^{\top}$ and θ_i is the angle of vector i measured counterclockwise from the positive horizontal axis. With a specified crank angle θ_2, Equation (2) can be readily solved for θ_3 and θ_4. Note that the hip angle is $\phi_h = \theta_4 - \theta_b$, where θ_b is the seat recline angle. In addition, the knee angle is $\phi_h = \pi - \theta_3 + \theta_4 - \tan^{-1}(R_6/R_5)$.

The time derivative of Equation (2) generates the velocity equation,

$$\begin{bmatrix} 0 & -1 \\ 1 & 0 \end{bmatrix} (R_2\dot{\theta}_2\vec{\theta}_2 - R_3\dot{\theta}_3\vec{\theta}_3 - R_4\dot{\theta}_4\vec{\theta}_4) = 0. \tag{3}$$

As with the position analysis, the angular velocities of the lower leg $\vec{\theta}_3$ and upper leg $\vec{\theta}_4$ are determined from Equation (3).

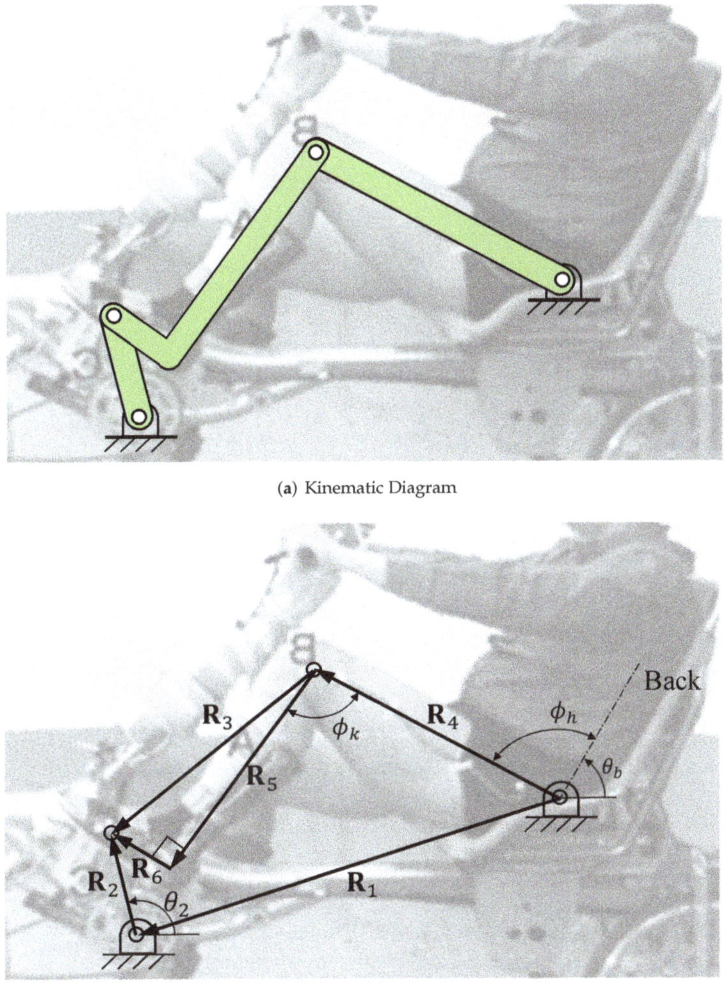

(**a**) Kinematic Diagram

(**b**) Vector Loop

Figure 2. Kinematic model for the TRT mechanism, adapted from Ref. [23], 2014 Elsevier Ltd.

Szecsi's experimental work [21] imposed a constant angular velocity (cadence) of $\dot{\theta}_2$ = 60 rpm on the cyclist pedaling. Using a gear reduction of λ = 1.31 between the crank sprocket and the driving wheel, the linear velocity of the tricycle was determined by using a wheel diameter of D_w = 66 cm, to be $\overline{v}_t = \dot{\theta}_2 D_w/(2\lambda)$ = 5.71 km/h, which is consistent with the average speed of cyclists at the Cybathlon in 2016 [14]. Given $\dot{\theta}_2$ and the TRT dimensions, a position, velocity, and acceleration analysis are performed. Determining the relationship of the knee angle ϕ_k and hip angle ϕ_h to the crank angle θ_2 is required to utilize the experimental data on the alternate six-bar linkage designs.

Free-body diagrams of the thigh, cnemis/foot, and crank are shown in Figure 3. A quasi-static model was deemed appropriate as the angular velocities and accelerations of $\mathbf{R}_3$ and $\mathbf{R}_4$ are low. As a consequence, inertial forces are ignored when generating the free-body diagrams. Further, the cycle-averaged power will be used as a comparison parameter between the TRT and the alternative six-bar designs. Over one cycle of the crank, the potential and kinetic energies related to the leg segment masses, would sum to zero and would have no effect on the cycle-averaged power. From Figure 3, the force and moment equilibrium equations become

$$\begin{aligned} F_{C_x} R_4 \sin\theta_4 - F_{C_y} \cos\theta_4 &= M_h + M, \\ F_{C_x} R_3 \sin\theta_3 - F_{C_y} R_3 \cos\theta_3 &= M_k, \\ F_{C_x} R_2 \sin\theta_2 - F_{C_y} R_2 \cos\theta_2 - M_A &= 0. \end{aligned} \tag{4}$$

The hip-joint moment M_h and knee-joint moment M_k as a function of crank angle θ_2 serves as input for the force analysis. Given the link kinematics $\theta_i, i = 1, \ldots, 4$, from Equations (2) and (4) become a square system of F_{C_x}, F_{C_y}, and M_A. As such, a statically equivalent torque at the crank center, M_A, is generated a function of θ_2.

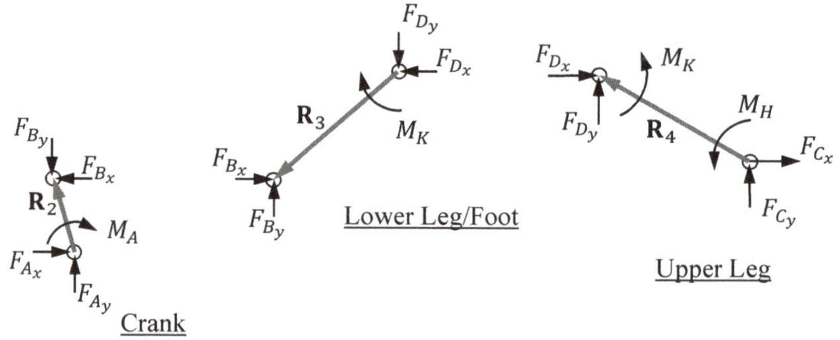

Figure 3. Free-body diagrams for the TRT force model.

The experimental data of Figure 1 represents the average of several FES cyclists, each with different leg lengths. To generate a model that represents an average rider, R_4, R_5, and R_6, were altered until a simulated cycle-averaged power was consistent with Szecsi's experiments. The average leg lengths for the P1P2 group are R_4 = 51.5 cm, R_5 = 51.5 cm, and R_6 = 6.9 cm. The average leg lengths that represent the P1P3 group are R_4 = 54.2 cm, R_5 = 54.2 cm, and R_6 = 6.9 cm.

The crank torque M_A is calculated with the P1P2 and P1P3 knee and hip moments from Figure 1 for $0\,^\circ \leq \theta_2 < 360\,^\circ$ using a constant angular velocity of the crank. The instantaneous power is calculated through Equation (1) and overlayed on Szecsi's data in Figure 4.

Each SCI FES cyclist in Szecsi's experiments produced an average of 30 Watts per cycle of the crank. Presuming that each leg equally contributes to the power generation, the single-leg TRT model is expected to produce a cycle-average of 15 W. With representative rider dimensions selected as R_4 = 52.0 cm, R_5 = 52.0 cm, and R_6 = 6.9 cm, the P1P2 group produced $\overline{P}_{P1P2}$ =14.7 W per leg whereas the P1P3 produced $\overline{P}_{P1P3}$ = 16.2 W.

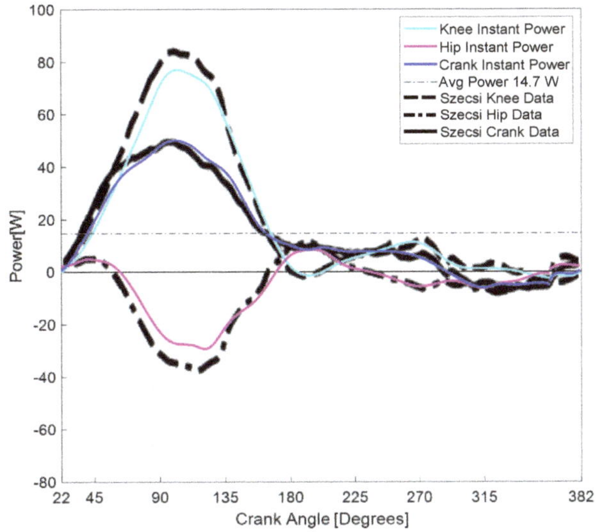

(**a**) P1P2 group

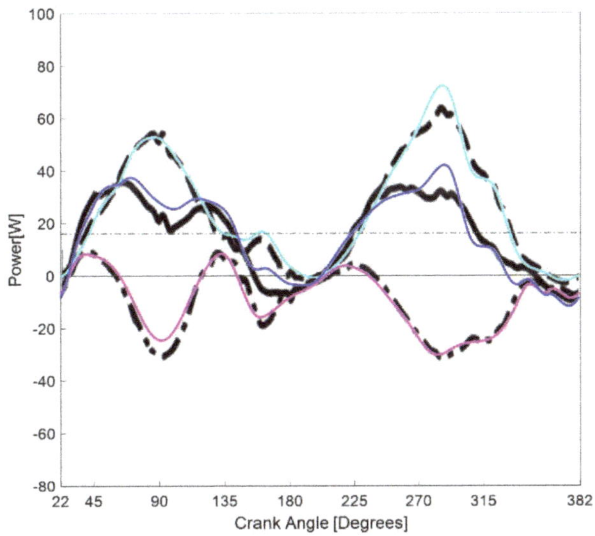

(**b**) P1P3 Group

Figure 4. The color curves represent the power calculated by the TRT model at the crank center attributed to the knee (cyan), hip (magenta) and total (blue). The calculated data overlays the black curves that represent the power reported by Szecsi.

4. Joint Moment Transformation

The measured joint torques (i.e., joint moments) shown in Figure 1 are provided by Szesci as a function of crank angle. Those torques are used as inputs to the TRT power model of Section 3. Alternative tricycle linkage designs will exhibit significantly different leg kinematics. In order to utilize Szesci's measurements when exploring alternative designs, the joint torque dependency must be transformed.

Yoon et al. [26] show that joint torque is a function of three inputs: joint angle, joint motion, and muscular contraction speed. The preferred FES cyclist torque data would be available as a function of these dependencies. Without that experimental data, the approach taken in this work combines the three inputs. Joint torque is mapped from crank angle to joint angle and joint motion (i.e., joint extension and flexion). Additionally, the motion of the hip and knee joints are prevented from exceeding those experienced in the TRT cyclist model. This constraint is present not because the leg is incapable of moving beyond these ranges, but because of the lack of data for modeling purposes.

The TRT model provides the joint angle as a function of crank angle, which is combined with Szesci's torque as a function of crank angle. In that way, a continuous curve of joint torque as a function of joint angle is produced and shown in Figure 5. The angular velocity of each joint is used to distinguish joint extension or flexion, which affects joint torque as the motion is derived from stimulating different muscle groups. A positive (or zero) joint angular velocity is identified as extension. Conversely, a negative angular joint velocity is identified as flexion. Extension and flexion are denoted in Figure 5 as alternate line styles.

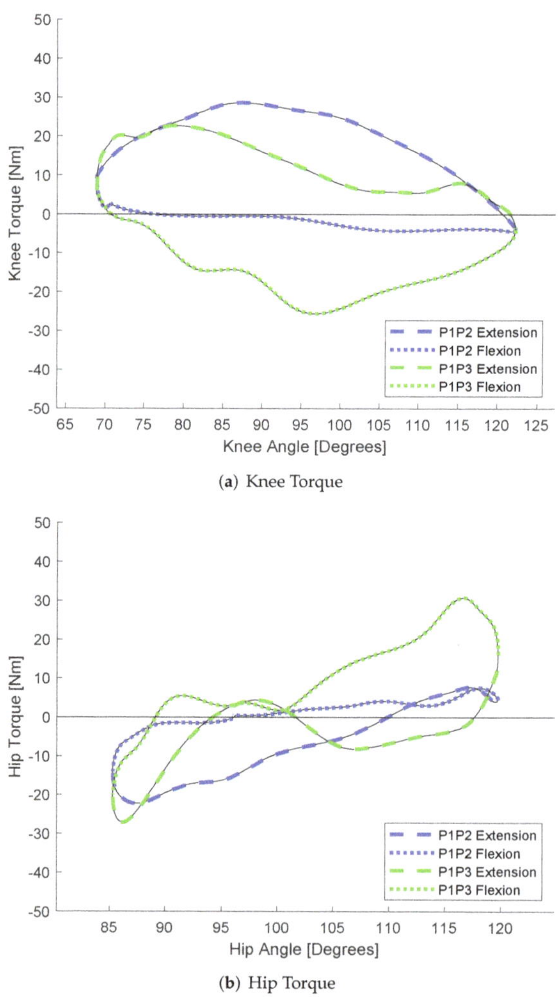

(a) Knee Torque

(b) Hip Torque

Figure 5. Joint moments as a function of relative joint angles.

5. Alternate Tricycle Models

Two alternative six-bar linkage designs for the recumbent tricycle drive train are described below.

5.1. The CDT Model

When including the thigh and cnemis as links in the kinematic diagram, the Coupler-Driver Tricycle (CDT) forms a Stephenson III six-bar mechanism. The vector diagram for modeling the rider and CDT interaction is shown in Figure 6b. As with the TRT in Figure 2b, the upper leg $\mathbf{R}_4$ serves as the oscillating input link. The cnemis $\mathbf{R}_5$ and foot $\mathbf{R}_6$ are constrained by a boot to define link $\mathbf{R}_3$. Note that vectors $\mathbf{R}_8$, $\mathbf{R}_9$, and $\mathbf{R}_{11}$ define a rigid triangle that acts as a coupler for the four-bar defined by $\mathbf{R}_2$, $\mathbf{R}_7$, $\mathbf{R}_9$, and $\mathbf{R}_{10}$. The rider's foot, affixed to a point on the triangular coupler, follows a cyclic motion corresponding to oscillations of the upper and lower leg. In addition, note that θ_2 is not the actual rotation of the drive wheel as there can be a gear reduction λ inserted between the crank and the drive wheel due to the chain and sprockets.

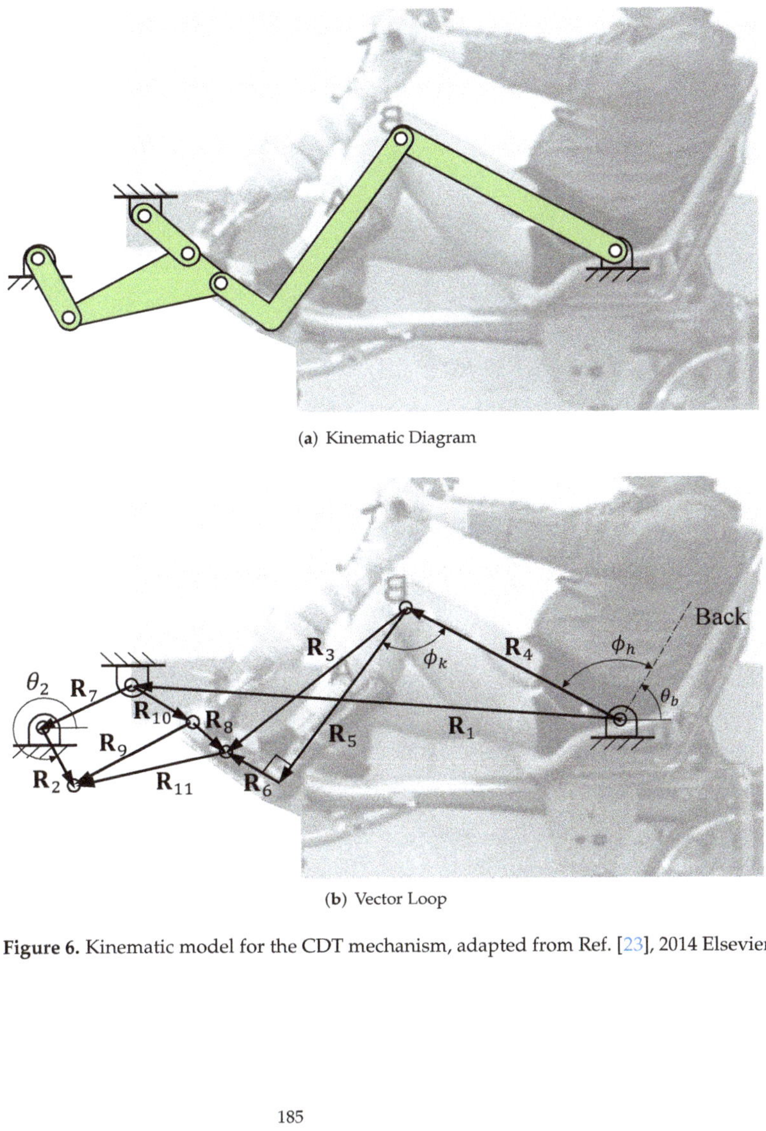

Figure 6. Kinematic model for the CDT mechanism, adapted from Ref. [23], 2014 Elsevier Ltd.

The vector loops for the CDT are

$$R_1\vec{\theta}_1 + R_{10}\vec{\theta}_{10} + R_8\vec{\theta}_8 - R_3\vec{\theta}_3 - R_4\vec{\theta}_4 = 0, \tag{5}$$

$$R_2\vec{\theta}_2 - R_9\vec{\theta}_9 - R_{10}\vec{\theta}_{10} + R_7\vec{\theta}_7 = 0. \tag{6}$$

where the nomenclature R_i and $\vec{\theta}_i$ is adopted from Equation (2). In addition, $\theta_8 = \theta_9 + \gamma$, where $\gamma = cos^{-1}[(R_8^2 + R_9^2 - R_{11}^2)/(2R_8R_9)]$. Taking time derivatives of the loop closure gives the CDT linkage velocity equations,

$$\begin{bmatrix} 0 & -1 \\ 1 & 0 \end{bmatrix}(R_{10}\dot{\theta}_{10}\vec{\theta}_{10} + R_8\dot{\theta}_8\vec{\theta}_8 - R_3\dot{\theta}_3\vec{\theta}_3 - R_4\dot{\theta}_4\vec{\theta}_4) = 0, \tag{7}$$

$$\begin{bmatrix} 0 & -1 \\ 1 & 0 \end{bmatrix}(R_2\dot{\theta}_2\vec{\theta}_2 - R_9\dot{\theta}_9\vec{\theta}_9 - R_{10}\dot{\theta}_{10}\vec{\theta}_{10}) = 0. \tag{8}$$

Similar to the TRT model, a quasi-static model was used to calculate the torque production at the crank. The corresponding free-body diagrams are shown in Figure 7. The dynamic effects of the leg and the linkage masses are neglected by acknowledging the low angular velocities and accelerations experienced through one cycle of the crank. Over the course of one cycle, all potential and kinetic energies sum to zero. As in the TRT, the mass of each link was ignored as the concern of this study is the average power produced at the crank. The equilibrium equations from Figure 7 are

$$\begin{aligned} F_{C_x}R_4\sin\theta_4 - F_{C_y}\cos\theta_4 &= M_h + M_k, \\ F_{C_x}R_3\sin\theta_3 - F_{C_y}R_3\cos\theta_3 &= M_k, \\ -F_{C_x} + F_{F_x} + F_{E_x} &= 0, \\ -F_{C_y} + F_{F_y} + F_{E_y} &= 0, \\ -F_{E_x}R_8\sin\theta_8 + F_{E_y}R_8\cos\theta_8 - R_{11}\sin\theta_{11}F_{E_x} + R_{11}\cos\theta_{11}F_{E_y} &= 0, \\ F_{F_x}R_9\sin\theta_9 - F_{F_y}R_9\cos\theta_9 &= 0, \\ F_{F_x}R_2\cos\theta_2 - F_{F_y}R_2\sin\theta_2 - M_A &= 0. \end{aligned} \tag{9}$$

Thus, given the complete kinematics and values of $\theta_i, i = 1, \ldots, 11$, the quasi-static analysis of Equations (9) generates the statically equivalent torque at the crank M_A given the torque produced by the cyclist at the knee and hip joints, M_k and M_h, respectively.

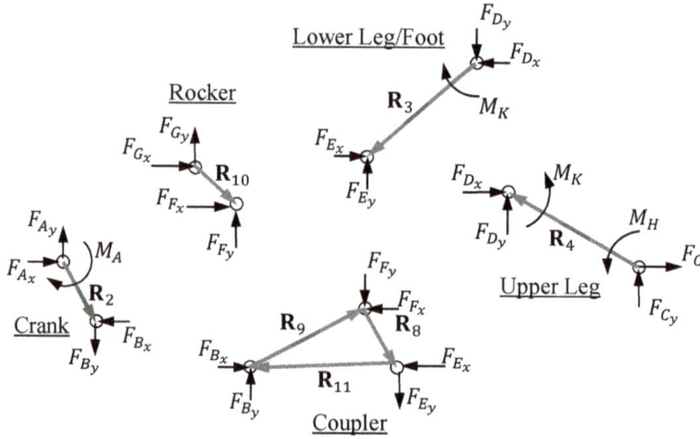

Figure 7. Free-body diagrams for the CDT mechanism.

5.2. The CRT Model

When including the thigh and cnemis as links in the kinematic diagram, the Crank-Rocker Tricycle (CRT) with the rider's legs forms a Watt II six-bar mechanism. The vector diagram for modeling the rider and CRT interaction is shown in Figure 8b. Note that vectors $\mathbf{R}_8$ and $\mathbf{R}_{10}$ are collinear and part of the same rigid body, acting together as the rocking input link for the four-bar defined by $\mathbf{R}_2$, $\mathbf{R}_7$, $\mathbf{R}_9$, and $\mathbf{R}_{10}$. The rider's foot oscillates, affixed to a point on the rocker, with corresponding oscillations of the thigh and cnemis. This system also has the gear reduction λ.

The collinearity of vectors $\mathbf{R}_8$ and $\mathbf{R}_{10}$ may appear to eliminate a potential design variable in the angle between them. As $\mathbf{R}_7$ may change in both the x and y component, the fixed pivot at the crank center could potentially be located well below or above the fixed pivot of the oscillating input link. The angle between $\mathbf{R}_8$ and $\mathbf{R}_{10}$ may then be introduced to rotate the entire $\mathbf{R}_2\mathbf{R}_7\mathbf{R}_9\mathbf{R}_{10}$ linkage and move $\mathbf{R}_7$ to an improved location with respect to the actual design of the tricycle. Rotating the entire linkage in this fashion does not change the forces and does not effect the power over a full rotation of the crank.

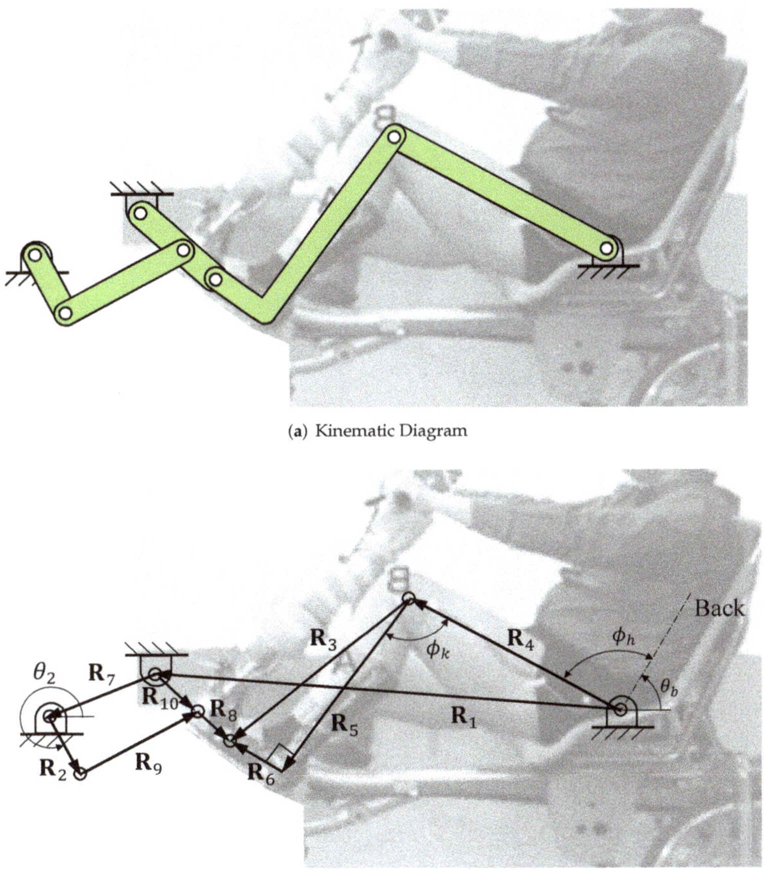

(a) Kinematic Diagram

(b) Vector Loop

Figure 8. Kinematic model for the CRT mechanism adapted from Ref. [23], 2014 Elsevier Ltd.

The vector loops for the CRT are

$$R_1\vec{\theta}_1 + R_{10}\vec{\theta}_8 + R_8\vec{\theta}_8 - R_3\vec{\theta}_3 - R_4\vec{\theta}_4 = 0, \tag{10}$$

$$R_2\vec{\theta}_2 + R_9\vec{\theta}_9 - R_{10}\vec{\theta}_8 + R_7\vec{\theta}_7 = 0. \tag{11}$$

where the nomenclature R_i and $\vec{\theta}_i$ is adopted from Equation (2). Taking time derivatives of the loop closure gives the CDT linkage velocity equations,

$$\begin{bmatrix} 0 & -1 \\ 1 & 0 \end{bmatrix} (R_{10}\dot{\theta}_8\vec{\theta}_8 - R_8\dot{\theta}_8\vec{\theta}_8 - R_3\dot{\theta}_3\vec{\theta}_3 - R_4\dot{\theta}_4\vec{\theta}_4) = 0, \tag{12}$$

$$\begin{bmatrix} 0 & -1 \\ 1 & 0 \end{bmatrix} (R_2\dot{\theta}_2\vec{\theta}_2 + R_9\dot{\theta}_9\vec{\theta}_9 - R_{10}\dot{\theta}_8\vec{\theta}_8) = 0. \tag{13}$$

The CRT free-body diagrams are shown in Figure 9, noting that those for bodies 2, 3 and 4 remain unchanged between the CDT and CRT. Furthermore, all assumptions about masses stay the same and the energies still sum to zero.

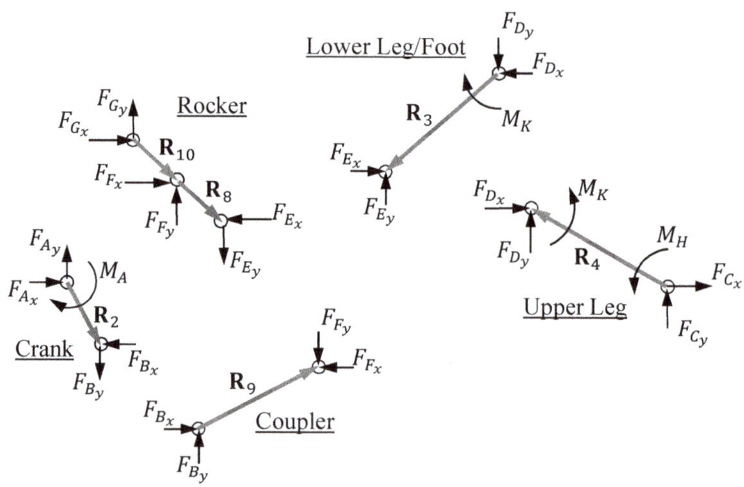

Figure 9. Free-body diagrams for the CRT mechanism.

The equilibrium equations for the CRT are

$$\begin{aligned} F_{C_x} R_4 \sin\theta_4 - F_{C_y} \cos\theta_4 &= M_h + M_k, \\ F_{C_x} R_3 \sin\theta_3 - F_{C_y} R_3 \cos\theta_3 &= M_k, \\ -F_{C_x} + F_{F_x} + F_{E_x} &= 0, \\ -F_{C_y} + F_{F_y} + F_{E_y} &= 0, \\ -F_{F_x}(R_8 \sin\theta_8 + R_{10}\sin\theta_{10}) + F_{F_y}(R_8\cos\theta_8 + R_{10}\cos\theta_{10}) & \\ -F_{E_x} R_8 \sin\theta_8 + F_{E_y} R_8 \cos\theta_8 &= 0, \\ F_{F_x} R_9 \sin\theta_9 - F_{F_y} R_9 \cos\theta_9 &= 0, \\ F_{F_x} R_2 \cos\theta_2 - F_{F_y} R_2 \sin\theta_2 - M_A &= 0. \end{aligned} \tag{14}$$

Given the complete kinematics and values of $\theta_i, i = 1, \ldots, 10$, Equations (14) represent a square system for F_{C_x}, F_{C_y}, F_{E_x}, F_{E_y}, F_{F_x}, F_{F_y}, and M_A. As with the other devices, the quasi-static analysis generates the statically equivalent torque at the crank M_A given the torque produced by the cyclist at the knee and hip joints, M_k and M_h, respectively.

6. Optimization

Both six-bar models have ten design parameters, and not the 14 stated in the Introduction because the parameters associated with the human leg cannot be varied. The ten parameters are collected as $\mathbf{x}$=[$\mathbf{R}_1$, R_2, R_6, $\mathbf{R}_7$,R_8, R_9, R_{10}, λ]. An objective function is posed to maximize the average power throughout the cycle $\overline{P}$. The optimization is formally posed as

$$\begin{aligned} \underset{\mathbf{x}}{\text{maximize}} \quad & \overline{P} \\ \text{subject to} \quad & \Im(\theta_j) = 0, \quad j = 1, \ldots, n_j \\ & v_t = 5.71\text{km/h}, \\ & R_2 + R_9 - R_7 - R_{10} \leq 0, \\ & L_2 \geq 2.5\text{cm}, \\ & L_{10} \geq 2.5\text{cm}, \\ & L_7 \leq 35\text{cm}, \\ & 69^\circ \leq \phi_k \leq 122^\circ, \\ & 85^\circ \leq \phi_h \leq 120^\circ. \end{aligned} \tag{15}$$

where n_j is the number of angles within the kinematic model and $\Im$ refers to the imaginary part of θ_j.

In order to determine $\overline{P}$ from Equation (1), the kinematic vector loop closure equations are solved throughout a cycle of $0 < \theta_2 < 360^\circ$. The kinematic loop closure for the CDT are given in Equations (5) and (6) and for the CRT in Equations (10) and (11). Subsequently, the force and moment equilibrium equations, given in Equations (9) for the CDT and Equations (14) for the CRT, are used to determine the reaction moment at the crank M_A that is used in Equation (1). The motivation for the optimization constraints posed in Equation (15) are provided below.

The loop closure equations must produce a linkage that can be assembled for every position of the crank. That is, $\Im(\theta_j) = 0, \quad j = 1, \ldots, n_j$. Additionally, the crank rotation must produce a linear velocity consistent with the TRT (v_t = 5.71 km/h),

$$\dot{\theta}_2 = \frac{2\lambda v_t}{D_w}, \tag{16}$$

Grashof's criteria is enforced to ensure full crank rotatability,

$$R_2 + R_9 - R_7 - R_{10} \leq 0, \tag{17}$$

where R_2 is required to be the shortest of the four links.

To align with the results generated during the TRT analysis, the hip joint was restricted to the range $69^\circ \leq \phi_k \leq 122^\circ$. Likewise, the knee joint was restricted to the range $85^\circ \leq \phi_h \leq 120^\circ$. As expected, test optimizations produced unusable mechanisms due to positioning of components and undesirable link lengths. As such, a final set of constraints on dimensions was introduced for the sake of practicality and manufacturability: the crank $R_2 \geq 2.5$ cm, the rocker $R_{10} \geq 2.5$ cm, the angle-to-pedal length $R_6 \leq 35.5$ cm, and crank center and rocker center locations constrained to lie within the envelope shown in Figure 10.

This nonlinear problem includes a significant number of inequality constraints. As such, a standard gradient-based optimization proved problematic. A Random Pursuit Optimization was used instead to identify optimal design values $\mathbf{x}^*$. The optimization initiates with a guess, $\mathbf{x}_0$, for the ten design parameters that satisfies all constraints. The corresponding average power over one full cycle, $\overline{P}_0$, is calculated for $\mathbf{x}^* = \mathbf{x}_0$. A new guess, $\mathbf{x}_1$, is generated from $\mathbf{x}_0$ through the introduction of small, random changes. That is, for the potential first optimization step, $i = 1$,

$$\mathbf{x}_i = \mathbf{x}^* + [\delta_1 \ldots \delta_j \ldots \delta_{10}]^{\mathsf{T}}, \tag{18}$$

where $-0.01 \leq \delta_j \leq 0.01$. The values of $\mathbf{x}_1$ are evaluated against the entire set of constraints. If satisfied, $\mathbf{x}_1$ is passed to the objective function and $\overline{P}_1$ is determined. If $\overline{P}_1 > \overline{P}_0$, the new optimal is $\mathbf{x}^* = \mathbf{x}_1$. A new guess, $\mathbf{x}_2$, is generated by making small, random changes to $\mathbf{x}_1$ according to Equation (18). If $\overline{P}_1 \leq \overline{P}_0$, the optimal design values remain unchanged and $\mathbf{x}^* = \mathbf{x}_0$. A new guess for $\mathbf{x}_1$ is generated by making a different set of random changes to $\mathbf{x}_0$.

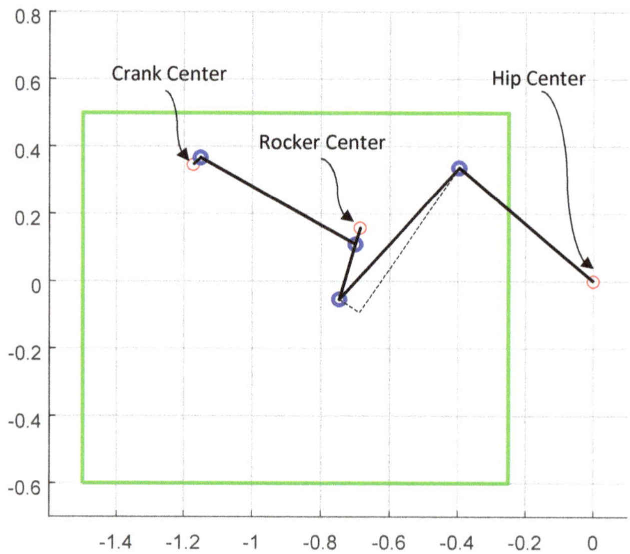

Figure 10. The design space shown in green bounds the region of usable crank center and rocker arm pivot locations for both the CRT and CDT designs, in units of meters.

Convergence was determined by limiting the number of iterations since the last improvement. Due to the randomness associated with altering the ten parameters, and to be confident that a design approaching the optimum has been obtained, the convergence criteria was established to be 10,000 consecutive iterations of parameters that satisfy the constraints without improving $\overline{P}$. Code to perform the aforementioned procedure was generated by the authors using MATLAB.

7. Results and Discussion

An optimization as described in the prior section was performed for both six-bar designs using the torque input for the P1P2 group and the P1P3 group. The resulting power and extreme joint angles encountered during motion for the TRT, CRT and CDT architectures are presented in Table 1. The $\mathbf{x}^*$ solution for the CRT design that generated the highest power output for the P1P2 group is designated R1, while $\mathbf{x}^*$ for the CRT design with the P1P3 group is designated R2. Similarly, D1 is $\mathbf{x}^*$ corresponding with the CDT design for the P1P2 group and D2 is $\mathbf{x}^*$ corresponding with the CDT design for the P1P3 group. Once the design was achieved, the cycle-averaged power output for both groups $\overline{P}_{P1P2}$ and $\overline{P}_{P1P3}$ was determined and are given in Table 1. The average power for both groups $\overline{P}_{av}$ is also shown.

Table 1. Designs R1 and R2 were selected from the CRT optimization results and designs D1 and D2 were selected from the CDT optimization results for further analysis to identify the best design for spinal cord injured FES cyclists. The powers listed are in watts (W). For reference, the TRT is shown in Figure 2, the CDT in Figure 6 and the CRT in Figure 8.

Design		$\overline{P}_{P1P2}$ (W)	$\overline{P}_{P1P3}$ (W)	$\overline{P}_{av}$ (W)	ϕ_k	ϕ_h
TRT		14.70	16.20	15.45	69–122°	85–120°
CRT	R1	23.98	22.79	23.39	76–110°	111–118°
	R2	21.36	29.00	25.18	73–112°	94–102°
CDT	D1	31.29	29.34	30.32	70–121°	106–118°
	D2	26.19	35.83	31.01	69–121°	92–109°

Note from Table 1 that all six-bar designs provide substantial increases in power output when compared to the TRT. Further, the R1 and D1 designs were optimized for the P1P2 group, yet the P1P3 group would generate comparable power with those tricycle designs. Conversely, the R2 and D2 designs were optimized for the P1P3 group, but the P1P2 group generates notably less power.

Designs R1, R2, D1, D2 are further evaluated to identify the best six-bar tricycle design based upon the following set of metrics: manufacturability, joint torque curve continuity, equal power distribution over the entire crank cycle, and percent increase in average power for each group.

7.1. Manufacturability

The dimensions of the TRT, R1, R2, D1 and D2 designs are provided in Table 2. The length R_6 represents the distance from the ankle to the center of the pedal. Since the cyclist's foot is placed in a boot, zero and negative values are permissible.

Table 2. Designs R1, R2, D1, D2 Optimized Design Parameter Results for further analysis to identify the best design for spinal cord injured FES cyclists. Link dimensions (and vectors) are in cm and λ is a dimensionless ratio.

Design		$\mathbf{R}_1$	R_2	R_6	$\mathbf{R}_7$	R_8	R_9	R_{10}	λ
TRT		$[-74, 30]^T$	15	7	-	-	-	-	1.31
CRT	R1	$[-70, 11]^T$	3	7	$[-50, 14]^T$	17	51	5	2.03
	R2	$[-62, 32]^T$	4	8	$[-49, 1]^T$	15	44	9	1.81
CDT	D1	$[-62, 25]^T$	16	10	$[-38, 9]^T$	14	34	22	2.10
	D2	$[-64, 28]^T$	15	5	$[-29, 13]^T$	7	32	19	2.04

Manufacturability considerations include several design criteria such as the distance from the hip center to crank center, length of the the rocker arm, height of the rocker arm pivot relative to rider's line of sight, estimated weight of the design, etc. All four design six-bar designs are deemed suitable and will yield an acceptable mechanical design.

7.2. Joint Torque Curve Continuity

The joint torque curves of Figure 5 are truncated when the kinematics of a six-bar linkage design requires the cyclist's knee or hip to move less than the TRT during a full pedaling motion. The discontinuity at the ends of the plots on Figure 11 are not biomechanically correct. Yet, these truncated curves are used in the CRT and CDT power modeling because the experimental data needed to generate continuous curves does not exist.

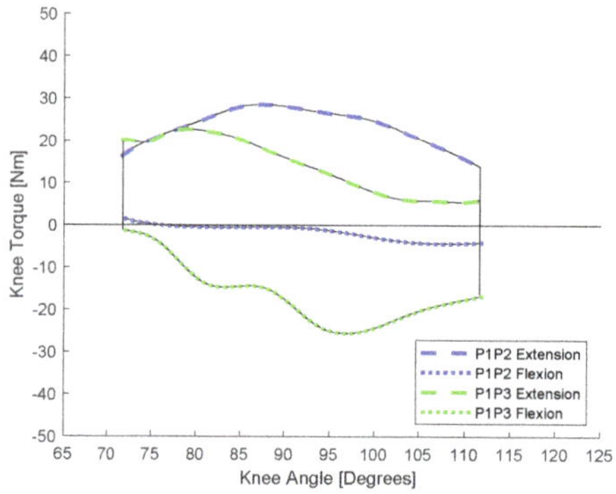

(a) Knee Torque for CRT Design R2

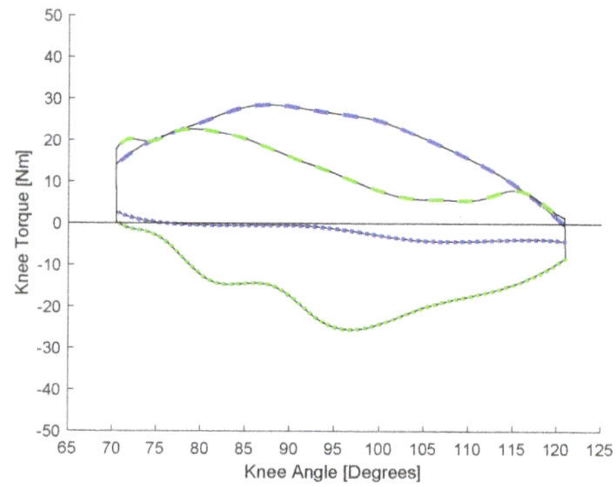

(b) Knee Torque for CDT Design D1

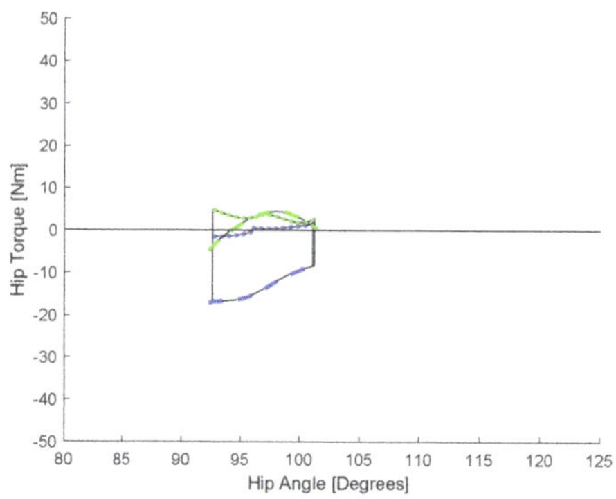

(c) Hip Torque for CRT Design R2

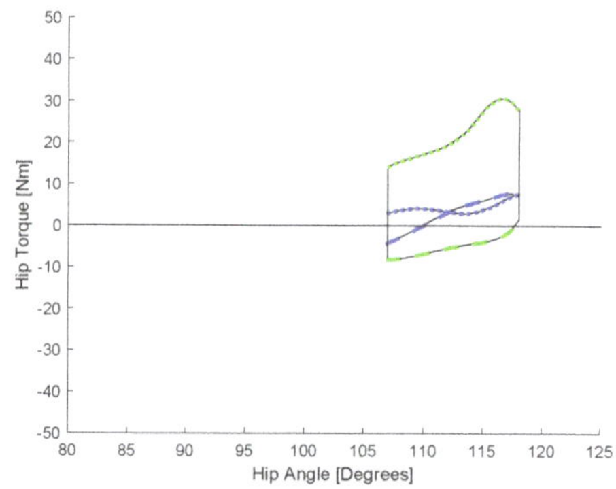

(d) Hip Torque for CDT Design D1

Figure 11. Joint torque curves for CRT design R2 and CDT design D1. Note that the axes limits are kept the same as Figure 5 for comparative purposes.

For example, notice that design R2 requires knee movement from 73° to 112°, whereas the TRT requires 69° to 122°. At the lower extreme of 73°, the knee joint torque is assumed to shift instantaneously from the extension curve with 20 Nm of torque to flexion curves with −1 Nm. A continuous joint torque curve would likely reduce the power generated as the sharp corners in Figure 11 would become rounded and decrease the torque produced at the joint limit extremes. As such, the power from the six-bar linkage models will be more accurate when the design requires joint movements that closely align with the TRT joint movements. The CDT designs D1 and D2 have more similar joint angle extremes when compared to the TRT. Accordingly, the CDT modeled power is more representative of the biomechanical output.

7.3. Power Distribution

An ideal tricycle design enables a cyclist to generate a constant level of power throughout the crank rotation, thereby creating smooth forward motion. However, Szesci [21] observed that FES cyclists generate uneven levels of power, exhibiting propulsive and recovery phases. Understanding that power fluctuations will occur, a favorable design will limit the instantaneous power variations and prevent negative values. When the power output becomes negative, power is extracted from the tricycle's kinetic energy and used to move the cyclist's legs. This transfer of power from tricycle to the cyclist's legs is referred to as an inactive zone, where the leg is not in a configuration that can be stimulated to produce positive work. The inactive zones create a choppy pedaling motion and may lead to the cyclist becoming locked at the TDC or BDC of the TRT crank. Video from the Cybathlon demonstrates this phenomena [15].

The instantaneous power curves for the R2 and D1 designs are given in Figure 12 for both the P1P2 and P1P3 groups. The curves in Figure 12a,c exhibit a high power peak in one phase of the crank because the P1P2 group riders generate almost all of their joint torques in this region. This single peak is consistent with the TRT instantaneous power with the P1P2 group shown in Figure 4a. Note that in the TRT inactive zones, the remainder of the power decreases in the second half of the cycle while becoming negative. In the CRT and CDT designs of Figure 12a,c, the power remains above or at zero throughout the remainder of the cycle.

The P1P3 group exhibits two power peaks as discussed in Section 3 and shown in Figure 4b. Accordingly, the R2 and D2 designs with the P1P3 group yield the highest power output. Additionally, the power curves in Figure 12b,d are not dominated by a single peak and are expected to produce smoother tricycle motion.

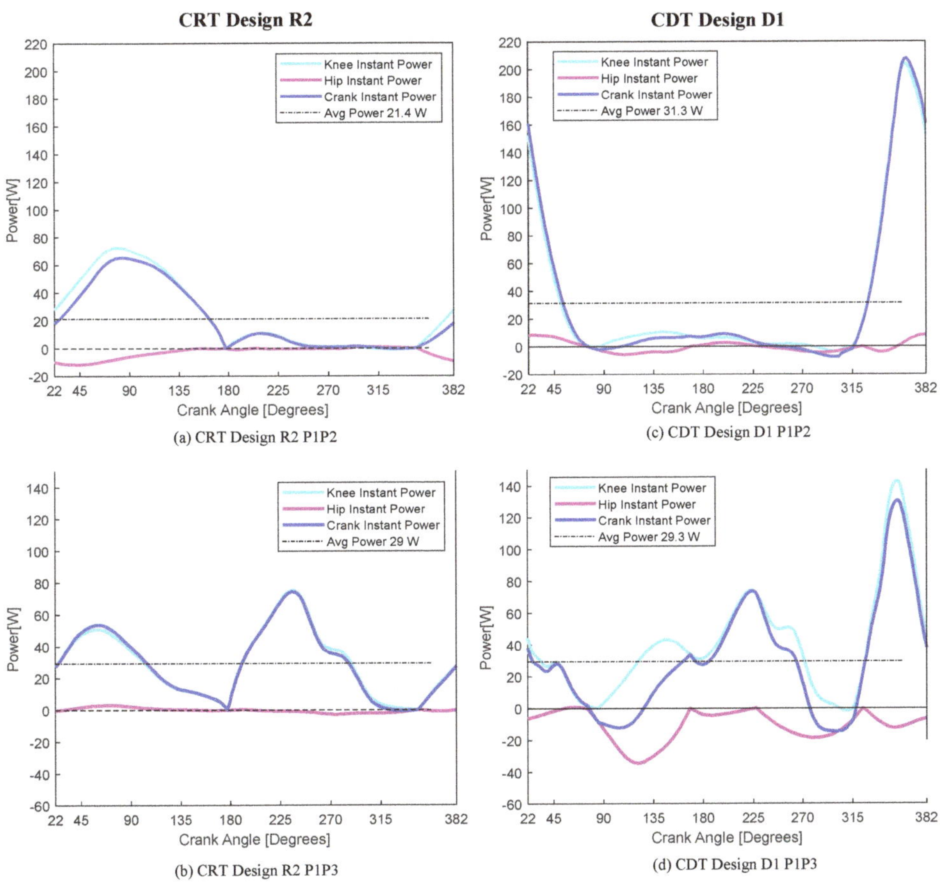

Figure 12. CRT instantaneous power curve at crank center for design R2 and CDT instantaneous power curve at crank center for design D1. Plot (**a**) shows the CRT design R2 instantaneous power curve for the P1P2 group. Plot (**b**) shows the CRT design R2 instantaneous power curve for the P1P3 group. Plot (**c**) shows the CDT design D1 instantaneous power curve for the P1P2 group. Plot (**d**) shows the CDT design D1 instantaneous power curve for the P1P3 group.

7.4. Design Recommendation

The tricycle design selected for further consideration should allow any FES rider of an unknown power group to effectively produce a significant power improvement above the TRT. The tricycle design should have high average power for both groups ($\overline{P}_{P1P2}$ and $\overline{P}_{P1P3}$), satisfy the manufacturability criteria, and biomechanically satisfy the additional constraints required for FES cycling. Design D1 produces comparable power in each group with a group average power of 30.32 W, which is nearly double the TRT. Design D1 achieves the increased power by altering the motion of the leg and maintaining the joint angle in regions of high toque to ensure that positive power is produced for the duration of the crank cycle. The optimization recognized that a significant portion of the hip range did not produce positive power at the crank, therefore the kinematics of the design D1 limits the hip from entering this region. Further, design D1 limits any antagonistic joint moments created at the hip through limiting its joint range. Lastly, the D1 design also nearly matches the knee joint angle ranges of the TRT. Therefore, the recommendation is that the biomechanical interaction of the cyclist and the Coupler Driver Tricycle, design D1 be investigated to confirm that an an FES cyclist is capable of producing the anticipated joint torque curves. A scaled kinematic sketch of the recommended Design D1 is shown in Figure 13.

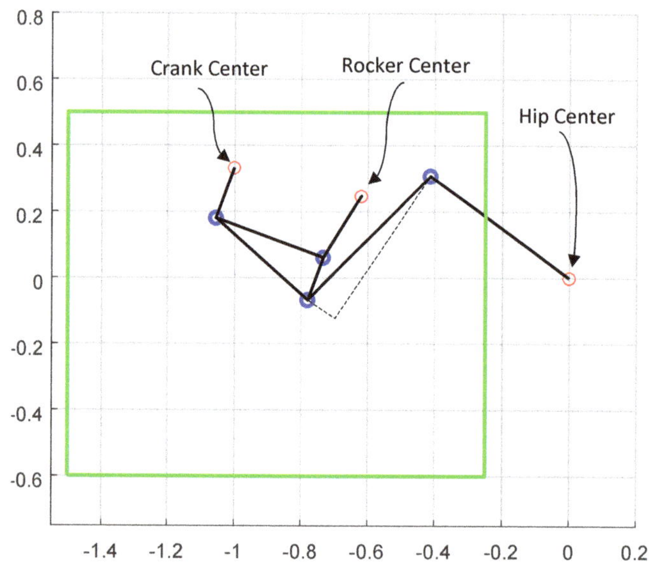

Figure 13. A scaled kinematic sketch of CDT Design D1, in meters.

8. Conclusions

This paper presents two alternative tricycle designs and modeling techniques for spinal cord injury functional electric stimulated cycling. Traditional cycling for AB subjects is vastly different from FES cycling. Average power production for FES cyclist is an order of magnitude less than for an AB cyclist. Additionally, the four power peaks observed in AB cycling are not seen in FES cycling. Therefore, alternative drive trains may have better force/power transmission to the driven wheel. The CRT tricycle design employs a four-bar architecture that allows the pedal, and therefore leg, to travel in a back-and-forth motion instead of the traditional circular motion of the pedal. The CDT tricycle design drives the leg in the same motion as the CRT while holding the angular constraint in a different position than the CRT. the optimization of these architectures lead to designs that improved the throughput power of P1P2 cyclists by as much as 102% and P1P3 cyclists by as much as 138% on the CDT architecture. These designs take advantage of joint angular ranges where cyclists can produce large amounts of torque. Furthermore, the back-and-forth motion more evenly distributes power throughout the cycle of the crank and thus reduces inactive zones. Increased power throughput to the driving wheel and a smoother pedaling cycle are believed to improve SCI FES cycling. This paper shows alternative tricycle architectures can improve mechanical power throughput of FES cycling. Further research is recommended to determine the biomechanical properties of the rider to supplement the mechanical models presented in this paper.

Author Contributions: Conceptualization, N.A.L., D.H.M., A.L.B. and A.P.M.; methodology, D.H.M. and A.P.M.; software, N.A.L. and A.L.B.; validation, N.A.L. and A.L.B.; formal analysis, N.A.L., D.H.M., A.L.B. and A.P.M.; investigation, N.A.L. and A.L.B.; data curation, N.A.L. and A.L.B.; writing—original draft preparation, D.H.M., A.L.B. and A.P.M.; writing—review and editing, N.A.L., D.H.M. and A.P.M.; visualization, N.A.L., D.H.M. and A.L.B.; supervision, D.H.M. and A.P.M.; project administration, D.H.M. and A.P.M. All authors have read and agreed to the published version of the manuscript.

Funding: This research received no external funding.

Data Availability Statement: Data is available upon request.

Acknowledgments: The authors are grateful for motivation and valuable guidance from Christine Azevedo-Coste.

Conflicts of Interest: The authors declare no conflict of interest.

References

1. Hull, M.; Jorge, M. A Method for Biomechanical Analysis of Bicycle Pedaling. *J. Biomech.* **1985**, *18*, 631–644. [CrossRef]
2. Chen, F.-C. Analysis and Verification of a Watt I Six-Bar Furniture Hinge Mechanism. *Proc. Inst. Mech. Eng. Part C J. Mech. Eng. Sci.* **2005**, *219*, 1107–1117. [CrossRef]
3. Soong, R.-C. An Adjustable Six-Bar Mechanism with Variable Input Speed for Mechanical Forming Presses. *Trans. Can. Soc. Mech. Eng.* **2008**, *32*, 453–466. [CrossRef]
4. Batayneh, W.; Al-araidah, O.; Malkawi, S. Biomimetic Design of a Single DOF Stephenson III Leg Mechanism. *Mech. Eng. Res.* **2013**, *3*, 43. [CrossRef]
5. Plecnik, M.; McCarthy, J.M.; Wampler, C.W. Kinematic Synthesis of a Watt I Six-Bar Linkage for Body Guidance. In *Advances in Robot Kinematics*; Springer: Cham, Switzerland, 2014; pp. 317–325.
6. Gezgin, E.; Chang, P.; Faruk, A. Synthesis of a Watt II six-bar linkage in the design of a hand rehabilitation robot. *Mech. Mach. Theory* **2016**, *104*, 177–189. [CrossRef]
7. Azam, F.I.; Rani, A.M.A.; Altaf, K.; Zaharin, H.A. Experimental and Numerical Investigation of a Six-Bar Linkage Application to Bellow Glovbe Valve for Compact Design. *J. Appl. Sci.* **2018**, *8*, 1980. [CrossRef]
8. Bazler, A.; Myszka, D.; Murray, A. The Redesign of a Recumbent Trucycle using a Crank Rocker Mechanism to Increase Power Throughput in FES Cycling. In Proceedings of the 2021 ASME International Design Engineering Technical Conference, Virtually, 17–19 August 2021.
9. Nussbaum, E.; Houghton, P.; Anthony, J.; Rennie, S.; Shay, B.; Hoens, A. Neuromuscular Electrical Stimulation for Treatment of Muscle Impairment: Critical Review and Recommendations for Clinical Practice. *Physiother. Can.* **2017**, *69*, 1–76. [CrossRef]
10. Agarwal, S.; Triolo, R.; Kobetic, R.; Miller, M.; Bieri, C.; Kukke, S.; Davis, J. Long-term user perceptions of an implanted neuroprosthesis for exercise, standing, and transfers after spinal cord injury. *J. Rehabil. Res. Dev.* **2003**, *40*, 241–252. [PubMed]
11. Chen, S.C.; Lai, C.H.; Chan, W.P.; Huang, M.H.; Tsai, H.W.; Chen, J.J.J. Increases in bone mineral density after functional electrical stimulation cycling exercises in spinal cord injured patients. *Disabil. Rehabil.* **2005**, *27*, 1337–1341. [CrossRef] [PubMed]
12. Berkelmans, R. FES Cycling. *J. Autom. Control* **2008**, *18*, 73–76. [CrossRef]
13. Salter, A.; Bagg, S.; Creasy, J.; Romano, C.; Romano, D.; Richmond, F.; Loeb, G. First clinical experience with BION implants for therapeutic electrical stimulation. *Neuromodulation Technol. Neural Interface* **2004**, *7*, 38–47. [CrossRef] [PubMed]
14. Cybthalon Official Website, Functional Electrical Stimulation Bike Race. Available online: https://cybathlon.ethz.ch/en/event/disciplines/fes (accessed on 3 February 2022).
15. Bergeron, V. FES Cycling Race Cybathlon 2016, 2018. Available online: www.youtube.com/watch?v=RWiN5ihf3og (accessed on 3 February 2022).
16. Sijobert, B.; Fattal, C.; Daubigney, A.; Azevedo-Coste, C. Participation to the first Cybathlon: An overview of the FREEWHEELS team FES-cycling solution. *Eur. J. Transl. Myol.* **2017**, *27*, 7120. [CrossRef] [PubMed]
17. Yamazaki, H.; Matsuda, A. Joint Torque Evaluation of Lower Limbs in Bicycle Pedaling. In Proceedings of the 11th Conference of the International Sports Engineering Association, Delft, The Netherlands, 11–14 July 2016.
18. Hull, M.; Jorge, M. Crank Inertial load has Little Effect on Steady-State pedaling Coordination. *J. Biomech.* **1996**, *29*, 1559–1567.

19. Gfohler, M.; Angeli, T.; Eberharter, T.; Lugner, P.; Mayr, W.; Hofer, C. Test Bed with Force-Measuring Crank for Static and Dynamic Investigations on Cycling by Means of Functional Electrical Stimulation. *IEEE Trans. Neural Syst. Rehabil. Eng.* **2001**, *9*, 169–180. [CrossRef] [PubMed]
20. Gfohler, M.; Lugner, P. Dynamic Simulation of FES-cycling: Influence of Individual Parameters. *IEEE Trans. Neural Syst. Rehabil. Eng.* **2004**, *12*, 398–405. [CrossRef] [PubMed]
21. Szecsi, J.; Straube, A.; Fornusek, C. A Biomechanical Cause of Low Power Production during FES Cycling of Subjects with SCI. *J. Neuroeng. Rehabil.* **2014**, *11*, 123. [CrossRef] [PubMed]
22. Szecsi, J.; Krause, P.; Krafczyk, S.; Brandt, T.; Straube, A. Functional output improvement in FES cycling by means of forced smooth pedaling. *Med. Sci. Sports Exerc.* **2007**, *39*, 764–780. [CrossRef]
23. Szecsi, J.; Straube, A.; Fornusek, C. Leg General Muscle Moment and Power Patterns in Able-Bodied Subjects during Recumbent Cycle Ergometry with Ankle Immobilization. *Med. Eng. Phys.* **2014**, *36*, 1421–1427. [CrossRef] [PubMed]
24. Ibrahim, K.; Gharooni, S.; Tokhi, M.; Massoud, R. Energy-Efficient FES Cycling with Quadriceps Stimulation. In Proceedings of the 13th Annal Conference of the International Function Electrical Stimulation Society, Freiburg, Germany, 21–15 September 2016.
25. Shan, G. Biomechanical Evaluation of Bike Power Saver. *Appl. Ergon.* **2008**, *39*, 45. [CrossRef]
26. Yoon, T.S.; Park, D.S.; Kang, S.W.; Chun, S.I.; Shin, J.S. Isometric and Isokinetic Torque Curves at the Knee Joint. *Yonsei Med. J.* **1991**, *32*, 33–43. [CrossRef] [PubMed]

robotics

MDPI

Article

Design of a Labriform-Steering Underwater Robot Using a Multiphysics Simulation Environment

Daniele Costa [1], Cecilia Scoccia [1], Matteo Palpacelli [1,*], Massimo Callegari [1] and David Scaradozzi [2]

[1] Department of Industrial Engineering and Mathematical Sciences, Polytechnic University of Marche, 60131 Ancona, Italy; d.costa@univpm.it (D.C.); c.scoccia@pm.univpm.it (C.S.); m.callegari@univpm.it (M.C.)
[2] Department of Information Engineering, Polytechnic University of Marche, 60131 Ancona, Italy; d.scaradozzi@univpm.it
* Correspondence: m.palpacelli@univpm.it

Abstract: Bio-inspired solutions devised for Autonomous Underwater Robots are currently investigated by researchers as a source of propulsive improvement. To address this ambitious objective, the authors have designed a carangiform swimming robot, which represents a compromise in terms of efficiency and maximum velocity. The requirements of stabilizing a course and performing turns were not met in their previous works. Therefore, the aim of this paper is to improve the vehicle maneuvering capabilities by means of a novel transmission system capable of transforming the constant angular velocity of a single rotary actuator into the pitching–yawing rotation of fish pectoral fins. Here, the biomimetic thrusters exploit the drag-based momentum transfer mechanism of labriform swimmers to generate the necessary steering torque. Aside from inertia and encumbrance reduction, the main improvement of this solution is the inherent synchronization of the system granted by the mechanism's kinematics. The system was sized by using the experimental results collected by biologists and then integrated in a multiphysics simulation environment to predict the resulting maneuvering performance.

Keywords: biomimetics; underwater robots; robotics; multibody systems; transmission systems; autonomous underwater vehicles

Citation: Costa, D.; Scoccia, C.; Palpacelli, M.; Callegari, M.; Scaradozzi, D. Design of a Labriform-Steering Underwater Robot Using a Multiphysics Simulation Environment. *Robotics* **2022**, *11*, 11. https://doi.org/10.3390/robotics11010011

Academic Editor: Charalampos P. Bechlioulis

Received: 22 November 2021
Accepted: 4 January 2022
Published: 7 January 2022

1. Introduction

The locomotion of aquatic animals has attracted the attention of biologists and engineers for a long time, and the last thirty years have witnessed a significant growth in the study of the comparative biomechanics of motion through water. The attempts to design machines, such as autonomous underwater vehicles (AUVs) capable of moving similar to biological swimmers are inspired by the superior performance of marine animals and fish in terms of both efficiency and maneuverability. Several prototypes of bio-inspired robots can thus be mentioned, whereas an extensive review is presented in [1]. The possibility of exploiting the swimming modes that fish have evolved over thousands of years requires insight into the fluid mechanic principles underlying aquatic animal locomotion. According to swim mechanics, propulsive thrust is generated from the momentum transfer due to the relative motion between the fish body and the surrounding water. Figure 1 illustrates the terminology used to identify morphological features commonly used in the literature. Most fish generate thrust by bending their bodies into a backward-moving propulsive wave that extends to their caudal fin, a type of swimming classified as Body and Caudal Fin (BCF) locomotion [1]. Other fish have developed alternative mechanisms that involve the use of their median and pectoral fins: these swimming modes are classified as Median and Paired Fin (MPF) locomotion. An estimated 15% of fish families employ non-BCF modes as their primary motion system, whereas those that rely on BCF modes for propulsion employ MPF modes for maneuvering and stabilization [2].

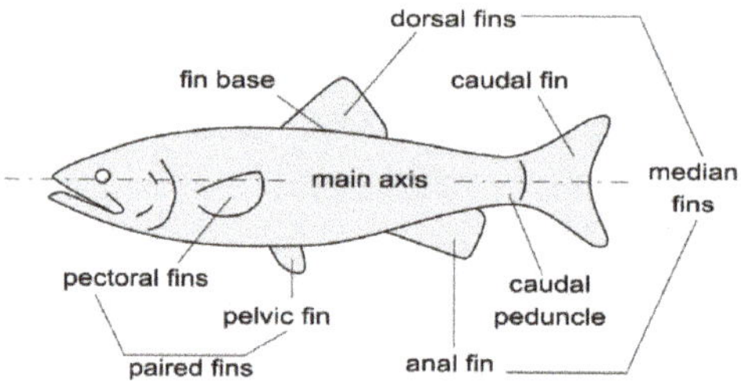

Figure 1. Fish morphological features commonly depicted in the literature.

In the last few years, the authors of this work designed, prototyped, and experimentally validated [3] a series of bio-inspired vehicles for both research and educational purposes: the robot in [4] is driven by an oscillating plate shaped such as a caudal fin and is hinged to the rigid forebody; this is the least efficient locomotion principle among the BCF swimming modes [1]; however, since the number of moving parts is very limited, the resulting system is inexpensive and easy to fabricate and seal. The search for higher propulsive efficiency has forced the authors to further improve their design by moving to the undulating tail presented in [5]: in the latter, the links of the transmission mechanism are driven by cam joints connected to a piecewise flexible shaft actuated by a single rotary actuator. The adopted swimming mode represents a compromise choice in terms of efficiency and maximum speed for an underwater robot propelled by a biomimetic thruster. However, the requirements of stabilizing a course and performing stationary turns have not been met yet. As a matter of fact, the robot in [5] is able to swim following a straight path, whereas it lacks any capability to adjust its course as well as steer, under both stationary and dynamic conditions. Therefore, before moving to the prototyping and testing phases, the authors aim to fulfill those necessities by improving their design, which is the focus of this paper. Particularly, the purpose of this work is to present a bio-inspired maneuvering device which exploits the momentum transfer mechanism of MPF locomotion to generate the steering torque. The system was designed so as to be integrated in the robot described in [5]; therefore, the final prototype will rely on BCF swimming modes for forward propulsion, whereas the pectoral fins will be deployed whenever a course correction is required. Ultimately, the concept that drove authors' research since their first work [4] is to exploit the best opportunities offered by nature to improve the speed and maneuverability of AUVs.

As stated before, MPF swimming modes are widespread among aquatic animals for maneuvering and stabilization; particularly, in labriform mode, the steering torque is generated through the pectoral fin motion. Two alternative oscillatory movement types were identified [6]: a rowing action based on drag forces and a flapping action due to lift generation similar to birds flying. In the former, the rowing action consists of two phases [6]: the power stroke of Figure 2a, where the fins move posteriorly at a high attack angle and high speed, and the recovery stroke of Figure 2b, where the fins are feathered at a near-zero angle of attack to reduce resistance while they are brought forward at low speed. On the contrary, in the lift-based mode displayed in Figure 2c, the propulsive force is always normal to the direction of the fin motion; as a result, no recovery stroke is necessary. According to [6,7], the drag-based method is more efficient at low speeds, whereas lift-based maneuvering requires higher velocities to generate significant forces. Due to the necessity to perform even stationary turn maneuvers, drag-based momentum transfer mechanism was chosen by the authors of this work.

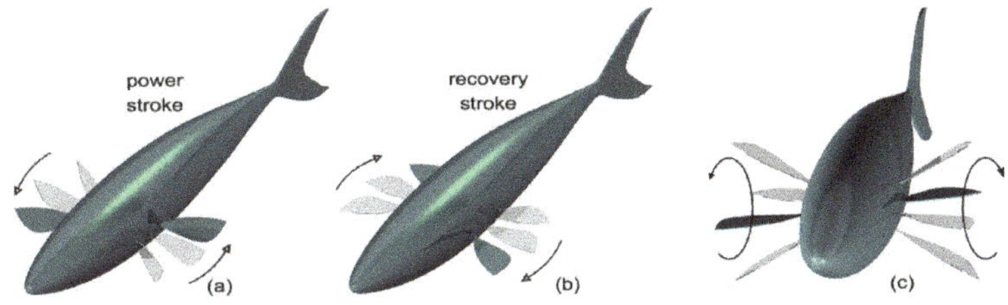

Figure 2. Pectoral fin kinematics during power strokes (**a**) and recovery strokes (**b**) in drag-based labriform locomotion; pectoral fin kinematics in lift-based labriform locomotion (**c**).

According to the literature, the potential for building stabilization and maneuvering devices based on pectoral fins motion has already been investigated by biologists [8,9] and researchers in the underwater robotic field: in [10], two concave-shape fins were connected to a servomotor arm sliding on a pair of parallel shafts fixed at the center of a pool. Due to the concave shape of the fins, the propulsive forces generated by the power strokes overcame the ones generated by the recovery strokes. The authors tested several fin-oscillations angles, whereas the optimum oscillation range was obtained for each power-to-recovery ratio in order to achieve the maximum thrust. Five values of the power-to-recovery ratio were tested, and the maximum thrust was obtained for a moderate ratio of the power-to-recovery stroke, i.e., 3:1. A similar driving system was adopted in [11–13]: in [11], each pectoral fin was driven by a couple of servomotors. The authors employed this architecture to obtain different types of motion: rowing, flapping, and hovering. The performance of the proposed solution was investigated experimentally by varying the fin flapping frequency and amplitude, and the optimal behavior was found as a function of the Strouhal number, a parameter commonly used to characterize flow–oscillation phenomena. In [12], the authors reduced the number of servomotors by designing the fin driving system in a symmetrical way. Experimental tests were performed to examine two different maneuvering conditions: the steering force was initially generated through asymmetrical tail undulations, whereas the pectoral fins were kept still in the feathering position; later, both the tail and the pectoral fins cooperated with the turn maneuver, improving the performance in terms of the turnabout radius and velocity. Turning characteristics were also measured in [13], where a robotic fish was driven by two degrees of freedom (DOF) pectoral fins. As in [12], the results of the numerical simulations and the experiments showed that the fastest turning speed was achieved when the robotic fish was cooperatively propelled by both the fins and the tail; the pectoral fins are capable by themselves to turn the robotic fish on the spot, but the resulting angular speed is the smallest. Finally, an active–passive solution was devised in [14,15], where the second servomotor was replaced by a flexible mechanism. Here, the authors presented a dynamic model for a robotic fish incorporating the proposed steering device. Experimental tests were performed by varying the geometric and kinematic parameters of the fin motion, whereas the flexibility effect was also investigated by using different compliant joints. The obtained measurements ultimately showed that maximum efficiency was achieved, as the Strouhal number was within the optimal range reported in the literature and was observed by researchers in the biological counterparts.

The robotic fish in [10–13] adopted drag-based labriform locomotion for stabilization and maneuvering. Here, the yawing–pitching oscillation of the pectoral fins was driven by a couple of servomotors connected in series: the first one, attached to the robot forebody, drives the rowing rotation around the yaw axis of the vehicle; whereas the second one, connected to the pectoral fin, allows the latter to trim its attitude and is used for feathering. Although this solution is simple to manufacture, the main disadvantages of a direct drive are the number of motors and the necessity to synchronize their rotations to generate the yawing–pitching oscillation presented in [6]. Position control is then necessary in order to comply with a non-linear function of time and to maintain the constant phase shift between the fin rotation components. Moreover, each servomotor must be accurately sealed, while watertight connections must be mounted both on the joint shafts and on the electrical cables, thus increasing the structure inertia, encumbrance, and possibility of failure.

On the basis of the aforementioned factors, the aim of the present work is to design a novel transmission system driven by a single rotary actuator that is capable of driving the pectoral fins of the robotic fish [5] while exploiting the drag-based momentum transfer mechanism of labriform locomotion. Aside from inertia and encumbrance reduction, the main improvement of this solution is the inherent synchronization of the system, meaning left and right fins as well as their individual rotation components. In fact, the phase shifts and the oscillation frequencies will remain constant by the mechanical constraints of the transmission system. Waterproofing issues will also be minimal because the section housing the pectoral fin mechanism will be fully flooded and only one actuator will need sealing.

2. Materials and Methods

The geometric and inertial features of the biomimetic robot designed in [5] were modeled after a mackerel (Scomber Scombrus) [16]. Fish belonging to the Scombridae family possess a streamlined body with a homocercal caudal fin and can achieve the highest speed among BCF swimmers. As stated before, fish relying on BCF modes for propulsion employ MPF modes for maneuvering and stabilization. Inspired by nature, the aim of the authors was to successfully replicate this behavior in an artificial device. The final assembly is a 2:1 scale of the biological mackerel: the robot is 800 mm long and its tail spans the last 300 mm.

2.1. Mechanics of Drag-Based Labriform Swimming and Blade-Element Theory

A complete analysis of the mechanics of drag-based *labriform* locomotion is beyond the scope of this work, and extensive reviews are presented in [17,18]. However, in order to size the transmission system devised in this paper, the following assumptions are necessary [6]:

1. Drag force is due to pressure drag only, whereas the effects of viscosity are negligible.
2. During the power stroke, the distal two-thirds of the fin are perpendicular to the horizontal plane.
3. The rowing rotation around the fish yaw axis is about 90 degrees.
4. At the end of the power stroke, as the fin moves forward, its distal two-thirds form a small angle with the horizontal plane (about 10–20 degrees).
5. The power stroke is about three times faster than the recovery stroke.
6. A blade-element approach is used to analyze pectoral fin mechanics in drag-based labriform swimming.

A schematic diagram showing pectoral fin positions and orientation during a beat cycle is shown in Figure 3a together with a typical blade-element during a power stroke, Figure 3b. Here, progressive numbers were used to show the consecutive positions of the fins during both power and recovery strokes. Blade-element theory is a mathematical process originally conceived to predict the behavior of propellers: their blades are broken down into small parts and the forces acting on each one of these elements are calculated. These forces are then integrated along the entire blade and over one revolution in order to compute the thrust and torque produced by the entire propeller. In the same way, in this paper, the pectoral fins were broken down into small parts along the span direction: the relative velocity components between each of these small blade-elements and the surrounding water were computed and used to calculate the resulting hydrodynamic forces. Finally, these forces were integrated along the fin span and over one beat cycle to obtain the produced work.

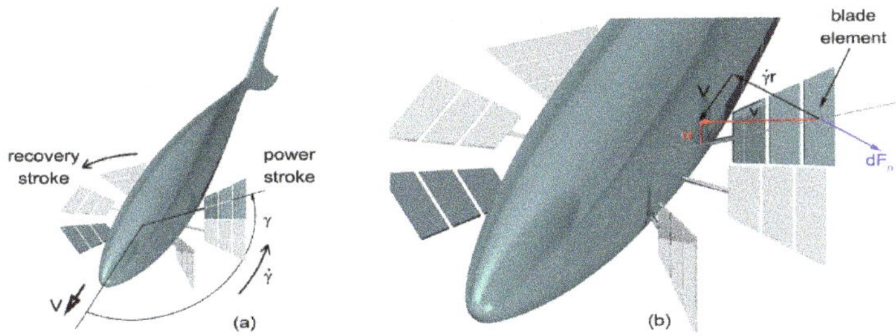

Figure 3. Geometry of power and recovery strokes: the fin was drawn thin during the power stroke and flat during the recovery stroke to show its perpendicular and parallel orientation with respect to the horizontal plane (**a**); blade-element kinematics [6] (**b**).

The normal and spanwise components of blade-element velocities v_n and v_s in Figure 3b are:

$$v_n = \dot{\gamma}r - V\sin\gamma \quad v_s = V\cos\gamma \tag{1}$$

where r is the distance from the base of the fin to the midpoint of a blade-element, V is the fish cruising velocity, whereas γ and its time derivative are the respective fin angular position and velocity with respect to the yaw axis. Therefore, the blade-element resultant relative velocity is:

$$v^2 = \dot{\gamma}^2 r^2 + V^2 - 2V\dot{\gamma}r\sin\gamma \tag{2}$$

The hydrodynamic angle of attack α in Figure 3b is computed as:

$$\tan\alpha = \frac{v_n}{v_s} = \frac{\dot{\gamma}r - V\sin\gamma}{V\cos\gamma} \tag{3}$$

The normal force dF_n and torque dM_n due to pressure drag acting on a blade-element are expressed as:

$$dF_n = \tfrac{1}{2}\rho v^2 dAC_n \quad dM_n = rdF_n \tag{4}$$

where ρ is the water density, dA is a blade-element wetted surface, and C_n is a normal force coefficient. This last parameter was experimentally measured in [6]: the normal force coefficient remains approximately constant at 1.1 when the attack angle varies between 40 and 140 degrees. However, when angle α is smaller than 40 degrees, the normal force coefficient progressively decreases to zero following a nonlinear function of the attack angle.

In labriform locomotion, the normal forces due to pressure drag are not the only mechanism of momentum transfer: added mass forces have almost the same impact, particularly during a power stroke. The added mass of a blade element can be computed as:

$$dm_a = \rho\pi(c/2)^2 dl \tag{5}$$

where c is the chord of the fin at the midpoint of an element and l is its length. Therefore, the added mass force and torque can be expressed as:

$$dF_a = \ddot{\gamma}r\rho\pi(c/2)^2 dl \quad dM_a = rdF_a \tag{6}$$

Therefore, the steering torque generated by the motion of the pectoral fins is computed as:

$$M_{PF} = \int_{FINS}(dM_n + dM_a) \tag{7}$$

2.2. Functional Design of the Transmission System

As stated before, as a fin approaches the end of a recovery stroke, it turns around its span axis so that its final two-thirds form a high angle with the horizontal plane, preparing for the upcoming power stroke. Similarly, at the end of a power stroke, the fin twists in the reverse direction to align itself with the horizontal plane, preparing for the upcoming recovery stroke. Supported by the evidence that the power stroke is about three times faster than the recovery motion, the authors decided to split the fin-beat cycle into four phases characterized by the same duration: phase 1 is the power stroke, while the recovery stroke consists of the series of the remaining three phases. Here, the yawing rotation is continuous whereas the pitching motion is intermittent. Specifically:

- Phase 1—the power stroke, Figure 4a: the pitching rotation is blocked, and the fin is kept perpendicular to the horizontal plane, while it spins about the yaw axis at high angular velocity.
- Phase 2—feathering recovery stroke, Figure 4b: the fin turns around its pitch axis until it is flat on the horizontal plane; at the same time, it rotates slowly about the fish yaw axis in the opposite direction with respect to the power stroke.
- Phase 3—main recovery stroke, Figure 4c: the rotation about the pitch axis is blocked and the fin is kept parallel the horizontal plane, while it continues to move slowly in the reverse direction with respect to the power stroke.
- Phase 4—unfeathering recovery stroke, Figure 4d: the fin turns about its pitch axis until it is perpendicular to the horizontal plane; at the same time, the fin slows down along the last part of the rotation about the yaw axis, preparing for the upcoming power stroke.

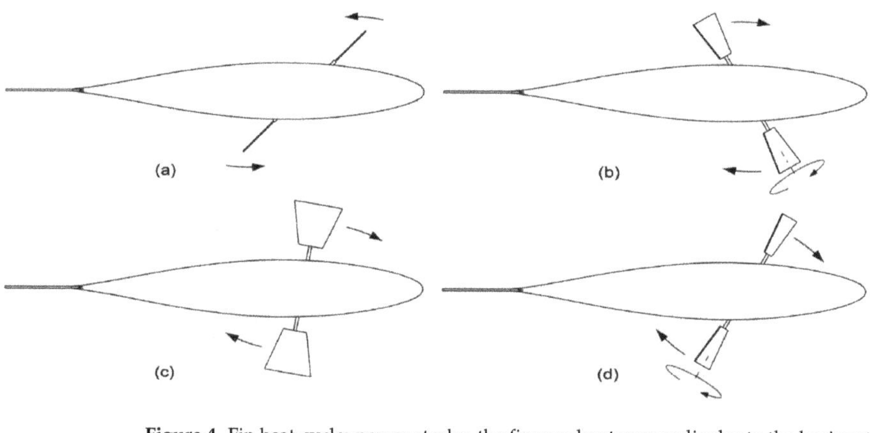

Figure 4. Fin beat-cycle: power stroke: the fins are kept perpendicular to the horizontal plane as they rotate counterclockwise about the fish yaw axis at high speed (**a**); feathering recovery stroke: the fins turn around their pitch axis, flattening on the horizontal plane, as they slowly rotate clockwise about the yaw axis (**b**); main recovery stroke: the fins are kept parallel to the horizontal plane as they continue to rotate clockwise about the yaw axis (**c**); unfeathering recovery stroke: the fins turn around their yaw axis until they are perpendicular to the horizontal plane, as they complete the rotation around the yaw axis, thus returning in the initial position and configuration (**d**).

Table 1 summarizes the kinematics of the pectoral fins based on the authors' design.

Table 1. Fin kinematics according to the phases of Figure 4.

-	Power Stroke		Recovery Stroke	
Time	**Time 1**	**Time 2**	**Time 3**	**Time 4**
Yaw rotation	Driven—fast	Driven—slow	Driven—slow	Driven—slow
Yaw angle γ [1]	45° -> 135°	135° -> 108.5°	108.5° -> 71.5°	71.5° -> 45°
Pitch rotation	Blocked	Driven	Blocked	Driven
Pitch angle φ	Trimmed at 90°	90° -> 0°	Trimmed at 0°	0° -> 90°

[1] The computation of the provided values is shown in Section 3.1.

As stated before, the transmission system presented in this paper drives the pectoral fins of a robotic fish by means of a single rotary actuator. The mechanism is composed of two parallel sub-systems, each of which is dedicated to one of the fin individual rotations: pitching and yawing. Both sub-systems are driven by the same input shaft, which is connected to the rotary actuator. Finally, the sub-systems output shafts recouple right before the fins to generate their composite rotation about the pitch and yaw axis.

Sub-system A converts the continuous rotation of the motor in the oscillation of the fins about the yaw axis, as shown in Figure 5a. The core of this solution is a quick-return mechanism, specifically an oscillating glyph: here, the oscillation amplitude was set to 90° to comply with the rowing amplitude of the fins in drag-based labriform locomotion. In this way, since the motor spins with constant velocity, the return rotation is three times faster than the forward rotation, perfectly matching the duration ratio between the power stroke and the recovery stroke. Furthermore, the fins are connected to the ends of the same shaft, which is perpendicular to the glyph output device: in this way, the fins rotate in the same direction, so the torques due to the normal forces have the same sign, summed up to generate a stronger steering moment.

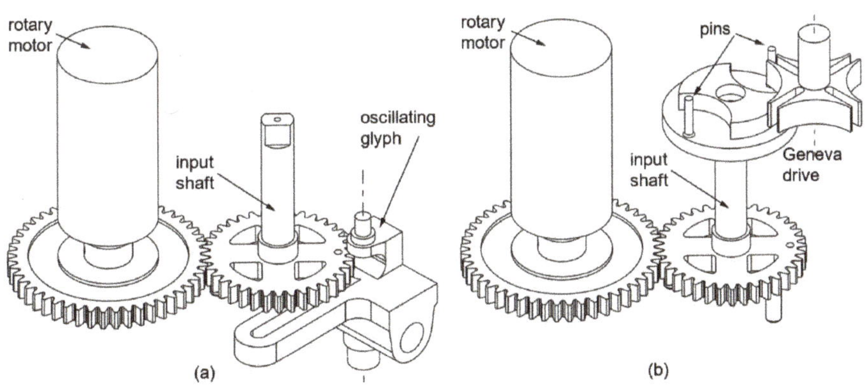

Figure 5. Sub-system A (**a**); Sub-system B (**b**).

Sub-system B generates discontinuous rotation of the fins about their pitch axis: the core of this device is an intermittent mechanism, namely, a Geneva drive. Here, the input rotation is the same as a sub-system A, meaning that is provided by the motor. However, since the fins need to trim twice in beat cycle, right before and after the power stroke, the Geneva system must be arranged with two pins on the driving wheel and four spokes on the driven wheel, as shown in Figure 5b.

3. Results

Figure 6 shows the complete assembly of the transmission system designed to drive the pectoral fins: here, the input shaft, which is connected to the motor by means of a gear joint, drives both the oscillating glyph (sub-system A) and the Geneva wheels (sub-system B). Shaft C, which is connected to the Geneva output device by means of a bevel gear joint, transmits the intermittent rotation of the driven wheel to fin shaft D through a third gear joint: in order to allow shaft C to follow glyph oscillations and maintain the required synchronization between sub-systems A and B, shaft C was divided in two symmetrical parts connected by a homokinetic joint. The presence of the latter is necessary to transmit the driving torque while preserving angular freedom and a constant rotation velocity. In this paper, the authors employed Double Cardan joints in homokinetic configuration [19]. Homokinetic Double Cardan joints require a centering element that maintains equal angles between the driving and driven shafts, as specified by the "homokinetic plane" arrangement: particularly, in order to produce a constant-velocity transmission, the centering element in the coupling must share the same plane of symmetry with the one between the input and output shafts, generally called the "homokinetic plane"; this plane must also contain the intersection of the axis of the joints forming the Double Cardan [19]. Figure 7 shows the arrangement proposed in this paper, where the homokinetic plane coincides with the symmetry plane of each Double Cardan joint. Moreover, the centering element is composed of two parts that may slide, relative to each other, by means of a prismatic joint. In this way, the length of the centering element is passively adjusted as shaft C changes its configuration according to the glyph oscillations while maintaining its mid-section coincident with the symmetry plane of the coupling and fulfilling the mentioned requirement of constant rotational velocity.

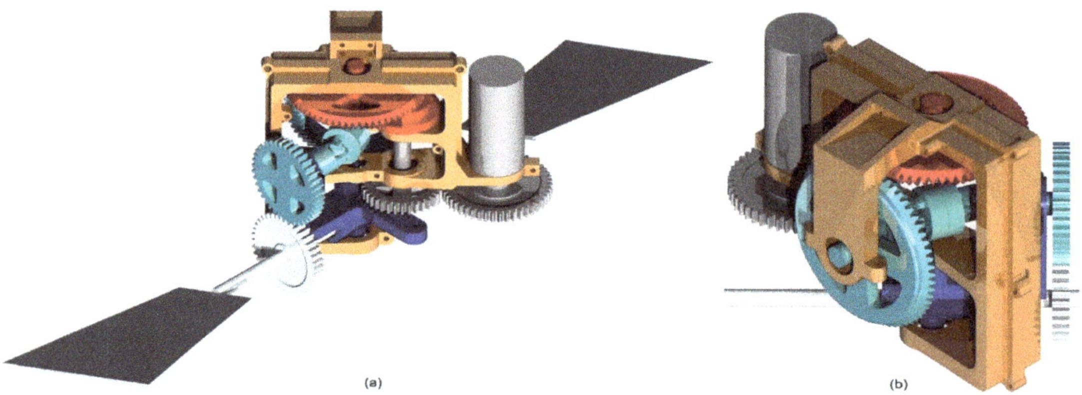

Figure 6. Transmission mechanism and pectoral fin complete assembly: frame (orange), rotary motor and input shaft (grey), sub-system A (dark blue), sub-system B (red), shaft C and Double Cardan joints (cyan), fin shaft D (pearl); front view (**a**), back view (**b**).

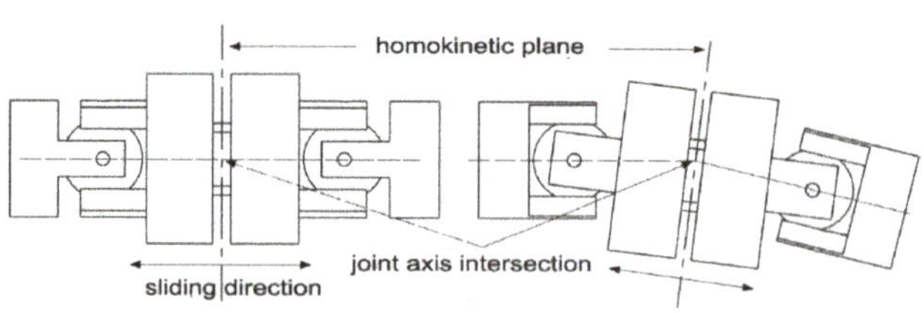

Figure 7. Double Cardan joint assembly.

3.1. Preliminary Analysis of the Transmission Mechanism

Figure 8 shows the configuration of the oscillating glyph and the pectoral fins at the beginning of the power stroke. Here, length L measures 90 mm. Furthermore, according to blade-element theory [6], the trapezoidal fins were divided in three parts, each of which is 30 mm long. Moreover, the fins have a constant taper ratio, where the outermost blade element is twice as broad as the innermost, measuring 100 mm at its tip. The relative velocities of the midpoints of each blade elements, as well as their attack angles α, are then expressed by Equation (1), which was expanded here to account for the left and right fin:

$$\begin{array}{cc} v_{n,B} = \dot{\gamma} r - V \sin\gamma & v_{n,F} = \dot{\gamma} r + V \sin\gamma \\ v_{s,B} = V \cos\gamma & v_{s,F} = -V \cos\gamma \\ \tan\alpha_B = \frac{\dot{\gamma} r - V \sin\gamma}{V \cos\gamma} & \tan\alpha_F = \frac{\dot{\gamma} r + V \sin\gamma}{-V \cos\gamma} \end{array} \quad (8)$$

where subscripts B and F refer to the respective right, backward-moving fin, and to the left, forward-moving fin.

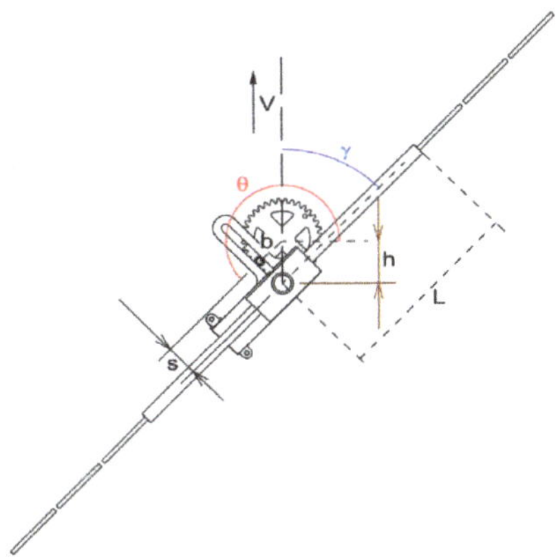

Figure 8. Oscillating glyph at the beginning of a power stroke.

The fin angular position γ and its time derivative during a fin-beat cycle can be easily computed as a function of motor rotation θ and its constant angular velocity ω:

$$\begin{array}{c} \tan\gamma = -\frac{h + b \sin\theta}{b \cos\theta} \\ s^2 = b^2 + h^2 + 2bh \sin\theta \\ \dot{\gamma} = \omega \frac{b}{s} \cos(\theta + \gamma) \end{array} \quad (9)$$

where s, b, and h are the geometric features of the oscillating glyph shown in Figure 8.

Finally, Figure 9 shows both the fin angular position γ and its pitch orientation ϕ driven by the Geneva mechanism as a function of motor rotation θ during a fin-beat cycle.

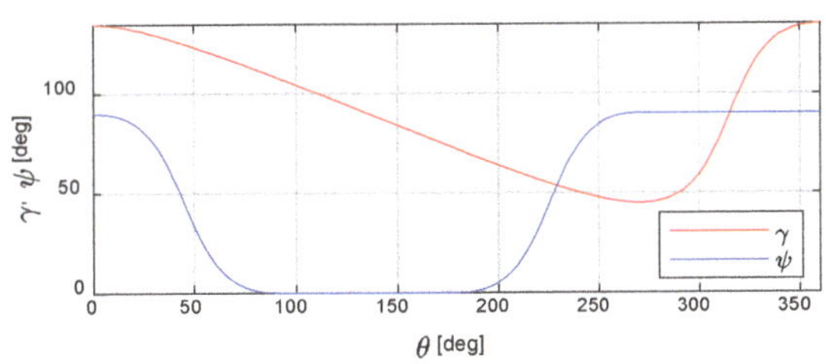

Figure 9. Fin angular position and orientation in a fin beat-cycle.

By combining Equations (8) and (9), the attack angles can be computed at the midpoint of the fin blade-element during a power stroke when the fish swimming velocity varies from zero to 0.5 BL/s. Figure 10 shows that α has negative values at the beginning of the power stroke, when the fin angular velocity is rising from zero, as seen in Angelfish swimming [6]. As the fin angular velocity continues to rise, the attack angle quickly closes to the range [40–130°] where the normal force coefficient C_n is constant at 1.1, as stated in

Section 2.1; alternatively, when α is less than 40°, C_n can be approximated by the following Equation [6]:

$$C_n = \mathrm{k} \sin \alpha \quad \mathrm{k} = 2.5 \tag{10}$$

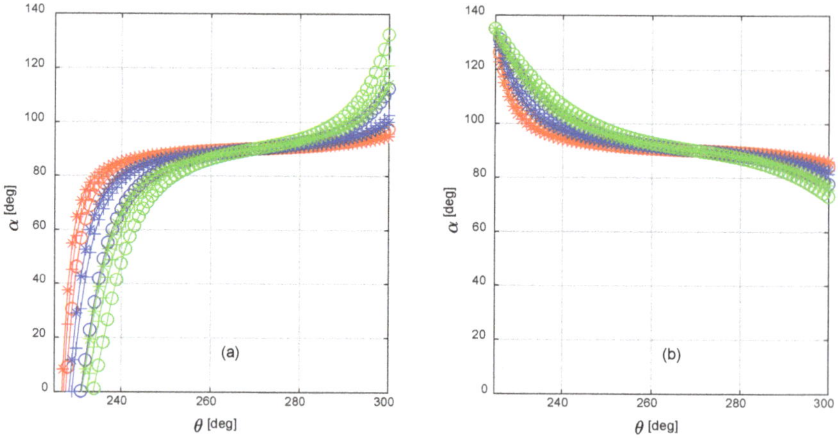

Figure 10. Attack angle α during a power stroke: right fin (**a**) and left fin (**b**); the values were computed when the swimming velocity V was 0.125 BL/s (red), 0.25 BL/s (blue), and 0.5 BL/s (green). The markers on the curves refer to the innermost (circle), median (plus sign), and outermost (asterisk) blade-element.

These results will be used to calculate the steering torque in the upcoming multibody analysis.

3.2. Multibody Analysis

In order to predict the performance of the transmission mechanism presented in this paper, the authors employed the multiphysics platform developed and validated in previous works as a sizing and prediction tool devised to verify and optimize its design [3,20]. The backbone of this platform is a multibody model of the investigated robot: as a matter of fact, multibody techniques are particularly suited to a large class of bio-inspired underwater robots due to their multi-joint rigid architecture. Additionally, the hydrodynamic loads, obtained through experimental investigations or predicted by means of computational fluid dynamic analysis, can be easily integrated in the aforementioned dynamic model in order to allow the analysis to account for their effect on the solution.

Figure 11a shows a simple biomimetic robot, composed of a cylindrical forebody and a plane caudal fin, swimming in the surrounding water. Reference frame Σ_b, O_b–$x_b y_b z_b$ is attached to the robot, whereas vector $\nu_1 = [u\ v\ w]^{\mathrm{T}}$ expresses the velocity of the origin O_b in the body frame Σ_b; similarly, its angular velocity is represented by vector $\nu_2 = [p\ q\ r]^{\mathrm{T}}$ [21].

In this paper, the authors narrowed their analysis on plane motion: therefore, according to the Newton–Euler formulation, the dynamics equations can be formulated as:

$$\begin{aligned} m(\dot{u} - vr) &= X + F_T \\ m(\dot{v} + ur) &= Y + F_L \\ I_z \dot{r} &= N + M_T + M_{PF} - x_{CF} F_L \end{aligned} \tag{11}$$

where m is the robot total mass and I_z is its z principal moment of inertia, computed under the hypothesis that frame Σ_b is coincident with the body central frame; finally, length x_{CF} represents the distance between the frame origin O_b and the revolute joint connecting the caudal fin to the rigid forebody. The terms X, Y, and N on the right side of Equation (11) represent the hydrodynamic loads applied to a multibody system moving in the surrounding fluid. In order to accurately compute those terms, the uncompressible flow

Navier–Stokes equations should be solved. Nevertheless, if the velocities are sufficiently low, most of the hydrodynamic effects have no significant influence on the resulting motion; furthermore, if the body features three planes of symmetry, the terms X, Y, and N can be linearized and replaced by the simplified expressions presented in the following equation:

$$\begin{aligned} m(\dot{u} - vr) &= X + F_T = -X_{\dot{u}}\dot{u} - Y_{\dot{v}}vr - X_{u|u|}u|u| + F_T \\ m(\dot{v} + ur) &= Y + F_L = -Y_{\dot{v}}\dot{v} + X_{\dot{u}}ur - Y_{v|v|}v|v| + F_L \\ I_z\dot{r} &= N + M_T + M_{PF} = -N_{\dot{r}}\dot{r} - N_{r|r|}r|r| + M_T + M_{PF} \end{aligned} \tag{12}$$

where the subscripts with the time derivative of the velocity components refer to added mass loads coefficients, while the velocity subscripts identify damping coefficients [21,22]. Table 2 collects the expressions of the respective terms for a cylinder with radius R, length L, and mass m. Regarding the damping coefficients, it might be noteworthy to point out that linear damping due to skin friction was neglected in Equation (12). The significance of linear and quadratic damping was measured in [23]: as a matter of fact, linear damping should be always accounted for under station-keeping conditions as well as for marine crafts operating in waves, whereas neglecting its effect is a good assumption for maneuvering. Nevertheless, since the steering device presented in this paper was designed even for low-speed maneuvers, the latter assumption will be experimentally investigated in future work.

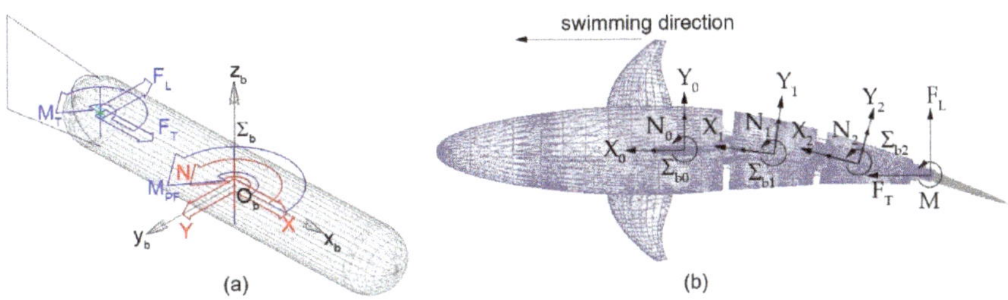

Figure 11. Cylindrical body subject to the hydrodynamic loads (red) and to the propulsive/maneuvering forces generated by the caudal and pectoral fins (blue) (**a**); robotic fish subject to the hydrodynamic and propulsive loads applied to the tail sections and to the rigid forebody (**b**).

Table 2. Added mass and damping coefficient for a cylinder with radius R, length L, and mass m [22] (Reprinted from Ref. [5]).

$X_{\dot{u}}$	$Y_{\dot{v}}$	$N_{\dot{r}}$	$X_{u\|u\|}$	$Y_{v\|v\|}$
$0.1m$	$\pi\rho R^2L$	$\pi\rho R^2L^3/12$	$\rho A_f c_{D,f}/2$	$\rho A_l c_{D,l}/2$
$N_{r\|r\|}$	A_f	A_l	$c_{D,f}$	$c_{D,l}$
$\rho A_l c_{D,l} L^3/16$	πR^2	2RL	0.5	[0.8–1.2]

Figure 11b shows the multibody model of a robotic fish composed of a rigid forebody hinged to a three-joint tail linkage ending with a caudal fin. Body reference frames $\Sigma_{b,i}$ ($i = 0 \ldots 3$) are attached to the rigid bodies of the model; the 0 index identifies the robot forebody while non-zero indexes refer to the tail links. When moving from the single body shown in Figure 11a to the multibody system of Figure 11b, it is necessary to apply the hydrodynamic loads of Equation (12) to the individual bodies of the assembly; then, the resulting system can be solved to compute the robot dynamics.

Figure 12 shows the multibody model of the carangiform swimming robot designed in [5]. As stated before, the robot head and the tail links, except the caudal fin, are subject to the hydrodynamic loads expressed in Equation (12). Regarding the propulsive forces and torque applied to the caudal fin as a result of tail undulation, they were predicted in [5] by means of computational fluid dynamic analysis; their expression is the following:

$$\begin{aligned} F_T &= K_T \rho {A_0}^2 f^2 Bc \sin(2\pi f t + \varphi_T) \\ F_L &= K_L \rho {A_0}^2 f^2 Bc \sin(2\pi f t + \varphi_L) \\ M_T &= K_M \rho {A_0}^2 f^2 Bc^2 \sin(2\pi f t + \varphi_M) \end{aligned} \quad (13)$$

where A_0, B, and c are the translation amplitude, span, and mean chord of the caudal fin, f is the tail undulation frequency, φ_T, φ_L, φ_M are phase constants, whereas K_T, K_L, and K_M are the caudal fin force and torque coefficients.

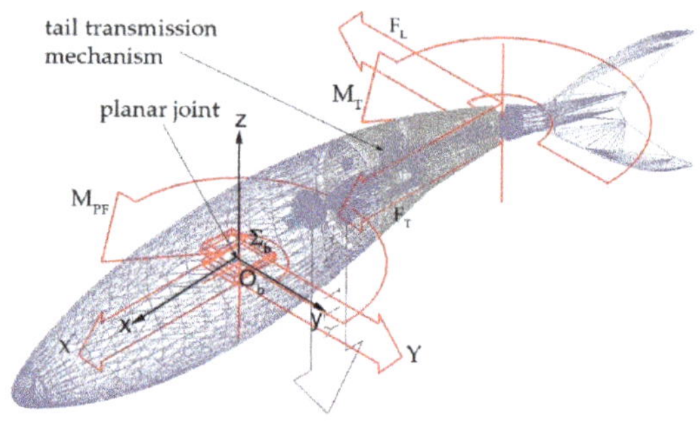

Figure 12. Multibody model of the underwater robot.

Finally, M_{PF} is the steering moment generated by pectoral fin undulations and expressed by Equation (7). According to blade-element theory, the resulting torque was computed as the sum of moments acting on the fin blade parts: particularly, the relative velocity v in Equations (1) and (8) was calculated in their respective centers of mass; then, the normal force due to pressure drag in Equation (4) was approximated as:

$$F_{n,BE} = \frac{1}{2} \rho {v_{BE}}^2 A_{BE} C_{n,BE} \quad (14)$$

where subscript BE identifies the three blade-elements into which the fins were split. The torque due to the normal forces $F_{n,BE}$ was thus computed as the product of Equation (14) multiplied by the distance of the blade-element center of mass from the fin yaw rotation axis.

The normal force coefficient C_n was set constant at 1.1 during the power strokes, as stated in Section 3.1, and zero during the recovery strokes when both fins were flat on the robot horizontal plane due to the feathering motion. In this way, the steering action in the multibody system is limited to the power strokes as in labriform locomotion.

In order to investigate the maneuvering performances of the robotic fish modeled by the multibody system of Figure 12, the dynamic equations were solved by using MSC Adams View. The transmission mechanism presented in this paper was assembled in the fish forebody section, and the steering torque generated by pectoral fin motion was added to the system. In the dynamic analysis, the tail undulation frequency was suitably set in order to obtain a steady cruising condition with forward speed V limited to the range of interest equal to [0–1 BL/s]. Finally, the angular velocity of the motor driving the pectoral fin mechanism was varied in the simulations, and the robot turnabout radius, expressed in robot BLs, was measured. Table 3 shows the obtained results.

Table 3. Turnabout radius [BL] predicted by the multibody simulations as a function of the fish swimming speed and motor frequency.

Motor Frequency	Swimming Speed			
	0.125 BL/s	**0.25 BL/s**	**0.50 BL/s**	**1 BL/s**
0.5 Hz	3.16	7.17	12.25	14.93
1 Hz	1.46	3.15	7.15	12.17
1.5 Hz	0.98	1.97	4.44	9.35
2 Hz	0.75	1.45	3.13	7.09
3 Hz	0.53	0.96	1.94	4.41

3.3. Mechanism Prototyping and Mechatronic Components

Figure 13 shows the physical prototype of the transmission system proposed in this paper. All parts, except for the Double Cardan joint, were drawn in a CAD environment by using Dassault CATIA and then printed by high-resolution stereolithography (SLA). The mechanism was then assembled to verify the quality of the manufacturing process, and minor issues have already been solved.

Figure 13. Physical prototype of the transmission system manufactured by SLA.

Future work will focus on the drawing of a CAD model of the vehicle major components, including tail and hull sections, which will also be printed by SLA. Regarding the mechatronic components, the transmission system designed in this paper will be driven by a 12-V DC brushed gearmotor with a quadrature encoder providing 64 counts per revolution. Feedback control on the constant angular velocity setting will be provided by the FPGA system of a National Instrument MyRIO Board, whereas the motor power will be supplied by a Sabertooth 2 × 32 motor driver. The MyRIO board will also run the Navigation, Guidance, and Control system designed and validated by the authors in [4,24].

4. Discussion

Several approximations were introduced throughout the modelling and design process of the labriform steering mechanism presented in this work: first, when calculating the normal force coefficient, the effect of viscosity—i.e., shear stresses—was neglected, whereas the drag forces were due to pressure drag only; however, the attack angle was greater than 40° during almost 90% of the power stroke, as shown in Figure 10. Under this condition, it is reasonable to assume that the flow separated from the fin rear surface, thus the drag force was almost entirely due to pressure drag, while skin friction can be neglected [25]. Second, the fins were modeled as a series of flat plates set at a high angle of attack with respect to the incident flow: although the innermost and the median blade-elements did not actually experience the flow regime of a three-dimensional flat plate in a freestream, because they were bounded by the neighboring elements, it could be shown that the outermost blade part was responsible for more than 80% of the steering torque. Therefore, the introduced approximation affects the simulations in an acceptable manner.

The maneuvering performance predicted by the multibody analysis, collected in Table 3, show that the bio-inspired mechanism proposed in this paper is capable of steering the robot with a turnabout radius comparable to its length when its swimming velocity is less than 0.5 BL/s. However, as the cruising speed rises, the efficiency of the drag-based propulsive system considerably drops, as seen in the biological counterparts. In order to overcome this limitation, the proposed system could be improved by inserting supplementary joints both on the vehicle forebody and on its tail, thus allowing the robot to perform fast C-shape turn maneuvers.

By means of the steering device proposed in this paper, the carangiform swimming robot in [5] is now capable to adjust its course and perform turn maneuvers at low velocity, which could be a necessity for an AUV employed for survey missions [26]. As a matter of fact, the biomimetic vehicle developed by the authors relies on BCF swimming modes to generate forward propulsion, mimicking the mackerel it was modeled after, whereas MPF locomotion is employed for maneuvering. It might be noteworthy to point out that Scombrids do not rely on the labriform swimming mode for steering. Nevertheless, since the aim of the authors was to allow the robot to perform efficient low-speed maneuvers, labriform locomotion was chosen. In the end, the intent of the whole project is not the construction of an artificial mackerel; that would be mimicking nature, which differs from biomimesis. The creation of a robotic fish that includes all the features of its biological counterpart remains a necessity when the purpose is to investigate the fluid mechanics of swimming; but in general, the aim of bio-inspired design is to apply the marine animal locomotion principle to solve the major design and control issues of underwater vehicles, such as power-efficient cruising and maneuverability.

5. Conclusions

Bio-inspired solutions devised for Autonomous Underwater Robots are currently investigated by researchers as a source of propulsive improvement. To address this ambitious objective, the authors designed a carangiform swimming robot, which represents a compromise choice in terms of efficiency and maximum speed. However, the requirements of stabilizing a course and performing turns were not fulfilled in their previous works. In this paper, the authors thus proposed a novel transmission mechanism capable of improving the vehicle maneuvering capacities. The system transforms the constant angular velocity of a single rotary actuator into the pitching–yawing rotation of the robot's pectoral fins. The individual rotation components are driven by a dedicated sub-system that recouples on the fin shaft to generate the composite rotation of the fins. In this way, the inherent synchronization of the system is maintained at a constant level by mechanism kinematics. When compared to a driving mechanism composed of multiple servomotors, the proposed solution reduces the effort of the control system, inertia, and encumbrance, whereas waterproofing issues are minimal because only one motor needs to be sealed.

The current project is now at the beginning of the prototyping phase. Future work will focus on the design adjustments necessary to perform C-shape maneuvers. At the same time, the multibody platform will undergo extensive improvements to simulate depth-changing maneuvers.

Author Contributions: Conceptualization, D.C.; methodology, M.P.; software, D.C.; validation, D.C., M.P. and C.S.; formal analysis, D.C.; investigation, D.C.; resources, M.C. and D.S.; data curation, M.P.; writing—original draft preparation, D.C.; writing—review and editing, M.C.; visualization, M.C.; supervision, D.S.; project administration, D.S. All authors have read and agreed to the published version of the manuscript.

Funding: This research received no external funding.

Institutional Review Board Statement: Not applicable.

Informed Consent Statement: Not applicable.

Data Availability Statement: Not applicable.

Conflicts of Interest: The authors declare no conflict of interest.

References

1. Scaradozzi, D.; Palmieri, G.; Costa, D.; Pinelli, A. BCF swimming locomotion for autonomous underwater robots: A review and a novel solution to improve control and efficiency. *Ocean Eng.* **2017**, *130*, 437–453. [CrossRef]
2. Sfakiotakis, M.; Lane, D.; Davies, J. Review of fish swimming modes for aquatic locomotion. *IEEE J. Ocean. Eng.* **1999**, *24*, 237–252. [CrossRef]
3. Costa, D.; Palmieri, G.; Scaradozzi, D.; Callegari, M. Experimental Validation of a Bio-Inspired Thruster. *J. Dyn. Syst. Meas. Control* **2021**, *143*, 081004. [CrossRef]
4. Costa, D.; Palmieri, G.; Palpacelli, M.C.; Callegari, M.; Scaradozzi, D. Design of a bio-inspired underwater vehicle. In Proceedings of the 2016 12th IEEE/ASME International Conference on Mechatronic and Embedded Systems and Applications (MESA), Auckland, New Zealand, 29–31 August 2016; pp. 1–6. [CrossRef]
5. Costa, D.; Palmieri, G.; Palpacelli, M.C.; Scaradozzi, D.; Callegari, M. Design of a Carangiform Swimming Robot through a Multiphysics Simulation Environment. *Biomimetics* **2020**, *5*, 46. [CrossRef]
6. Blake, R.W. The mechanics of labriform locomotion I. Labriform locomotion in the angelfish (*Pterophyllum eimekei*): An analysis of the power stroke. *J. Exp. Biol.* **1979**, *82*, 255–271. [CrossRef]
7. Vogel, S. *Life in Moving Fluids: The Physical Biology of Flow*; Princeton University Press: Princeton, NJ, USA, 2020.
8. Kato, N.; Furushima, M. Pectoral fin model for maneuver of underwater vehicles. In Proceedings of the Symposium on Autonomous Underwater Vehicle Technology, Monterey, CA, USA, 2–6 June 1996.
9. Westneat, M.W.; Walker, J.A. Applied aspects of mechanical design, behavior, and performance of pectoral fin swimming in fishes. In Proceedings of the Special Session on Bio-Engineering Research Related to Autonomous Underwater Vehicles, Durham, NH, USA, 10–11 September 1997. Available online: https://apps.dtic.mil/sti/pdfs/ADA330550.pdf#page=157 (accessed on 21 November 2021).
10. Naser, F.A.; Rashid, M.T. Design, modeling, and experimental validation of a concave-shape pectoral fin of labriform-mode swimming robot. *Eng. Rep.* **2019**, *1*, e12082. [CrossRef]
11. Sitorus, P.E.; Nazaruddin, Y.Y.; Leksono, E.; Budiyono, A. Design and implementation of paired pectoral fins locomotion of labriform fish applied to a fish robot. *J. Bionic Eng.* **2009**, *6*, 37–45. [CrossRef]
12. Zhong, Y.; Li, Z.; Du, R. The design and prototyping of a wire-driven robot fish with pectoral fins. In Proceedings of the 2013 IEEE International Conference on Robotics and Biomimetics (ROBIO), Shenzhen, China, 12–14 December 2013.
13. Li, Z.; Ge, L.; Xu, W.; Du, Y. Turning characteristics of biomimetic robotic fish driven by two degrees of freedom of pectoral fins and flexible body/caudal fin. *Int. J. Adv. Robot. Syst.* **2018**, *15*, 1729881417749950. [CrossRef]
14. Behbahani, S.B.; Tan, X. Bio-inspired flexible joints with passive feathering for robotic fish pectoral fins. *Bioinspiration Biomim.* **2016**, *11*, 036009. [CrossRef] [PubMed]
15. Singh, N.; Gupta, A.; Mukherjee, S. A dynamic model for underwater robotic fish with a servo actuated pectoral fin. *SN Appl. Sci.* **2019**, *1*, 659. [CrossRef]
16. Borazjani, I.; Sotiropoulos, F. Numerical investigation of the hydrodynamics of carangiform swimming in the transitional and inertial flow regimes. *J. Exp. Biol.* **2008**, *211*, 1541–1558. [CrossRef] [PubMed]
17. Drucker, E.G.; Walker, J.A.; Westneat, M.W. Mechanics of pectoral fin swimming in fishes. *Fish Physiol.* **2005**, *23*, 369–423.
18. Lauder, G.V.; Jayne, B.C. Pectoral fin locomotion in fishes: Testing drag-based models using three-dimensional kinematics. *Am. Zool.* **1996**, *36*, 567–581. [CrossRef]
19. Fischer, I.S.; Paul, R.N. Kinematic Displacement Analysis of a Double-Cardan-Joint Driveline. *J. Mech. Des.* **1991**, *113*, 263–271. [CrossRef]
20. Costa, D.; Palmieri, G.; Callegari, M.; Scaradozzi, D. Multi-body Analysis of a Bio-inspired Underwater Robot. In *Advances in Italian Mechanism Science*; Mechanisms and Machine Science; Springer: Berlin/Heidelberg, Germany, 2019; Volume 68.
21. Antonelli, G. *Underwater Robots: Motion and Force Control of Vehicle-Manipulator Systems*; Springer: Berlin/Heidelberg, Germany, 2003.
22. Fossen, T.I. *Marine Control System-Guidance, Navigation and Control of Ships, Rigs and Underwater Vehicles*; Marine Cybernetics: Trondheim, Norway, 2002.
23. Fossen, T.I. *Handbook of Marine Craft Hydrodynamics and Motion Control*; John Wiley & Sons: Hoboken, NJ, USA, 2011.
24. Costa, D.; Palmieri, G.; Palpacelli, M.-C.; Panebianco, L.; Scaradozzi, D. Desing of a Bio-inspired Autonomous Underwater Robot. *J. Intell. Robot. Syst.* **2018**, *91*, 181–192. [CrossRef]

25. Prandtl, L.; Tietjens, O.G.; Den Hartog, J.P. *Applied Hydro-And Aeromechanics*; Dover: New York, NY, USA, 1957; Volume 35.
26. Scaradozzi, D.; Conte, G.; De Capua, G.P.; Sorbi, L.; Luciani, C.; De Cecco, P.G.; Sorci, A. Innovative technology for studying growth areas of Posidonia oceanica. In Proceedings of the 2009 IEEE Workshop on Environmental, Energy, and Structural Monitoring Systems, Crema, Italy, 25 September 2009; pp. 71–75.

robotics

MDPI

Article

Modal Kinematic Analysis of a Parallel Kinematic Robot with Low-Stiffness Transmissions

Paolo Righettini *, Roberto Strada and Filippo Cortinovis

Department of Engineering and Applied Sciences, University of Bergamo, 24044 Dalmine (BG), Italy; roberto.strada@unibg.it (R.S.); filippo.cortinovis@unibg.it (F.C.)
* Correspondence: paolo.righettini@unibg.it

Abstract: Several industrial robotic applications that require high speed or high stiffness-to-inertia ratios use parallel kinematic robots. In the cases where the critical point of the application is the speed, the compliance of the main mechanical transmissions placed between the actuators and the parallel kinematic structure can be significantly higher than that of the parallel kinematic structure itself. This paper deals with this kind of system, where the overall performance depends on the maximum speed and on the dynamic behavior. Our research proposes a new approach for the investigation of the modes of vibration of the end-effector placed on the robot structure for a system where the transmission's compliance is not negligible in relation to the flexibility of the parallel kinematic structure. The approach considers the kinematic and dynamic coupling due to the parallel kinematic structure, the system's mass distribution and the transmission's stiffness. In the literature, several papers deal with the dynamic vibration analysis of parallel robots. Some of these also consider the transmissions between the motors and the actuated joints. However, these works mainly deal with the modal analysis of the robot's mechanical structure or the displacement analysis of the transmission's effects on the positioning error of the end-effector. The discussion of the proposed approach takes into consideration a linear delta robot. The results show that the system's natural frequencies and the directions of the end-effector's modal displacements strongly depend on its position in the working space.

Keywords: parallel manipulator; robot modeling and simulation; robot design

Citation: Righettini, P.; Strada, R.; Cortinovis, F. Modal Kinematic Analysis of a Parallel Kinematic Robot with Low-Stiffness Transmissions. *Robotics* **2021**, *10*, 132. https://doi.org/10.3390/robotics10040132

Academic Editor: Raffaele Di Gregorio

Received: 15 November 2021
Accepted: 8 December 2021
Published: 10 December 2021

1. Introduction

Parallel robots are Parallel Kinematics Machines (PKM) which have been investigated for a long time, as demonstrated by some key papers [1–3]. The parallel kinematic configuration gives to these robotic systems a high stiffness/inertia ratio. This characteristic is due to closed kinematic chains, where some members are mainly subject to axial loads; hence, these links can be designed to be slender and light without the risk of decreasing the system's overall stiffness. Moreover, the parallel kinematic configuration allows the positioning of the driving motors on the fixed frame of the robot, reducing, therefore, the moving masses [4,5].

The linear delta kinematic configuration [3] is characterized by linear axes instead of the rotary ones used within the well-known classical Delta configuration [1,2]. As a result, for this configuration, the working volume is spatially extended along a direction whose length depends only on the length of the linear axes. Applications in the industrial field requiring high speed and working volume extended along a direction can use linear delta robots designed with a suitable transmission that on one hand permits high linear speed and on the other has low inertia. This configuration is suitable for the primary packaging station of a production line or for transferring production goods between two lines. Linear transmissions based on a timing belt are an interesting solution due to the low mass, high speed, long travel length and low cost, especially if compared with other kinds of linear transmissions, such as ball-screws.

It should be moreover remarked that the overall performance of the system depends not only on the kinematic structure of the manipulator, but also on the characteristics of transmissions between motors and joints; hence, in order to achieve the expected system performance, the synthesis requires a synergistic design approach [6].

Despite the attractive characteristics of a belt transmission from the inertial point of view, it is also a well-known intrinsically deformable device that introduces significant flexibility between motors and driven joints. Several researchers have investigated belt transmissions for many years under several aspects. Some authors defined a methodology for the estimation of the characteristic parameters of a linear belt drive, considering lumped parameter modeling, mainly for control purpose [7]. Other authors focused their attention on the acoustic radiation of a timing belt due to its vibrations. Such dynamic behavior was investigated by means of a numerical multi-body model [8]. As far as modeling is concerned, many authors developed numerical models of timing belts, taking into consideration also the interaction between the belt and pulley [9–11]. Other recent works also focused on the influence of the tensioner on the dynamic behavior of the belt [12,13], while other papers dealt with new issues concerning the usage of oval pulleys in belt transmissions [14,15]. As confirmation of the researcher's interest in the dynamic behavior of timing belts, there is also an interesting review on the methodologies for modeling and analysis of the axial and torsional vibration of these kinds of transmissions [16].

Regarding the dynamics of a parallel kinematics robot, many works can be found in the literature where the modal analysis of the system is performed, but only taking into consideration the robotic system and not the contribution due to the transmissions' stiffness. In this context, some authors performed the modal analysis, developing the models by means of a commercial software [17–20], while other authors developed their own model [21–24], sometimes with the aim to investigate the low-order dynamics of the system [25]; in all these cases, nothing about the transmission is taken into consideration.

In the literature, papers focusing on both the robotic system and on the transmission can be found. However, they generally concern the definition of control algorithms for the suppression of vibrations [26,27], the estimation of the system's parameters [28], or the definition of the positioning error in the workspace [29,30].

Our research is related to investigating the dynamic vibration behavior of parallel kinematics robots, where the mechanical transmissions have lower stiffness than the one of the parallel kinematic structure. This approach is quite different from the methodologies detailed in the previously cited works. In particular, we are interested in the overall system, transmissions included, focusing on the influence of the transmissions' stiffness on the natural frequencies and the mode shapes of the parallel robot.

The main contribution of our paper is to highlight, taking into consideration also the transmissions and their stiffness, how the modes of vibration of the belt transmissions' carriages, which depend on the mass distribution within the system and on the coupling effect due to the parallel kinematic chains, are reflected to the end-effector as a function of its position within the workspace. The results are the end-effector's modal displacements' directions due to the system's characteristics. Moreover, these results are a function of the end-effector position in the workspace. For this reason, we have named our approach "Modal Kinematic Analysis".

The paper takes into consideration a 3 d.o.f. linear delta produced by the Italian company Mechatronics and Dynamic Devices s.r.l. [31]. This device has linear belt transmissions on which we have already carried out some experimental activities, showing the influence of the belt transmissions' stiffness on the dynamics of the overall system [32]. As previously mentioned, the stiffness of the parallel kinematic part is higher than the rigidity of the belts; this is why we have modeled the parallel kinematic part as being rigid, considering the belt transmissions to be the only deformable elements. Regarding the transmissions, we already highlighted that we are not interested in analyzing specific aspects, such as acoustic emissions, due to the belt's vibration or the contact between the belt and pulley. Moreover, in the case of a linear drive system that moves a carriage with a mass significantly higher than the belt's mass, the latter can be neglected; hence, the transmission can be modeled as the lumped stiffnesses, one for each part of the belt, changing with the configuration of the transmission, i.e., changing with the carriages positions.

2. Materials and Methods

2.1. The Parallel Robot under Analysis

The parallel robot on which we have applied our "Modal Kinematic Analysis" approach is a 3 d.o.f. parallel kinematic machine produced by the Italian company Mechatronics and Dynamic Devices s.r.l. [31] that is characterized by the typical linear delta kinematic configuration.

Figures 1 and 2 present the linear delta, whose main characteristics are the following:

- The three linear axes are parallel.
- Each linear axis is composed of a linear belt transmission. The belt is a HTD-5 characterized by a 15 mm width and a specific stiffness (or stiffness per unit of length and unit of width) of $k_{sp} = 2.42 \times 10^6$ N/m.
- The system is driven by means of brushless motors, characterized by a nominal torque of 0.7 Nm; a maximum velocity of 10,000 rpm; and a rotor inertia of 0.017×10^{-3} kg·m^2.
- The motors are connected to the driven pulley of the linear belt transmission by means of a planetary gearbox characterized by a reduction ratio equal to 10.
- The maximum axis stroke is 1.2 m.
- The distance l_r between axes is 200 mm.
- The length l_d of the links connecting the carriages to the end-effector is 400 mm.

The constraints between the links and, respectively, the carriages and the mobile platform are realized by means of universal joints, highlighted in blue for the carriages and in red for the mobile platform in Figure 3. Hence, each link is composed of two rods realizing a four bar linkage with a parallelogram configuration.

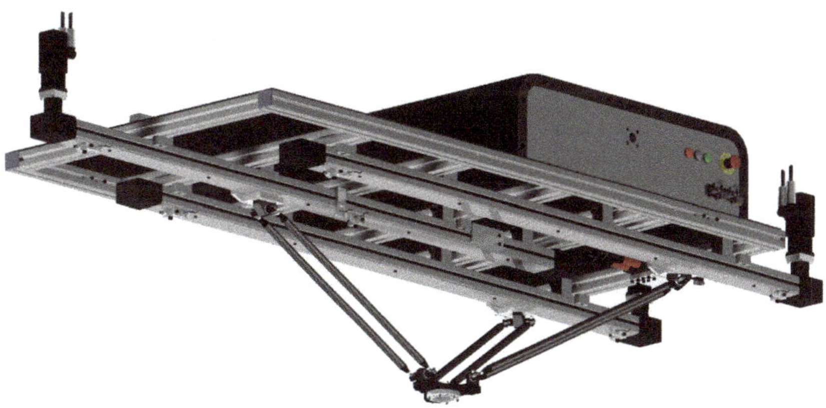

Figure 1. Linear delta under analysis.

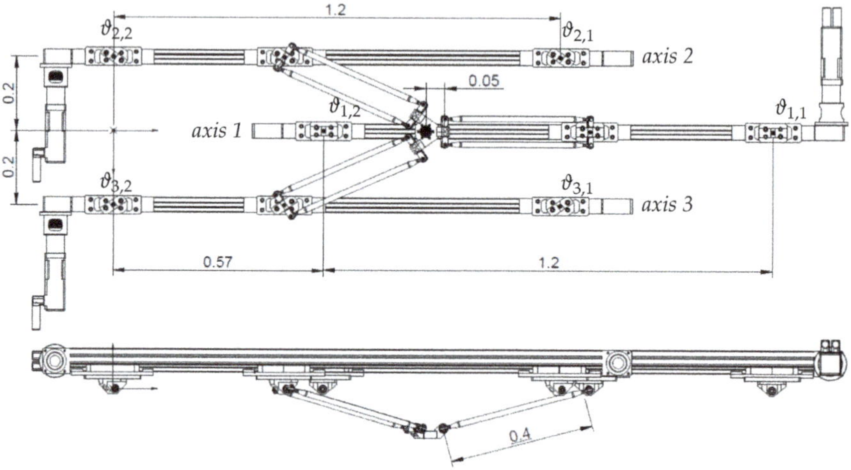

Figure 2. Diagram of the linear delta.

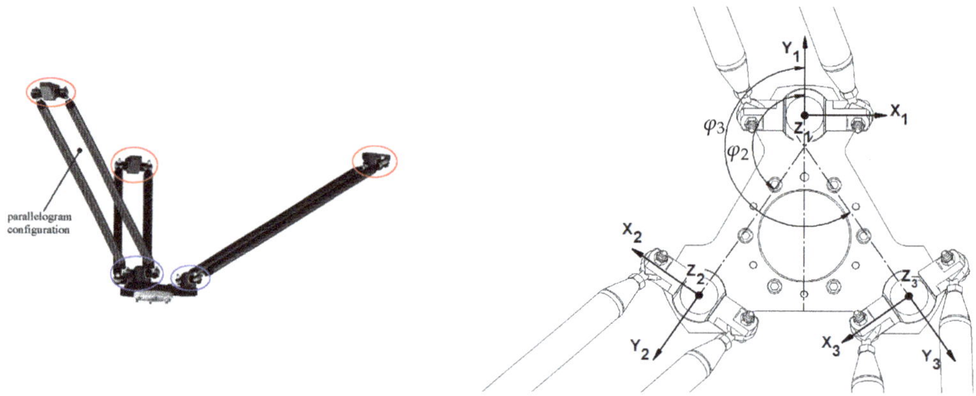

Figure 3. On the left: connection links and joints. On the right: local coordinate systems on the mobile platform.

2.2. Kinematics and Dynamics of the Parallel Kinematic Part

This section presents the model for the kinematics and the dynamics of the parallel kinematic part of the manipulator. This part has very high stiffness and low weight due to the carbon fiber composite material of the rods. For these reasons, it is possible to say that the flexibility of the parallel kinematic part can be considered negligible with respect to the compliance of the linear belt transmissions. This subsystem can be thus modeled as being composed of rigid bodies, whereas the only flexible elements are the transmissions. Moreover, the belt's mass is negligible with respect to the carriages mass; accordingly, we neglect the inertial contributions of the belt, and simply model its stiffness using configuration-dependent lumped parameters.

The rigid structure of the manipulator is responsible not only for the dynamic coupling between the linear axes, but also for the overall mass distribution of the system; the rigid-body kinematics and dynamics of this subsystem—composed of the mobile platform, of the distal links and of the trucks—must, therefore, be thoroughly accounted for.

The kinematics of the manipulator are largely analogous to those of the Delta robot designed by Reymond Clavel and analyzed, for example, in [1,2]; as such, this mechanical structure results in a mobile platform with three translational degrees of freedom [3]. Figure 2 shows the reference frame with respect to which the end effector's position is measured, while Figure 3 depicts three sets of local coordinates useful to completely describe the kinematics of the parallel bar linkages.

The angular positions $\varphi_j, j = 1,2,3$ of the universal joints that connect the distal rods to the platform are also shown in Figure 3. The free coordinates x_1, x_2 and x_3 that govern the kinematics of the manipulator are those relative to the longitudinal displacement of the three trucks. Given the geometric parameter l_r that defines the distance between two adjacent linear guides, the truck positions $\boldsymbol{e}_j$ are trivially expressed as functions of x_j:

$$\boldsymbol{e}_1 = \begin{bmatrix} x_1 \\ 0 \\ 0 \end{bmatrix}, \boldsymbol{e}_2 = \begin{bmatrix} x_2 \\ -l_r, \\ 0 \end{bmatrix}, \boldsymbol{e}_3 = \begin{bmatrix} x_3 \\ l_r, \\ 0 \end{bmatrix}. \tag{1}$$

The direct and inverse functions between these quantities and the center point $\boldsymbol{p}$ of the moving platform must furthermore be determined. The chief relationships that allow to do so are the vector loop equations written for each kinematic chain:

$$\boldsymbol{e}_j + \boldsymbol{l}_{d,j} + (\boldsymbol{p} - \boldsymbol{c}_j) = \boldsymbol{p} \quad \forall j \in [1,2,3]. \tag{2}$$

In Equation (2), $\boldsymbol{l}_{d,j}$ is the vector pointing from the j^{th}truck $\boldsymbol{e}_j$ to the position $\boldsymbol{c}_j$ of the j^{th}platform universal joint constraint, while the difference $(\boldsymbol{p} - \boldsymbol{c}_j)$ is a constant vector entirely determined by the geometry of the moving platform.

The vector loop equations, whose geometric meaning is depicted in Figure 4, must be satisfied for every assembled configuration of the robot. To exactly enforce them when the truck position vectors $\boldsymbol{e}_j$ are known, it is sufficient to intersect three spheres of radius l_d and centered in the points,

$$\boldsymbol{s}_j = \boldsymbol{e}_j + (\boldsymbol{p} - \boldsymbol{c}_j); \tag{3}$$

indeed, the intersection point coincides with the admissible $\boldsymbol{p}$, from which the constraint positions $\boldsymbol{c}_j$ can be then found.

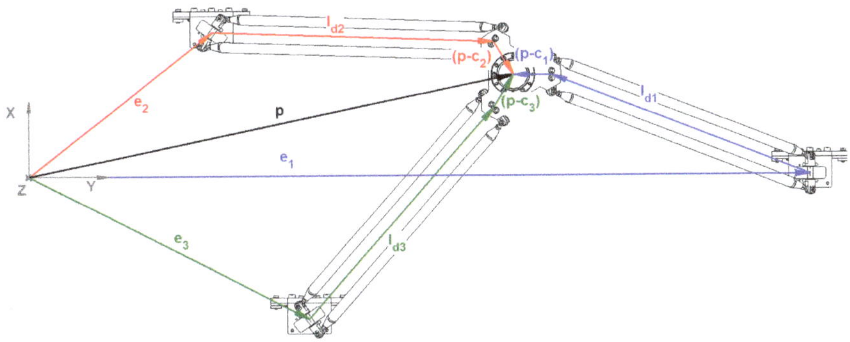

Figure 4. Vector loops for the three kinematic chains.

On the other hand, if $\boldsymbol{p}$ is given the points $\boldsymbol{c}_j$ can be immediately determined; the intersection of the sphere of radius l_d and center $\boldsymbol{c}_j$ with the segment generated by the j^{th}linear guide allows then the determination of $\boldsymbol{e}_j$ (and x_j) that satisfy the vector loop equations.

Other positional quantities of interest are the centers of mass of the distal linkages and the Euler angles describing the rotation of the linkage rods. Since the two rods of each linkage are constrained to perform the same motion, the kinematics of the entire distal link can be abstracted to those of its center-line.

Given that each distal rod is a slender, axially symmetric body, the position $\boldsymbol{d}_j$ of the center of mass of the j^{th}linkage can be expressed as a weighted average of the constraint and carriage position.

$$\boldsymbol{d}_j = \frac{W_c \boldsymbol{c}_j + W_t \boldsymbol{e}_j}{W_c + W_t}, \tag{4}$$

where W_c is the weight of the universal joint and W_t is the weight of the truck.

A physically meaningful description of the distal link rotation must take into account the actual configuration of the universal joint connecting each rod pair to the end-effector. With reference to Figure 5, where the red dashed line represents the direction of the generic link j, the most straightforward solution involves a joint allowing a rotation of the rod pair first around the local x-axis by an angle α_j, and then around the rotated local y$'$-axis by an angle β_j. The third Euler angle γ_j around the local z$''$-axis, directed as the distal rod itself, is kinematically constrained to be equal to zero.

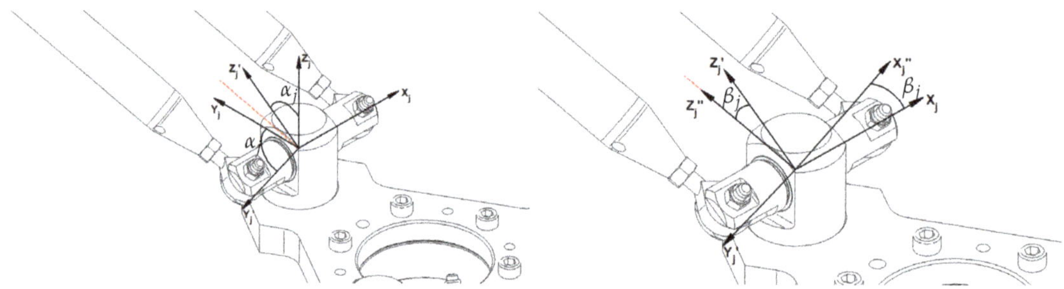

Figure 5. Euler angle for the definition of generic link j orientation.

Given the vector

$$\boldsymbol{u}_j = \boldsymbol{e}_j^{[j]} - \boldsymbol{c}_j^{[j]} \tag{5}$$

expressed in the jth frame of reference, these angles are easily computed as:

$$\alpha_j^{[j]} = \arctan\left(\frac{-u_{j,2}}{u_{j,3}}\right) \tag{6}$$

$$\beta_j^{[j]} = \arcsin\left(\frac{u_{j,1}}{||\boldsymbol{u}_j||}\right) \tag{7}$$

$$\gamma_j^{[j]} = 0\,. \tag{8}$$

Additionally, the rotation matrix $\boldsymbol{R}_{d,j}^{[j]}$ of the distal link can be constructed from the Euler angles through standard formulas for the composition of elementary rotations.

As the position kinematics of the platform are described as a sequence of geometric rather than analytical operations, the Jacobian analysis must also be developed in the same vein. The first goal is to determine the Jacobian matrix relating the time derivative of the truck coordinates to the velocity of the mobile platform.

To do so, the formula for the velocity kinematics of rigid bodies can be applied to the three kinematic chains connecting the platform to the base. Recalling that the platform can only translate and therefore $\dot{\boldsymbol{c}}_j = \dot{\boldsymbol{p}}$ the following equations can be written:

$$\dot{\boldsymbol{p}} = \dot{\boldsymbol{e}}_1 + \boldsymbol{\omega}_1 \wedge (\boldsymbol{p} - \boldsymbol{e}_1) \tag{9}$$

$$\dot{\boldsymbol{p}} = \dot{\boldsymbol{e}}_2 + \boldsymbol{\omega}_2 \wedge (\boldsymbol{p} - \boldsymbol{e}_2) \tag{10}$$

$$\dot{\boldsymbol{p}} = \dot{\boldsymbol{e}}_3 + \boldsymbol{\omega}_3 \wedge (\boldsymbol{p} - \boldsymbol{e}_3)\,. \tag{11}$$

The yet unknown distal angular velocities ω_j can be eliminated from Equations (9) and (11) by multiplying each side by the quantity $(p - e_j)^\top$. In matrix notation:

$$\begin{bmatrix} (p-e_1)^\top \\ (p-e_2)^\top \\ (p-e_3)^\top \end{bmatrix} \dot{p} = \begin{bmatrix} (p-e_1)^\top \dot{e}_1 \\ (p-e_2)^\top \dot{e}_2 \\ (p-e_3)^\top \dot{e}_3 \end{bmatrix}. \tag{12}$$

The truck velocities can then be expressed in terms of the time derivatives of the free coordinates by highlighting the Jacobian matrices $D_{e,j}$ of e_j with respect to $x = \begin{bmatrix} x_1 & x_2 & x_3 \end{bmatrix}^\top$:

$$\dot{e}_1 = \overbrace{\begin{bmatrix} 0 & 0 & 0 \\ 1 & 0 & 0 \\ 0 & 0 & 0 \end{bmatrix}}^{D_{e,1}} \dot{x}, \quad \dot{e}_2 = \overbrace{\begin{bmatrix} 0 & 0 & 0 \\ 0 & 1 & 0 \\ 0 & 0 & 0 \end{bmatrix}}^{D_{e,2}} \dot{x}, \quad \dot{e}_3 = \overbrace{\begin{bmatrix} 0 & 0 & 0 \\ 0 & 0 & 1 \\ 0 & 0 & 0 \end{bmatrix}}^{D_{e,3}} \dot{x}, \tag{13}$$

where $\dot{x} = \begin{bmatrix} \dot{x}_1 & \dot{x}_2 & \dot{x}_3 \end{bmatrix}^\top$. By substitution of Equation (13) in Equation (12), the following terms can be highlighted:

$$D_l \dot{p} = D_r \dot{x}. \tag{14}$$

More explicitly,

$$D_l = \begin{bmatrix} (p-e_1)^\top \\ (p-e_2)^\top \\ (p-e_3)^\top \end{bmatrix}, \; D_r = \begin{bmatrix} (p-e_1)^\top D_{e,1} \\ (p-e_2)^\top D_{e,2} \\ (p-e_3)^\top D_{e,3} \end{bmatrix}. \tag{15}$$

The platform Jacobian matrix can finally be expressed as $D_p = D_l^{-1} D_r$.

Each Jacobian matrix related to the motion of the center of mass of the jth distal linkage is then easily derived:

$$D_{d,j} = \frac{W_c D_p + W_t D_{e,j}}{W_c + W_t}. \tag{16}$$

Given that both positions and velocities of the notable points of the structure are made explicit, it is also possible to determine analytically the time derivative of the platform Jacobian matrix as a function of x and $\dot{x}$:

$$\dot{D}_p = D_l^{-1}(\dot{D}_r - \dot{D}_l D_p). \tag{17}$$

To compute $\dot{D}_r$ and $\dot{D}_l$, it is enough to differentiate, with respect to time, each element of D_r and D_l, which can be easily done thanks to the clear geometric meaning of these matrices.

The Jacobian analysis of the distal link rotations might then be performed. Equations (6)–(8) can be expressed in vector form as follows:

$$\psi_j = f(u_j(x)), \tag{18}$$

where $\psi_j = \begin{bmatrix} \alpha_j & \beta_j & \gamma_j \end{bmatrix}^\top$.

The function f can be easily differentiated with respect to the components of u_j to yield the matrix D_f. On the other hand, the Jacobian matrix of u_j with respect to x is

$$D_{u,j} = R_{f,j}^\top (D_{e,j}^{[j]} - D_p{}^{[j]}), \tag{19}$$

where $R_{f,j}$ is the constant rotation matrix due to φ_j. It follows that

$$\dot{\psi}_j = D_f D_{u,j} \dot{x} = D_{\psi,j} \dot{x}. \tag{20}$$

A linear relationship $\omega_j^{[j]} = \Psi_j \dot{\psi}_j$ subsists between the angular velocity of the distal rod and the time derivative of its Euler angles, with Ψ_j being defined as

$$\Psi_j = \begin{bmatrix} 1 & 0 & \sin(\beta_j) \\ 0 & \cos(\alpha_j) & -\cos(\beta_j)\sin(\alpha_j) \\ 0 & \sin(\alpha_j) & \cos(\beta_j)\cos(\alpha_j) \end{bmatrix}. \tag{21}$$

As a result the angular velocity of the distal rods can be expressed as

$$\omega_j^{[j]} = \Psi_j D_{\psi,j} \dot{x}. \tag{22}$$

Finally, in the absolute frame of reference,

$$\omega_j = R_{f,j} \Psi_j D_{\psi,j} \dot{x} = D_{\omega,j} \dot{x}. \tag{23}$$

Since the kinematics of the rigid subsystem is expressed as a function of x and $\dot{x}$, its kinetic energy can be written as

$$T_{ss} = \frac{1}{2} \dot{x}^\top \left[D_p{}^\top m_p D_p + \sum_{j=1}^{3} \left(D_{e,j}^\top m_t D_{e,j} + D_{d,j}^\top m_d D_{d,j} + D_{\omega,j}^\top I_{d,j} D_{\omega,j} \right) \right] \dot{x}. \tag{24}$$

The mass matrix M_{ss} of the manipulator, which appears inside the parentheses of Equation (24), is thus composed of terms related to the translations of the mobile platform; the translations of the trucks; the translations of the center of mass of each distal linkage; and the rotations of the distal links. Accordingly, the masses m_t, m_d and m_p of the trucks, the distal linkages and the platform appear alongside the inertia matrix $I_{d,j}$ of the distal links. These quantities are set for the following numerical analyses as

- $m_t = 1\,\text{kg}$,
- $m_d = 0.7\,\text{kg}$,
- $m_p = 1.5\,\text{kg}$.

Assuming $W_t = W_c$, the distal link inertia is calculated in the principal and barycentric frame as

$$\tilde{I}_{d,j} = m_d \frac{l_d^2}{12} \begin{bmatrix} 1 & 0 & 0 \\ 0 & 1 & 0 \\ 0 & 0 & 0 \end{bmatrix}. \tag{25}$$

Conversely, the gravitational potential of the subsystem is related only to the mobile platform and to the center of mass of the distal links, as only these are allowed to move along the vertical direction. Indicating as g the gravitational constant, the gravitational potential U_g can be written as:

$$U_g = -g \begin{bmatrix} 0 & 0 & 1 \end{bmatrix} \left(m_p p + \sum_{j=1}^{3} m_d d_j \right). \tag{26}$$

The gradient of U_g can be computed straightforwardly as follows:

$$\nabla U_g = -g \begin{bmatrix} 0 & 0 & 1 \end{bmatrix} \left(m_p D_p + \sum_{j=1}^{3} m_d D_{d,j} \right). \tag{27}$$

2.3. Belt Transmission Dynamics

As already mentioned, the mass of the belts can be neglected; therefore, the kinetic energy associated to the belt transmission is determined by the rotational inertia not only of the pulleys, but more significantly of the three motors and gearboxes, which are assumed to rotate rigidly with the actuated pulley.

To properly describe the system, six additional coordinates $\vartheta_{j,1}$ $\vartheta_{j,2}$ (with $j = 1, 2, 3$) associated to the rotation of the pulleys are required. Figure 6 shows a diagram of the timing belt in which the stiffnesses and the coordinates are highlighted for the jth axis. Which pulley is driven by the motor is shown in Figure 2.

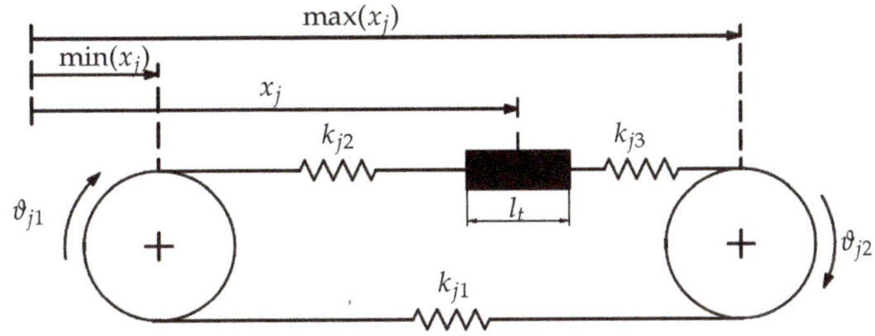

Figure 6. Diagram of the jth linear belt transmission.

Denoting as $I_{p,j,k}$ the rotational inertia of the pulley described by the coordinate $\vartheta_{j,k}$, the kinetic energy of the three transmissions can be expressed in the form

$$T_t = \frac{1}{2}\sum_{j=1}^{3}\left(I_{p,j,1}\dot{\vartheta}_{j,1}^2 + I_{p,j,2}\dot{\vartheta}_{j,2}^2\right), \tag{28}$$

By introducing the array of pulley coordinates

$$\boldsymbol{\vartheta}^\top = \begin{bmatrix}\vartheta_{1,1} & \vartheta_{2,1} & \vartheta_{3,1} & \vartheta_{1,2} & \vartheta_{2,2} & \vartheta_{3,2}\end{bmatrix} \tag{29}$$

and by suitably collecting the mass terms, the kinetic energy of the transmissions can be expressed in matrix notation as

$$T_t = \frac{1}{2}\dot{\boldsymbol{\vartheta}}^\top \boldsymbol{M}_t \dot{\boldsymbol{\vartheta}}\,. \tag{30}$$

In Equation (30) the transmission mass matrix $\boldsymbol{M}_t$ is a 6-by-6 diagonal and constant matrix. For the subsequent analyses, pulleys having the following characteristics are considered:

- Radius $R_p = 35\,\text{mm}$,
- Inertia around the rotation axis $I_{p,j} = 130 \times 10^{-6}\,\text{kg}\cdot\text{m}^2$.

Furthermore, the motor and gearbox inertia—suitably projected on the machine side through the gearbox reduction rate—is added to the inertia of the actuated pulleys.

3. Overall Dynamics

The total kinetic energy of the system can then be expressed as:

$$T(\boldsymbol{x}, \boldsymbol{\vartheta}, \dot{\boldsymbol{x}}, \dot{\boldsymbol{\vartheta}}) = T_{ss}(\boldsymbol{x}, \dot{\boldsymbol{x}}) + T_t(\boldsymbol{\vartheta}, \dot{\boldsymbol{\vartheta}})\,. \tag{31}$$

Introducing the array of the free coordinates of the overall system as

$$q = \begin{bmatrix}\boldsymbol{x}\\ \boldsymbol{\vartheta}\end{bmatrix}, \tag{32}$$

the complete mass matrix can be easily assembled:

$$M = \begin{bmatrix} M_{ss} & 0 \\ 0 & M_t \end{bmatrix}. \tag{33}$$

The overall mass matrix is a function of the trucks' positions. In a similar way, the mechanical stiffness of each transmission changes with the position of the truck, as this defines the free lengths of the three segments of which the timing belt is composed. As already highlighted in Section 2.1, the specific stiffness k_{sp} quantifies the stiffness of the timing belt for each unit length and unit width of the belt itself, and as such, it is one of the parameters commonly specified within the product datasheet. For a given configuration, the stiffness coefficients of the belt sections are therefore computed according to the following equation:

$$k_{j,k} = \frac{w_b k_{sp}}{l_{f,j,k}}, \tag{34}$$

in which w_b is the belt width, and $l_{f,j,k}$ is the free length of the considered belt segment.

More explicitly, the stiffness coefficients for each belt transmission are computed as

$$k_{j,1} = w_b k_{sp} (\max(x_j) - \min(x_j))^{-1} \tag{35}$$

$$k_{j,2} = w_b k_{sp} \left(x_j^* - \min(x_j) - \frac{1}{2} l_t \right)^{-1} \tag{36}$$

$$k_{j,3} = w_b k_{sp} \left(\max(x_j) - x_j^* - \frac{1}{2} l_t \right)^{-1}, \tag{37}$$

where x_j^* represents the undeformed j^{th} truck coordinate, l_t is the length of the truck, and $\min(x_j)$ and $\max(x_j)$, shown in Figure 6, are fixed structural parameters that correspond to the maximum and minimum displacements achievable by an ideally dimensionless cart.

If large rigid motions of the system are considered, the elastic actions do not admit a potential function, due to the non-constant stiffness coefficients. However, if small displacements around a given undeformed configuration are to be investigated, the stiffness variations can be neglected, and an elastic potential for each belt can be determined as

$$U_{el,j} = -\frac{1}{2} \left(k_{j,1} (R_p (\Delta\vartheta_{j,2} - \Delta\vartheta_{j,1}))^2 + k_{j,2} (\Delta x_j - R_p \Delta\vartheta_{j,2})^2 + k_{j,3} (R_p \Delta\vartheta_{j,1} - \Delta x_j)^2 \right). \tag{38}$$

Here, the variables $\Delta\vartheta_{j,1}$, $\Delta\vartheta_{j,2}$ and Δx_j are to be interpreted as displacements around the undeformed configuration $\vartheta_{j,1}^*$, $\vartheta_{j,2}^*$, x_j^*:

$$\Delta\vartheta_{j,1} = \vartheta_{j,1} - \vartheta_{j,1}^* \tag{39}$$

$$\Delta\vartheta_{j,2} = \vartheta_{j,2} - \vartheta_{j,2}^* \tag{40}$$

$$\Delta x_j = x_j - x_j^* . \tag{41}$$

The potential function of the free system can then be obtained by summation of the gravitational and elastic potentials:

$$U(q) = U_g(x) + \sum_{j=1}^{3} U_{el,j}(q). \tag{42}$$

The dynamic model describes a semi-definite system, which therefore can undergo rigid motions. Indeed, the motors apply torques to the system and do not set the position of the actuated pulleys. The equilibrium condition of the system, for a given position of the end-effector, can, however, be evaluated by introducing static torques $\tau_{st,1}$, $\tau_{st,2}$, $\tau_{st,3}$ applied to the actuated pulleys in order to keep the end-effector in the desired configuration.

The external generalized forces acting on the generalized coordinates of the system can therefore be expressed in vector notation as

$$\boldsymbol{Q} = \begin{bmatrix} 0 & 0 & 0 & \tau_{st,1} & 0 & 0 & 0 & \tau_{st,2} & \tau_{st,3} \end{bmatrix}^{\top}. \tag{43}$$

It should be noted that the vector $\boldsymbol{Q}$ defined in (43) is consistent with the location of the motors reported in Figure 2, and thus properly distinguishes the actuated degrees of freedom (namely $\vartheta_{1,1}$, $\vartheta_{2,2}$ and $\vartheta_{3,2}$) from the passive ones. In more detail, the position of the static torques within vector $\boldsymbol{Q}$ corresponds to the position of the driven pulleys' rotational angles in vector $\boldsymbol{q}$. The other elements of vector $\boldsymbol{Q}$ are equal to zero because they correspond to the position of the non-actuated coordinates.

A static equilibrium configuration should satisfy the condition

$$\nabla U(\boldsymbol{q}) + \boldsymbol{Q} = 0\,. \tag{44}$$

The solution of the set of Equation (44) yields the static equilibrium configuration around which we develop the modal analysis. In particular, the static torques and the pulleys' rotations are the unknowns of the problem, while the displacements of the trucks, which alone define the position of the mobile platform, are fixed. This choice is motivated by the need to express the several results of the modal analysis as functions of the mobile platform coordinates, as this can shed additional insight also on the functional characteristics of the robotic device.

4. Configuration-Dependent Modal Analysis

The modal analysis around a given equilibrium configuration $\boldsymbol{q}_{eq}$ can be performed by substituting into Lagrange's equations the approximated kinetic energy and potential functions.

In particular,

$$\widetilde{T}(\dot{\boldsymbol{q}}) = T(\boldsymbol{q}_{eq}, \dot{\boldsymbol{q}}) \tag{45}$$

$$\widetilde{U}(\boldsymbol{q}) = U(\boldsymbol{q}_{eq}) + \nabla U(\boldsymbol{q}_{eq})(\boldsymbol{q} - \boldsymbol{q}_{eq}) + \frac{1}{2}(\boldsymbol{q} - \boldsymbol{q}_{eq})^{\top}\boldsymbol{H}_{U}(\boldsymbol{q}_{eq})(\boldsymbol{q} - \boldsymbol{q}_{eq}), \tag{46}$$

where $\boldsymbol{H}_U$ is the Hessian matrix of the total potential.

Lagrange's equations assume the form

$$\frac{d}{dt}\frac{\partial \widetilde{T}}{\partial \dot{\boldsymbol{q}}} - \frac{\partial \widetilde{U}}{\partial \boldsymbol{q}} = \boldsymbol{Q}\,. \tag{47}$$

Remembering the equilibrium conditions, Equation (47) can be written in the familiar matrix notation,

$$\boldsymbol{M}\ddot{\boldsymbol{q}} + \boldsymbol{K}\boldsymbol{q} = 0\,, \tag{48}$$

in which the stiffness matrix is simply defined as $\boldsymbol{K} = -\boldsymbol{H}_U$.

It might be observed that the Hessian matrix associated to U_{el}, can be analytically obtained by double differentiation, while the one relative to the gravitational potential U_g is computed using the following property of the time derivative of the Jacobian matrices—chiefly $\boldsymbol{D}_p$—that govern the gradient ∇U_g,

$$\dot{\boldsymbol{D}}_p(\boldsymbol{x}, \dot{\boldsymbol{x}}) = \sum_{j=1}^{3}\frac{\partial}{\partial x_j}\boldsymbol{D}_p\dot{x}_j\,. \tag{49}$$

Evaluating $\dot{\boldsymbol{D}}_p(\boldsymbol{x}, \delta_{jk})$, with δ_{jk} being Kronecker's Delta, the partial derivatives $\frac{\partial}{\partial x_j}\boldsymbol{D}_p$ can be recovered, from which, in turn, the Hessian matrix of the gravitational potential can be straightforwardly constructed.

The mass and stiffness matrices are constant for each position $\boldsymbol{q}_{eq}$, taken into account inside the investigated workspace. For each position, an eigenvalue problem for the matrix $\boldsymbol{M}^{-1}\boldsymbol{K}$ is then set up and solved in order to highlight the configuration-dependent natural frequencies and modal vectors of the system. As expected, the first three modes represent the rigid motions of the system, while the remaining six are proper vibration modes.

Although a detailed workspace analysis of the manipulator is outside the scope of this work, the investigation refers to a significant working plane selected by the Jacobian's condition number of the rigid subsystem. The Jacobian's condition number of the chosen plane is more uniformly distributed and has the lower mean value, i.e., 3.5. Figure 7 shows the selected plane at the position $z = -245$ mm. An equally distributed sampling of this plane defines the investigated end-effector's pose, giving the equilibrium positions $\boldsymbol{q}_{eq}$.

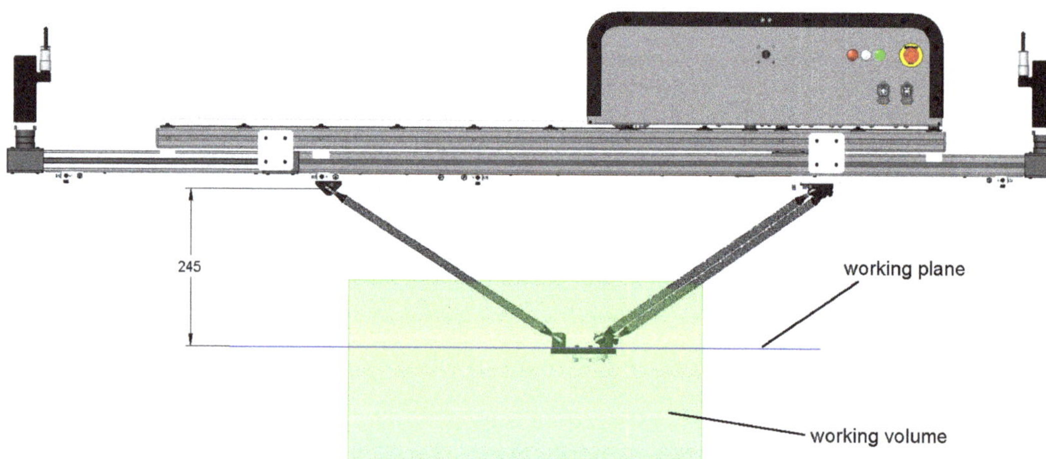

Figure 7. Reference plane position within the working volume (represented in green).

5. Results and Discussion

This section presents the results of the modal analysis developed on the end-effector's positions belonging to the plane defined in the previous section. The software package selected for the implementation of the system's model, and for the development of the modal analysis for different end-effector's positions is MathWorks® MATLAB®, owing to its matrix processing and data visualization capabilities. The first results reported are the natural frequencies for each of the six modes of vibration (the remaining three, related to the rigid motion of the system, are not included); $f_1 - f_6$ are the natural frequencies (expressed in Hz) of the modes 1 to 6, sorted in ascending order. In particular, Figure 8 shows how the natural frequencies change by changing the position(x_p, y_p) of the end-effector on the reference plane. The red markers are positioned along a line corresponding to the direction of the central axis (axis 1) at points, within the useful working area (boundary effects excluded), where the natural frequencies have their maximum and minimum value. The black markers qualitatively indicate the position of the three motors.

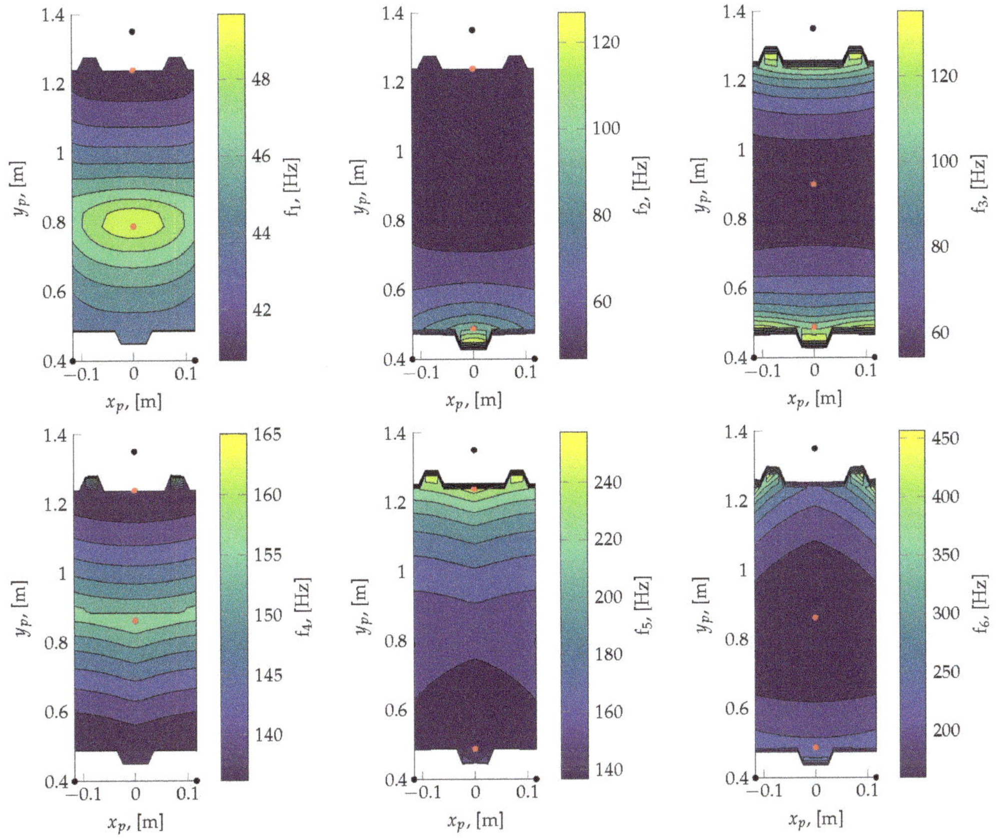

Figure 8. Natural frequencies.

As expected, all the graphs are symmetric with respect to the system's central linear axis (axis 1), while the change of the natural frequencies along the y axis depends on the specific mode of vibration. Regarding the first mode, the graph shows that the frequency has its maximum value located in the lower half of the reference plane; starting from this point, moving toward the upper or the lower end of the working plane (i.e., increasing of decreasing coordinate y_p), the frequency decreases, reaching its minimum value at the upper end. For the second mode, minimum and maximum values correspond, respectively, to the upper end and to the lower end of the working area. Moving toward lower y_p coordinate values, the frequency increases with a gradient more significant in the lower half of the area. The third mode has the frequency's minimum value at approximately the middle of the working area, and the frequency increases, both increasing and decreasing coordinate y_p. The gradient of the frequency becomes quite significant near the ends of the area. For the fourth mode, the behavior is almost symmetric also along the y direction, with the maximum value at the upper end and the minimum value in the middle. The fifth mode is characterized by a progressive increase of the frequency from the lower end to the higher one. The sixth mode has the minimum value located approximately in the middle of the area, and the maximum value at the lower end. Again, the frequency increases, both increasing and decreasing the position y_p, but with a gradient higher in the upper half rather than in the lower one.

However, the information concerning the natural frequencies is not enough to completely understand the system's behavior. As a matter of fact, even if we know the frequency of vibration corresponding to a specific mode, we do not know how the system vibrates according to the mode. So we do not know which is the influence of that specific mode on the behavior of the end-effector. In order to reach this goal, the attention must be focused on the modal vectors and on their projection on the end-effector's coordinates; this is the key point. The values of the modal vectors for the positions corresponding to the minimum and maximum values of the frequency, taken along the symmetry axis of the plane within the workspace area not affected by boundary effects, are summarized for each mode of vibration in Tables 1 and 2. The truck translations are divided by the pulley radius to compare non-homogeneous linear and angular coordinates fairly.

Table 1. Modal vector for each natural frequency, at the point of minimum.

Mode	f, [Hz]	x_1/R_p	x_2/R_p	x_3/R_p	$\vartheta_{1,1}$	$\vartheta_{2,1}$	$\vartheta_{3,1}$	$\vartheta_{1,2}$	$\vartheta_{2,2}$	$\vartheta_{3,2}$
1	40.873	-6.77×10^{-15}	-3.19×10^{-1}	3.19×10^{-1}	-9.69×10^{-15}	-1.37×10^{-1}	1.37×10^{-1}	-8.31×10^{-15}	1.00	−1.00
2	46.854	-4.04×10^{-4}	-2.54×10^{-2}	-2.54×10^{-2}	-1.50×10^{-2}	-5.02×10^{-1}	-5.02×10^{-1}	-1.50×10^{-2}	1.00	1.00
3	54.286	-1.89×10^{-2}	-6.88×10^{-3}	-6.88×10^{-3}	1.00	-5.35×10^{-2}	-5.35×10^{-2}	3.94×10^{-2}	2.50×10^{-1}	2.50×10^{-1}
4	136.237	-4.90×10^{-5}	-9.67×10^{-7}	-9.67×10^{-7}	7.36×10^{-2}	-3.35×10^{-5}	-3.35×10^{-5}	−1.00	2.37×10^{-6}	2.37×10^{-6}
5	136.835	2.17×10^{-5}	7.78×10^{-4}	7.78×10^{-4}	-5.46×10^{-5}	−1.00	−1.00	7.79×10^{-4}	4.15×10^{-2}	4.15×10^{-2}
6	159.823	-5.25×10^{-5}	-1.23×10^{-3}	-1.23×10^{-3}	-4.13×10^{-4}	1.00	1.00	1.92×10^{-2}	-2.40×10^{-2}	-2.40×10^{-2}

Table 2. Modal vector for each natural frequency, at the point of maximum.

Mode	f, [Hz]	x_1/R_p	x_2/R_p	x_3/R_p	$\vartheta_{1,1}$	$\vartheta_{2,1}$	$\vartheta_{3,1}$	$\vartheta_{1,2}$	$\vartheta_{2,2}$	$\vartheta_{3,2}$
1	49.692	5.33×10^{-1}	-5.35×10^{-3}	-5.35×10^{-3}	−1.00	5.05×10^{-3}	5.05×10^{-3}	3.63×10^{-2}	2.06×10^{-2}	2.06×10^{-2}
2	126.906	-2.14×10^{-17}	1.17×10^{-2}	-1.17×10^{-2}	1.45×10^{-16}	-6.46×10^{-1}	6.46×10^{-1}	-6.19×10^{-16}	−1.00	1.00
3	135.238	-7.71×10^{-4}	-2.68×10^{-2}	-2.68×10^{-2}	2.68×10^{-3}	2.65×10^{-1}	2.65×10^{-1}	-2.37×10^{-2}	1.00	1.00
4	157.310	8.51×10^{-4}	5.57×10^{-6}	5.57×10^{-6}	2.55×10^{-2}	9.86×10^{-3}	9.86×10^{-3}	−1.00	-2.75×10^{-4}	-2.75×10^{-4}
5	250.799	4.92×10^{-19}	6.90×10^{-4}	-6.90×10^{-4}	-3.36×10^{-16}	−1.00	1.00	-2.27×10^{-17}	9.76×10^{-3}	-9.76×10^{-3}
6	263.843	1.16×10^{-3}	2.78×10^{-5}	2.78×10^{-5}	8.66×10^{-3}	-1.10×10^{-4}	-1.10×10^{-4}	−1.00	-1.74×10^{-4}	-1.74×10^{-4}

In both Tables 1 and 2, it can be seen that the motion of the free pulley is dominant within the modal vectors of modes 4, 5, and 6; on the other hand, the component associated to the actuated pulleys is the most significant for modes 1, 2 and 3. This rather clean distinction between lower and higher frequency modes is also reflected in the magnitude of the displacements of the trucks; indeed the maximum absolute truck displacement over modes 1, 2 and 3 is at least one order of magnitude greater than that of modes 4, 5 and 6.

The kinematic modal analysis results are also represented from the working space point of view, highlighting, for different positions on the reference plane, the directions of the end-effector's displacement as a consequence of the projection of the modal vector components related to the carriages' displacement. In other words, they represent the projection of the modal vectors after suitable normalization on the mobile platform using the Jacobian matrix $\boldsymbol{D}_p$. The direction of the end-effector's displacements are represented, for modes 1 to 6, in Figures 9–14.

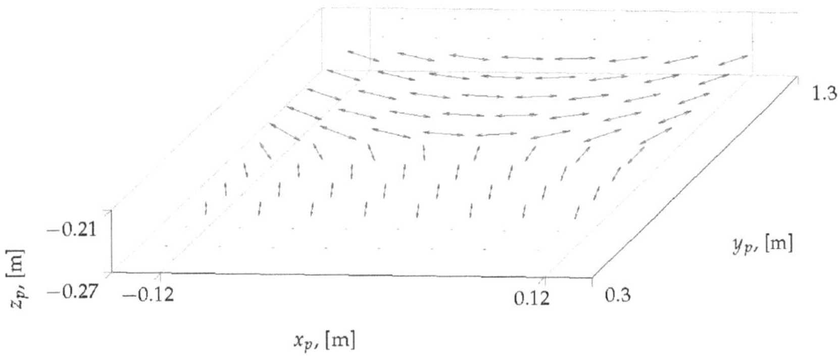

Figure 9. Displacement's direction of the end-effector for mode 1.

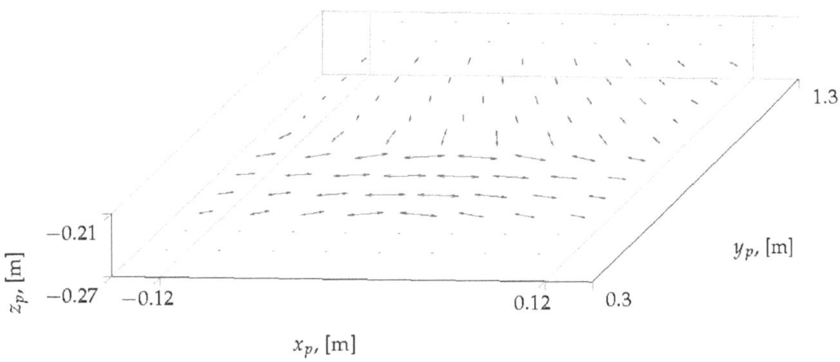

Figure 10. Displacement's direction of the end-effector for mode 2.

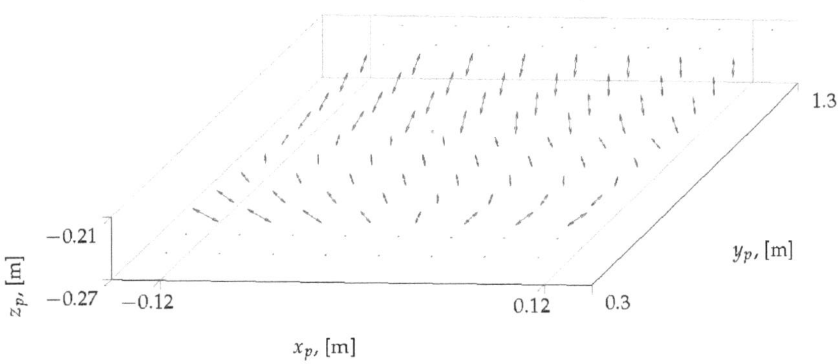

Figure 11. Displacement's direction of the end-effector for mode 3.

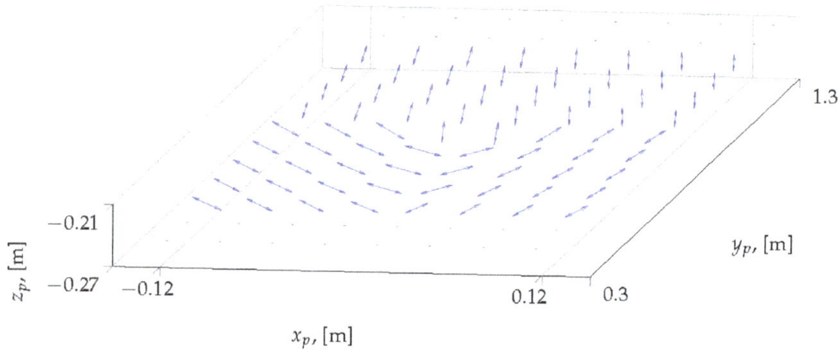

Figure 12. Displacement's direction of the end-effector for mode 4.

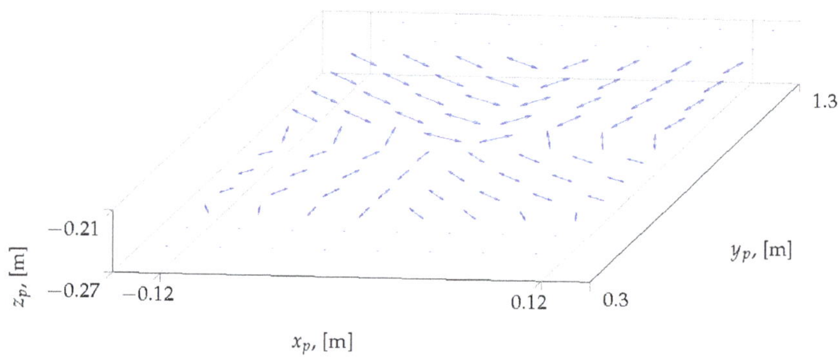

Figure 13. Displacement's direction of the end-effector for mode 5.

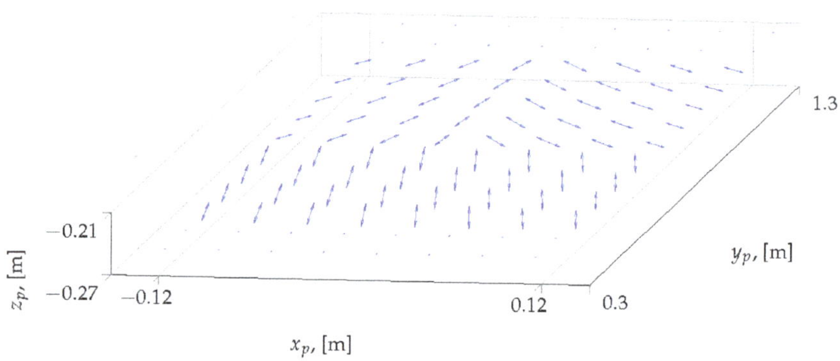

Figure 14. Displacement's direction of the end-effector for mode 6.

From the analysis of these diagrams, it is evident that the direction of vibration of the end-effector, due to a specific mode of vibration, strongly depends on its position within the reference plane. For all the modes, the displacement directions are symmetric with respect to the central axis, and they progressively change by changing the y_p coordinate. Figures 15–20 represent, in another way, the same modal vectors, focusing on values of in-plane and out-of-plane displacements rather than on the directions of vibration, using the same modal vector normalization for all the figures.

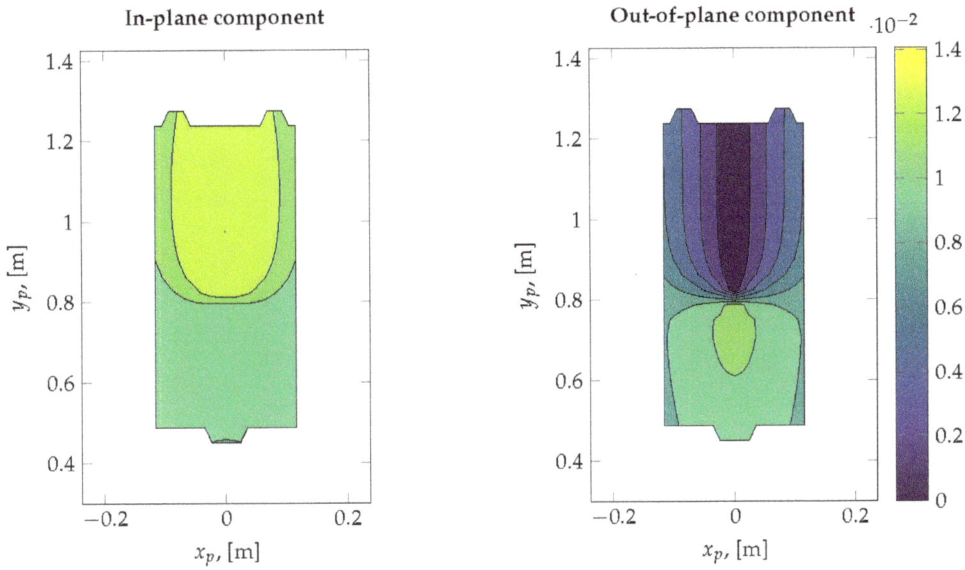

Figure 15. In-plane and out-of-plane components of the first modal vector projected on the mobile platform.

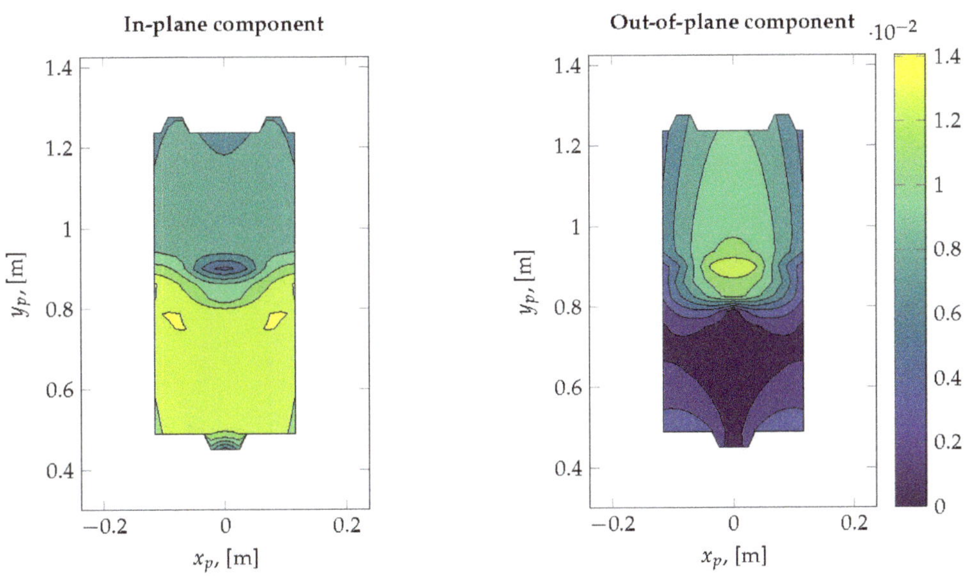

Figure 16. In-plane and out-of-plane components of the second modal vector projected on the mobile platform.

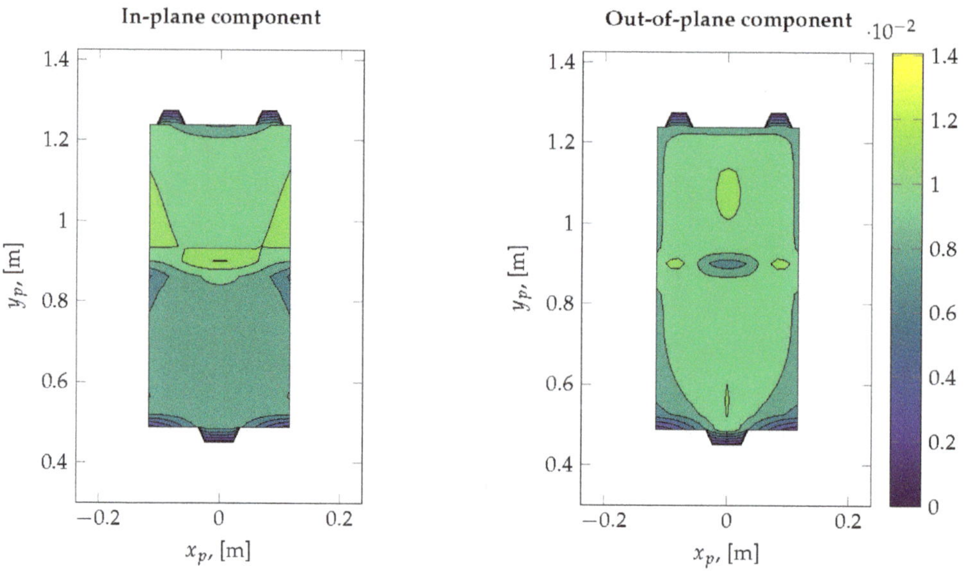

Figure 17. In-plane and out-of-plane components of the third modal vector projected on the mobile platform.

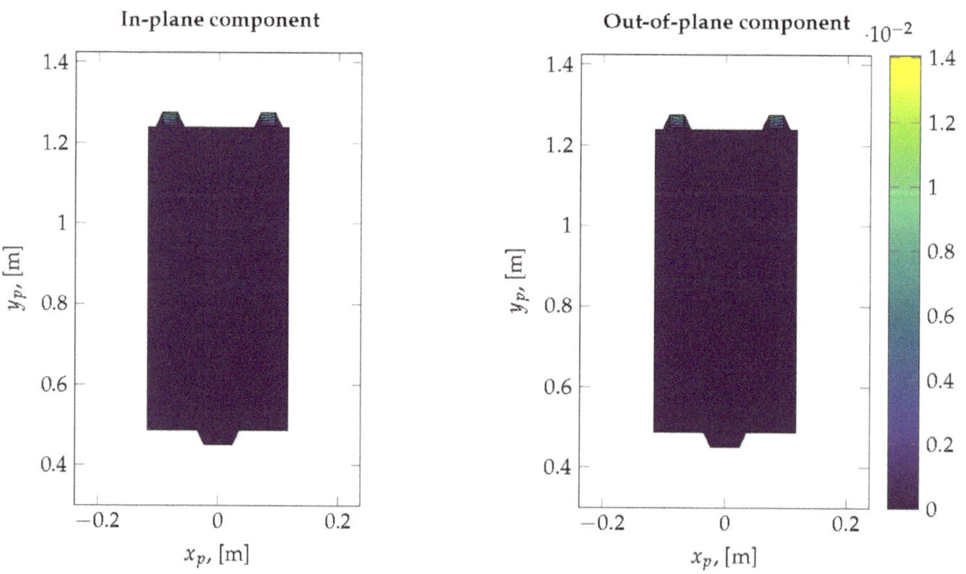

Figure 18. In-plane and out-of-plane components of the fourth modal vector projected on the mobile platform.

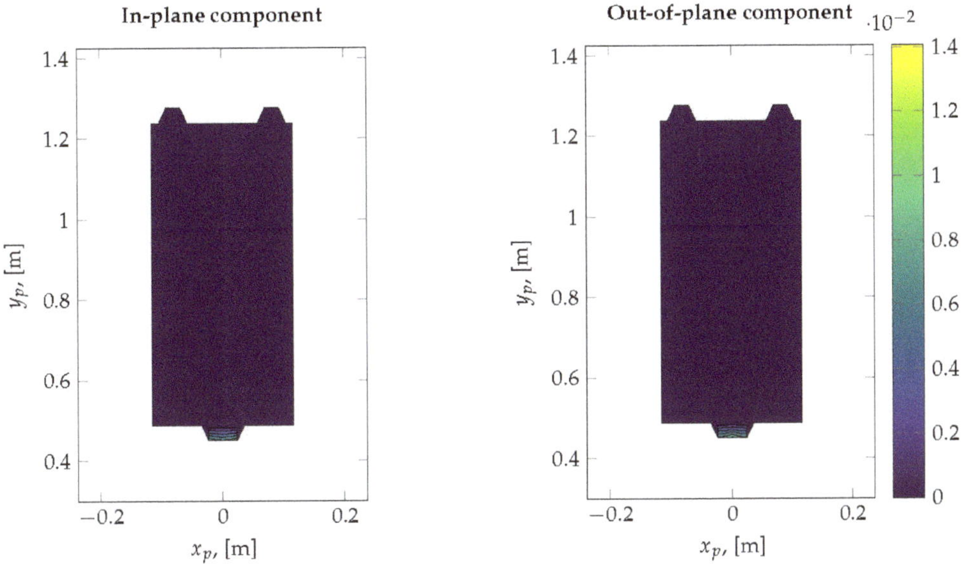

Figure 19. In-plane and out-of-plane components of the fifth modal vector projected on the mobile platform.

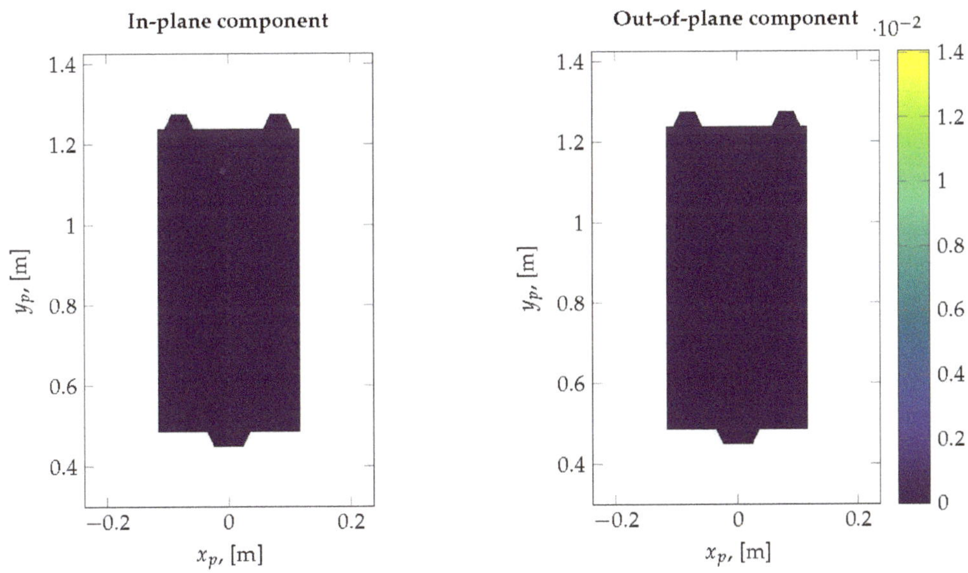

Figure 20. In-plane and out-of-plane components of the sixth modal vector projected on the mobile platform.

The figures on the left represent the displacement in the $x - y$ plane, while the figures on the right represent the displacement in the z direction.

This approach highlights the first-order effects of the truck displacements on the displacement of the end-effector. The in-plane component was defined and represented as $\sqrt{\delta p_1^2 + \delta p_2^2}$, where δp_1 and δp_2 are the first two components of the modal vector projected on the end-effector, while the absolute value of the vertical out of plane component δp_3 is represented separately. The vibrational displacements of the mobile platform are represented using the same scale, allowing a quantitative comparison between the effects of different modes.

In particular, Figures 15–17 are related to the first three modes and show that, for these modes, the direction's displacements of the end-effector, both along the reference plane and along the vertical direction, are significant (please note that the color bars of the graphs have all the same scale). Moreover, the shape of the diagrams is clearly symmetric with respect to the central axis. As far as the y direction is concerned, Figure 15, which refers to the first mode of vibration, shows that the upper half of the working plane, the one nearer to motors 1 and 2, is characterized by values of in-plane displacements lower than the ones corresponding to the lower half-plane, the one nearer to motor 1. Hence, displacements increase as the y_p coordinate increases. For the out-of-plane components graph, the behavior is opposite: as y_p coordinate increases, the displacement decreases. The displacement's maximum value is located in a small area just below the half of the plane. Moreover, it should be noted that along the symmetry axis, there is a position near to the middle of the area where the vertical displacement has a sudden change.

As regards the second mode of vibration, Figure 16 shows an opposite behavior with respect to Figure 15, for both the in-plane and out-of-plane displacements. The in-plane displacement decreases as the y_p coordinate increases, with the minimum value located just above the central position, and the maximum one located in small areas on opposite sides just below the center of the plane. On the contrary, the vertical displacement increases as the y_p coordinate increases, and there is an area around the middle of the plane where it has its maximum value and where a sudden change occurs.

For the in-plane displacement related to the third mode of vibration (Figure 17), the working plane is again divided in two equal parts: the lower one is characterized by values lower than the upper part. Moreover, no sudden changes are highlighted, just small areas where the values slightly change with respect to the values of the respective half area. The out-of-plane component is quite uniformly distributed along the plane, with a central area where the values increase and some areas are located approximately near the ends of the plane's horizontal and vertical middle lines where the values decrease.

As regards the displacements related to the fourth, fifth and sixth modes, respectively depicted in Figures 18–20, the displacements, both in-plane and out-of-plane, of the end-effector are clearly negligible. This confirms on the entire working plane that—as already seen for some notable configurations analyzed in Tables 1 and 2—the truck displacements associated to modes 4, 5 and 6 are comparatively smaller, and thus do not generate appreciable motions at the end-effector.

6. Conclusions

The proposed approach is applied to a parallel kinematic manipulator with driven joints characterized by low mechanical stiffness. It outlines the effects of the stiffness of the transmission, of the mass distribution and of the coupling between the joints due to the parallel kinematic chains on the end-effector's vibration direction and magnitude. The mathematical approach considers the masses of all the elements of the parallel kinematic part, the system's actual configuration, i.e., that it is subject to a motion control algorithm, where the motors are torque controlled, and the transmissions' stiffness, leading to a 9 d.o.f. system. The direction of vibration of the end-effector, which depends on the dynamic configuration of the system, is represented by the modal vector calculated from the equations of the system's model, linearized around the investigated points of the workspace.

The discussed results show that the magnitude and direction of the modal displacements at the end-effector, evaluated in suitable points of the working space, are influenced by the configuration-dependent mass distribution and transmission's stiffness.

The method, applied to a linear delta manipulator, highlights that the effects of the different vibration modes can be effectively compared, considering the amplitude of displacement of the end-effector in the $x - y$ plane or z-direction.

Moreover, the discussion of the results outlines that the displacement's direction of the end-effector changes along the working area as a function of the kinematic and transmission configuration. For the same vibration mode (i.e., the first), in some zones of the working space, the end-effector is subject to vibration in the vertical direction, while in other zones, the displacement's direction is predominantly horizontal. In the same way, the frequency associated with the mode varies in the workspace. The proposed approach constitutes a useful support for the system's design in evaluating the end-effector vibration direction for a given vibration mode, in choosing the working area within the plane, in selecting the proper motion laws, and in synthesizing the control system.

Finally, the proposed approach introduces, in an effective and computationally efficient way, the mass distribution and the coupling effects of the parallel kinematic part as a function of the investigated end-effector positions, thanks to the used mathematical approach. Lastly, our work could open the way toward its implementation in the cases of robots where the stiffness of the parallel kinematic part cannot be neglected.

Author Contributions: Conceptualization, P.R.; methodology, P.R.; software, F.C.; validation, P.R. and R.S.; investigation, P.R., R.S. and F.C.; writing—original draft preparation, P.R., R.S. and F.C.; supervision, P.R.; writing—review and editing, P.R., R.S. and F.C. All authors have read and agreed to the published version of the manuscript.

Funding: The APC was funded by the University of Bergamo.

Institutional Review Board Statement: Not applicable.

Informed Consent Statement: Not applicable.

Data Availability Statement: Not applicable.

Conflicts of Interest: The authors declare no conflict of interest.

References

1. Clavel, R. DELTA: A fast robot with parallel geometry. *Robotica* **1988**, *8*, 105–109.
2. Rey, L.; Clavel, R. *The Delta Parallel Robot*; Boër, C.R., Molinari-Tosatti, L., Smith, K.S., Eds.; Springer: London, UK, 1999.
3. Bourim, M.; Clavel, R. *The Linear Delta: Developments and Applications*; Springer: Munich, Germany, 2010.
4. Merlet, J. *Parallel Robots*; Kluwer Academic Publishers: Oxford, UK, 2000.
5. *Parallel Kinematic Machines: Theoretical Aspects and Industrial Requirements*; Boër, C.R.; Molinari-Tosatti, L.; Smith, K.S. (Eds.) Springer: Berlin/Heidelberg, Germany, 1999.
6. Righettini, P.; Lorenzi, V.; Zappa, B.; Ginammi, A.; Strada, R. Design and optimization of a PKM for micromanipulation. *Appl. Mech. Mater.* **2015**, *799*, 1088–1095. [CrossRef]
7. Nevaranta, N.; Parkkinen, J.; Lindh, T.; Niemelä, M.; Pyrhönen, O.; Pyrhönen, J. Online Estimation of Linear Tooth Belt Drive System Parameters. *IEEE Trans. Ind. Electron.* **2015**, *62*, 7214–7223. [CrossRef]
8. Poonia, S.; Singh, A.; Singh, J.; Sharma, S.; Kumar, N. *Noise Problem Resolution and Sound Quality Improvement of Valve Timing Belt in 4 Cylinders PFI Gasoline Engine*; SAE: Warrendale, PA, USA, 2019.
9. Belyaev, A.; Eliseev, V.; Irschik, H.; Oborin, E. Contact of two equal rigid pulleys with a belt modelled as Cosserat nonlinear elastic rod. *Acta Mech.* **2017**, *228*, 4425–4434. [CrossRef]
10. Oborin, E.; Vetyukov, Y.; Steinbrecher, I. Eulerian description of non-stationary motion of an idealized belt-pulley system with dry friction. *Int. J. Solids Struct.* **2018**, *147*, 40–51. [CrossRef]
11. Vetyukov, Y.; Oborin, E.; Krommer, M.; Eliseev, V. Transient modelling of flexible belt drive dynamics using the equations of a deformable string with discontinuities. *Math. Comput. Model. Dyn. Syst.* **2017**, *23*, 40–54. [CrossRef]
12. Long, S.; Wang, W.; Yue, X.; Zhang, C. Dynamic Modeling of the Belt Drive System with an Equivalent Tensioner Model. *J. Vib. Eng. Technol.* **2021**. [CrossRef]
13. Long, S.; Zhao, X.; Shangguan, W.B.; Zhu, W. Modeling and validation of dynamic performances of timing belt driving systems. *Mech. Syst. Signal Process.* **2020**, *144*, 106910. [CrossRef]
14. Zhu, H.; Zhu, W.; Hu, Y.; Wang, X. Periodic Response of a Timing Belt Drive System with an Oval Cogged Pulley and Optimal Design of the Pitch Profile for Vibration Reduction. *J. Comput. Nonlinear Dyn.* **2018**, *13*. [CrossRef]
15. Passos, S.; Manin, L.; Remond, D.; Sauvage, O.; Rota, L.; Besnier, E. Investigation on the Rotational Dynamics of a Timing Belt Drive including an Oval Driving Pulley. *J. Vib. Acoust.* **2021**, *143*, 051014. [CrossRef]
16. Zhu, H.; Zhu, W.; Fan, W. Dynamic modeling, simulation and experiment of power transmission belt drives: A systematic review. *J. Sound Vib.* **2021**, *491*, 115759. [CrossRef]

17. Kulkarni, V.; Chandrashekara, C.; Sethuram, D. Modal Analysis of 3-RRR SPM Model. In *Lecture Notes in Mechanical Engineering*; Springer: Singapore, 2022; pp. 673–678. [CrossRef]
18. La Mura, F.; Giberti, H.; Pirovano, L.; Tarabini, M. Theoretical and experimental modal analysis of a 6 PUS PKM. *ICINCO* **2019**, *2*, 276–283. [CrossRef]
19. Fiore, E.; Giberti, H. A montecarlo approach to test the modes of vibration of a 6-DoF parallel kinematic simulator. In *Shock & Vibration, Aircraft/Aerospace, Energy Harvesting, Acoustics & Optics*; Springer: Berlin/Heidelberg, Germany, 2017; pp. 315–323. [CrossRef]
20. Palmieri, G.; Martarelli, M.; Palpacelli, M.; Carbonari, L. Configuration-dependent modal analysis of a Cartesian parallel kinematics manipulator: Numerical modeling and experimental validation. *Meccanica* **2014**, *49*, 961–972. [CrossRef]
21. Ren, J.; Cao, Q. Dynamic modeling and frequency characteristic analysis of a novel 3-pss flexible parallel micro-manipulator. *Micromachines* **2021**, *12*, 678. [CrossRef] [PubMed]
22. Sayahkarajy, M. Mode shape analysis, modal linearization, and control of an elastic two-link manipulator based on the normal modes. *Appl. Math. Model.* **2018**, *59*, 546–570. [CrossRef]
23. Yang, C.; Li, Q.; Chen, Q. Natural frequency analysis of parallel manipulators using global independent generalized displacement coordinates. *Mech. Mach. Theory* **2021**, *156*, 104145. [CrossRef]
24. Dong, C.; Liu, H.; Huang, T.; Chetwynd, D. A screw theory-based semi-analytical approach for elastodynamics of the tricept robot. *J. Mech. Robot.* **2019**, *11*, 031005. [CrossRef]
25. Wu, L.; Dong, C.; Wang, G.; Liu, H.; Huang, T. An approach to predict lower-order dynamic behaviors of a 5-DOF hybrid robot using a minimum set of generalized coordinates. *Robot. Comput. Integr. Manuf.* **2021**, *67*, 102024. [CrossRef]
26. Hasegawa, A.; Fujimoto, H.; Takahashi, T. Robot Joint Angle Control Based on Self Resonance Cancellation Using Double Encoders. In Proceedings of the 2017 IEEE International Conference on Advanced Intelligent Mechatronics (AIM), Munich, Germany, 3–7 July 2017; pp. 460–465. [CrossRef]
27. Oaki, J.; Chiba, Y. Simple Physically Parameterized Observer for Vibration Suppression Control of SCARA Robot with Elastic Joints. *IFAC* **2017**, *50*, 6085–6092. [CrossRef]
28. Lee, T.; Choi, J.; Rhim, S. Dynamic analysis and parameter estimation of coupled three-link planar manipulator with flexible belt-drive system. *J. Mech. Sci. Technol.* **2015**, *29*, 981–988. [CrossRef]
29. Negahbani, N.; Giberti, H.; Fiore, E. Error Analysis and Adaptive-Robust Control of a 6-DoF Parallel Robot with Ball-Screw Drive Actuators. *J. Robot.* **2016**, *2016*. [CrossRef]
30. Negahbani, N.; Giberti, H.; Ferrari, D. A belt-driven 6-DoF parallel kinematic machine. In *Nonlinear Dynamics*; Springer: Berlin/Heidelberg, Germany, 2016; Volume 1. [CrossRef]
31. MDQUADRO srl | Advanced Mechatronic Solutions. Available online: https://www.mdquadro.com (accessed on 15 November 2021).
32. Righettini, P.; Strada, R.; Zappa, B.; Lorenzi, V. Experimental set-up for the investigation of transmissions effects on the dynamic performances of a linear PKM. *Mech. Mach. Sci.* **2019**, *73*, 2511–2520. [CrossRef]

robotics

MDPI

Communication

A Methodology for Flexible Implementation of Collaborative Robots in Smart Manufacturing Systems

Hermes Giberti [1,*], Tommaso Abbattista [2], Marco Carnevale [1], Luca Giagu [3] and Fabio Cristini [3]

1 Dipartimento di Ingegneria Industriale e dell'Informazione, Università degli Studi di Pavia, Via A. Ferrata 5, 27100 Pavia, Italy; marco.carnevale@unipv.it
2 ABB Striebel und John GmbH, 77880 Sasbach, Baden-Württemberg, Germany; tommaso.abbattista@de.abb.com
3 Dipartimento di Meccanica, Politecnico di Milano, Via La Masa 1, 20156 Milano, Italy; luca.giagu@mail.polimi.it (L.G.); fabio1.cristini@mail.polimi.it (F.C.)
* Correspondence: hermes.giberti@unipv.it

Abstract: Small-scale production is relying more and more on personalization and flexibility as an innovation key for success in response to market needs such as diversification of consumer preferences and/or greater regulatory pressure. This can be possible thanks to assembly lines dynamically adaptable to new production requirements, easily reconfigurable and reprogrammable to any change in the production line. In such new automated production lines, where traditional automation is not applicable, human and robot collaboration can be established, giving birth to a kind of industrial craftsmanship. The idea at the base of this work is to take advantage of collaborative robotics by using the robots as other generic industrial tools. To overcome the need of complex programming, identified in the literature as one of the main issues preventing cobot diffusion into industrial environments, the paper proposes an approach for simplifying the programming process while still maintaining high flexibility through a pyramidal parametrized approach exploiting cobot collaborative features. An Interactive Refinement Programming procedure is described and validated through a real test case performed as a pilot in the Building Automation department of ABB in Vittuone (Milan, Italy). The key novel ingredients in this approach are a first translation phase, carried out by engineers of production processes who convert the sequence of assembly operations into a preliminary code built as a sequence of robot operations, followed by an on-line correction carried out by non-expert users who can interact with the machine to define the input parameters to make the robotic code runnable. The users in this second step do not need any competence in programming robotic code. Moreover, from an economic point of view, a standardized way of assessing the convenience of the robotic investment is proposed. Both economic and technical results highlight improvements in comparison to the traditional automation approach, demonstrating the possibility to open new further opportunities for collaborative robots when small/medium batch sizes are involved.

Keywords: collaborative robots; small-scale production; skill-based programming

Citation: Giberti, H.; Abbattista, T.; Carnevale, M.; Giagu, L.; Cristini, F. A Methodology for Flexible Implementation of Collaborative Robots in Smart Manufacturing Systems. *Robotics* **2022**, *11*, 9. https://doi.org/10.3390/robotics11010009

Academic Editor: Oscar Reinoso Garcia

Received: 11 November 2021
Accepted: 31 December 2021
Published: 4 January 2022

1. Introduction

The increasing diversification of customer needs as well as greater regulatory pressure [1] add further costs and complexity to industrial operations. Companies throughout the world have embraced mass customization in an attempt to avoid these pitfalls and provide unique value to their customer in an efficient manner. Readily available information technology and flexible work processes permit the customization of goods or services for individual customers in high volumes and at a relatively low cost, but mass customization can still produce unnecessary cost and complexity. Customers can no longer be thought of as members of a homogeneous market grouping. In fact, the concept of markets also needs to be redefined as customization becomes more commonplace.

Instead of focusing on homogeneous markets and average offerings, mass customizers have identified the dimensions along which their customers differ in their needs. These points of common uniqueness reveal where every customer is not the same. In addition, it is at these points that traditional offerings, designed for average requirements, create gaps in customer satisfaction: the difference between a company offering and what each customer truly desires [2].

Due to this, automation is currently undergoing a strong process of evolution, keeping the pace with an economic context characterized by dynamic manufacturing systems: demand becomes more and more unstable, with batches of production shrunk and highly customized, requiring companies to be incredibly flexible in changing from one production to another quickly and cheaply. In such production systems, as well as in artisan enterprises, robotics penetration is rather low due to the limiting characteristics of traditional automation: rigid, fixed to a single process and demanding steady and high volumes, to repay the large investment costs and risks.

To meet these needs, scientists around the world are trying to implement modern solutions to traditional robot controllers aimed at defining control procedures to allow a safe interaction between human and robot [3] and advanced programming methods based on the physical interaction between human and robot (e.g., learning by demonstration [4], gesture and voice communication [5,6]).

Some experimental user interfaces are also being proposed for traditional industrial robots, since, due to the specificity of operations they can carry out, the need of adapting their control programs to the changing assortment is an important issue [7–9]. However, collaborative robots have certainly a great potential to satisfy the needs of this variable customer demand [10,11]. They are often used as traditional automation, i.e., fixed to a single process with the only advantage of not requiring safety cages and therefore lowering investment costs [12]. On the other hand, they could be exploited on assembly lines: dynamically adaptive to new production requirements and easily reconfigured and reprogrammed to the changes in production line [13]. The idea previously investigated by researchers, and further explored in this paper, is to transform these robots into tools that can be applied by non-expert robotics users into the desired production process when they are actually needed, according to the daily production agenda, in a fast and easy way.

Four main challenges arise when trying to implement a methodology to exploit cobots in the above-mentioned conditions; these are the starting points of this project:

- Expert engineers are needed for designing, programming and testing the robotic application.
- Programmers need to know the workstation requirements for developing the precise production process.
- The robot adoption in such production process conditions has to be technically justified, since small and varied batches are involved.
- Investment costs are certainly significant, and the economic return has to be assured.

Robot programming is recently evolving from the traditional written language to new more accessible directions. For example, kinesthetic teaching applied to a walk by demonstration approach simplifies a lot the workload required for programming a machine, and its path towards industry is on track, being regulated by ISO/TS 15066.

An interesting approach in the field of easy-programming research is the task-level layered framework, as described in [14]. A productive process, referred to as a task, is constituted by a sequence of standard, modular and parametric blocks named skills [15], arranged as a state machine architecture [16]. Skills may have different structures, implying a motion and a trajectory component as in [13], or have a slightly modified template as in [17–21]. In general, a skill is composed of a pre and post check phase, where preconditions and objectives are evaluated, and an execution phase in which elementary activities or component manipulation are performed. It must be remarked that most state-of-the-art task level programming still requires engineers or experts for creating a real industrial application, thus reducing the operative range of non-expert users. Some recent work is aimed at developing a framework that alleviates the problem of a high programming effort, e.g., exploiting the advantages of Learning by Demonstration, Learning by Programming and Learning by Interaction, combined [22].

Another relevant issue is the impossibility for programmers to standardize, in advance, the workstation, which would be needed for a reliable utilization of robots in the production process. Indeed, centering and positioning must be considered to guarantee the typical precision recognized in robotic applications. In [23] the robot can find the workstation center by means of its vision system, recognizing a QR code and using it for the correct positioning of the machine's Cartesian origin. Other calibration methods have been investigated by researchers, for example in [24] by applying hand-eye calibration or laser triangulation aided by image processing.

The last issues to be considered are related to the economic feasibility of the robotization process. Investing in robotics and automation requires great capital, usually repayable only if high and constant volumes are involved. The economic investment return can be guaranteed only by transforming robotics, typically rigidly utilized as a stable workstation, into a flexible tool that can be applied to different processes without involving significant costs or times.

The methodology proposed in this paper, named Interactive Refinement Programming (IRP), allows non-expert users to newly develop a production process by refining, step by step, built-in standard and parametric tasks in an easy way. The elements of novelty in this approach is a first translation phase carried out by engineers of production processes (in which the process to be carried out is translated into a robotic metalanguage based on skills), followed by an on-line setup and correction carried out by the non-expert user. Moreover, an economic analysis to generate a standardized way of assessing the convenience of the robotic investment is proposed. This also allows the production department to decide lot by lot the convenience of carrying out the production in a robotic or manual way.

First, a company department skilled in robotics, e.g., engineering of production processes, is responsible for a first translation of the production process from the manual activities, typically assigned to the operators, to a list of skills, thus creating a robotic metalanguage. This code structure is built by adopting pre-built skill blocks (which is a topic widely studied in the literature [17,18]) developed in our case by using the ABB code robot studio. At this stage the code is still not executable by a robot, since the created list of skills still needs an initialization and to be characterized by the actual process data. The process is then refined and finalized by a non-expert user into the real production process, by inserting the required parameters and correcting possible errors that may arise during its execution. From the point of view of industrial operation, this approach allows the shift of some operations typical of the production line department to the design stage. Each product can be conceived directly with its assembling procedure, and the robotic code structure can be directly given as an output of the design stage.

In production lines, non-skilled operators in the use of robots are enabled to provide input data to the pre-built code. The robot itself already has an inbuilt behavior that allows skill initialization. Moreover, the machine is able to drive a correction phase: even if each skill is standardly programmed, and the robot thus knows how the required activities or manipulations must be carried out, the specificities of each different workstation may cause some errors that make the task not properly executable. Built-in correction strategies are then preliminarily implemented, so that the machine can guide a non-expert user in this correction.

The paper is organized as follows: Section 2 gives an overview of the proposed methodology, of the exploited hardware and of the developed software. Section 3 details the Interactive Refinement Programming procedure. In Section 4 an economic analysis to evaluate a standardized way of assessing the robotic investment is proposed. The industrial test case and results are described in Section 5 and, finally, conclusions are drawn in Section 6.

2. Overview of the Proposed Methodology

The intrinsic properties of interaction and security of collaborative robots permit the flexibility needed in the reference context for this work, i.e., the possibility of being relocated and reused in an easy way. A widespread solution of cobots available on the market is the single-arm configuration with 6 or more Degrees of Freedom (DoF) and a payload on the order of 5–10 kg. An exception is made by the ABB YuMi, a double-arm, 7 + 7 DoF robot with a payload of 0.5 kg per arm. All the mentioned cobots present useful features that are exploitable to achieve the target of flexible implementation, such as the hand-guiding modality and the multi-modal interaction system (composed of a vision system, voice recognition and force/torque sensing), which are both very useful to interact with humans.

To obtain the required high level of flexibility, the workstation and all the related equipment (grippers, pliers or input and output systems) must be as general and standard as possible. A suitable solution consists in developing a general frame and some specific dowels per product, allowing for variation in as few components as possible when changing the production process. Obviously, the possibility of developing some specific components as well is still valid, but it would increase the implementation and setup times.

To increase system mobility, it could be useful to mount the robot on a cart, which can be manual or automatic, and to organize the shop floor into different fixed workstations which can be used both by human operators and robots independently. Each station should be of course provided with a proper blockage body and with a calibration system useful to correctly setting up the workstation itself. (This last part of activity is not carried out in the present work.)

As for the software, in order to obtain a tool that can be completely handled by a common robot-non-expert worker, the solution adopted is a pyramidal parametrized programming approach: a layered pyramidal framework composed of Primitives, Skills and Tasks, depicted in Figure 1:

- Primitives are the basic robot code command lines.
- Skills are blocks of code, composed by Primitives, that allow the robot to execute a specific operation.
- Tasks are logical and precise sequence of Skills used to allow the robot to complete an entire production process.

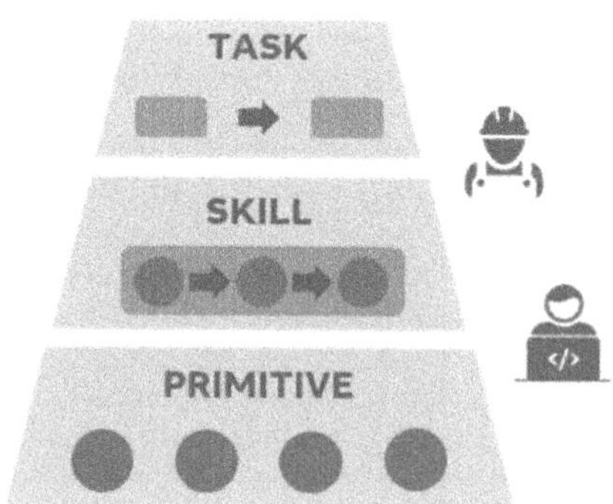

Figure 1. Layered pyramidal framework.

Each block of code, named Skill, represents a real activity or operation performed by the robot on the product to be manufactured. It is composed of a series of basilar command lines, called Primitives. The choice of creating a Skill equivalent to a production step is driven by robotics experts, in charge of developing Primitives and Skill libraries according to what the operator is used to doing during manual operations.

The structure of the considered Skills is divided into three main parts, reported in Figure 2.

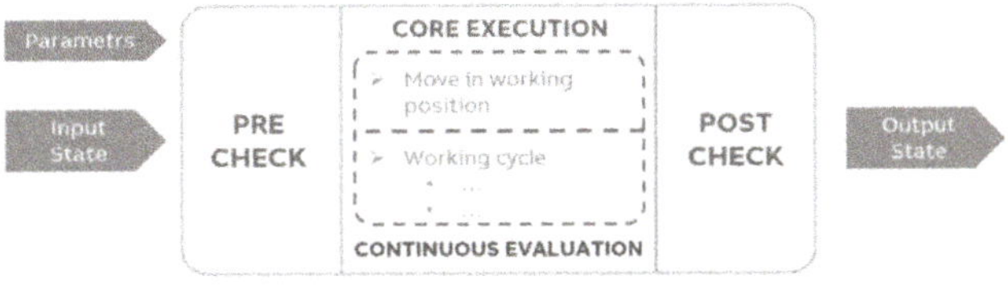

Figure 2. Structure of a standard and parametric skill.

- Pre-Check: the preliminary verification that has to be done before starting the execution of operations; the effective Execution can start only if all the Input States of the system have been verified.
- Core Execution: the parts that effectively execute the working cycle; the Core is always preceded by an initial movement to the working position, before starting the actual operation on the workpiece. In parallel to the Execution, a Continuous evaluation of the process in the workstation is carried out, and, consequently, certain methods to handle occurring errors are implemented.
- Post Check is performed when the working cycle is completed, in order to verify if the Core has correctly been executed and if the Output States are equal to the expected ones.

When a sequence of skills is created, each block, therefore, has Input and Output States which represents all the variables necessary for the execution of the current working cycle and for the execution of the following one.

Tasks, at the top of the layered pyramidal framework of Figure 1, are composed by connecting different Skills in sequence. The sequence can also present different branches, in a sort of state machine in which each block is executed when the value of the input state is correct. Different branches of the sequence can be thus executed on the basis of what has happened before.

The chosen Skill architecture is then very useful in order to keep what is happening under control and to avoid inconsistencies or mismatching errors. Input and Output States of each Skill, and thus Pre and Post Checks, are fundamental for the correct and congruent creation of a complete robot running code.

In the above-described scenario, the competencies required for creating a robotic application are split among different players. Robotics experts write the code for primitives and skill libraries, which are created after an analysis of all the involved production processes in the company, to make them as standardized as possible. The job of creating the application—generating a list of skills and thus creating a robotic metalanguage—is then assigned to the production process engineering department, who already possess the required competencies. Finally, shop floor non-expert users merely have to set input parameters for each skill during a Teaching Phase (which specifically relates each skill to the particular process), and to carry out, as a second step, a Correction phase to fix occurring errors. Non-expert users are helped in this phase by a software managing the entire teaching and correction procedure through a Graphical User Interface. These two phases will be described in detail in the following section.

Skill based programming has been extensively studied in the literature [17,18]. In this work, it has been adopted as an instrument to define and test an overall procedure accounting for every step in the flow of operations in an industrial environment, ultimately enabling the fine-tuning of a robotic cell by non-skilled operators.

3. Interactive Refinement Programming Procedure

This section describes the Interactive Refinement Programming (IRP) procedure, through which non-expert users can carry out the teaching and correction phases, ultimately leading to task execution. The main steps of the IRP are represented in Figure 3.

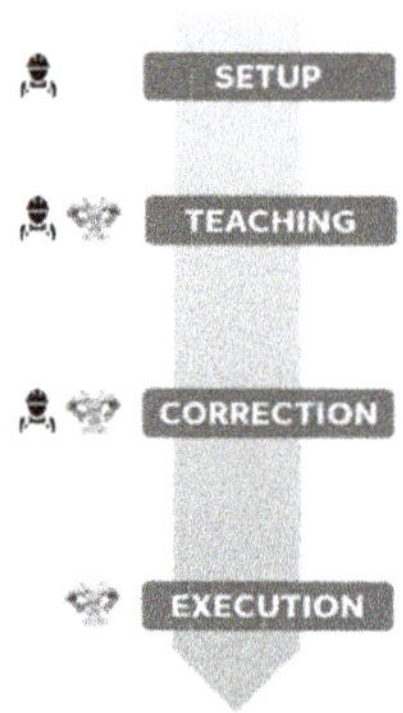

Figure 3. Steps of the Interactive Refinement Procedure (IRP).

3.1. Setup

Engineers in charge of designing and developing new products and manufacturing processes have to define the skill tree being executed based on operations traditionally done by hand, generating a list of skills in a robotic metalanguage. This sequence, at this stage, contains general purpose items that can still take shape accordingly to the context of the considered process. The task is not yet ready to be executed by the robot, since the inputs of each skill are still to be defined. This phase, named the Translation Phase, also includes the design and development of specific hardware or software components, in case specific adjustment of hardware or software, different from the general bundle, is required.

After the Translation Phase is completed, the operator in the production department receives the list with the sequence of Skills and the instructions to manage production (i.e., instructions and information for the operators on how to handle the workstation). The workstation can be set up and all the equipment mounted and verified according to the prescriptions in the list. When the station is ready, the setup phase is over. The operator uploads the sequence of Skills into a manager software in charge of driving the beginning of the Teaching phase for each Skill.

3.2. Teaching

If the manager software detects it is the first time a specific production process is implemented on a robot, a procedure to insert the precise input parameters for each skill is run. During this teaching phase, the user, guided by the manager software, exploits the interaction capabilities of the cobot to set the correct parameters needed for the execution of each skill. This procedure can be done in different ways, for example by using the keyboard and moving the robot into the correct positions using the teaching pendant (as in the present work), or by exploiting kinesthetic teaching (and thus moving manually the robot in the necessary positions) or, finally, by exploiting gestures [5] and voice recognition systems [6,25] or even more advanced interaction methods [26].

3.3. Correction

Once all the input parameters have been assigned, the robot contains an effectively runnable program. However, the execution could still find errors that should be detected in advance and corrected. Thus, a Test and Correction phase is performed: each operation is executed slower than the actual executing speed, so that the operator can supervise and intervene if any error or problem occurs. In such a circumstance, the program stops and the issue can be recovered through proper editing. Some typical problems are related to the arm movements (e.g., singularity points, wrong configurations, out of reach points or physical interference with other objects in the working area) or to the gripping/contact forces.

Some predefined corrective packages have been developed and preinstalled into the manager software to drive error recovering: this is the fastest and most straightforward procedure, which, however, requires all the possible occurring errors to be known in advance to permit the development of a pertinent correction procedure for each one. The list of correction packages developed is the following:

- Automatic Trajectory Generator.
- Waypoints Addition: the movement of the robot can be stopped and some passing points can be added to the path.
- Avoid Singularity.
- Lead by Demonstration: the operator drags the robot along the desired path between two points and the robot records the trajectory in order to re-execute it.
- Force routine: the robot closes the gripper repeatedly around an object, at each iteration increasing the force until the operator confirms that the object has been grabbed correctly.

When the Correction phase has been completed for each Skill, a low-speed test is performed in order to verify the correctness of the overall task.

3.4. Execution

At this point, the Task is ready for the Execution. The complete IRP procedure is obviously necessary only the first time that the process is implemented: the parameters are then saved and made available for the next time the same product will have to be manufactured. Nevertheless, each time a task has to be redeployed, the station must be recalibrated, and thus it is appropriate to initially perform a Test Execution and, eventually, again the Teaching and Correction steps. When everything is correct, the production cycle can finally start.

In manufacturing factories whose production is highly variable and whose products are customizable, it may occur that a production process for a customized product is slightly different from the process of the basic product. Thus, this solution allows for slight modification of an existing Task and for saving this as new one by adding some other skill branches to the state machine tree. This would maintain the desired grade of flexibility required by these kinds of production systems.

4. Economic Analysis

Investing in robotics requires high costs that should be paid back by the possibility of reducing labor hours, saving annual workforce costs or increasing production capacity.

In order to assess the economic convenience of installing a robotic cell, a deeper focus on the production activities of the department is needed. Every process should be analyzed to First evaluate its "robotic feasibility" (i.e., complexity of the operations to be carried out by the robot) and then its economic impact. The threshold value N_{thr} for the lot size that makes the adoption of robots convenient is computed by comparing the total robot setup time $T_{robot\ setup}$ (from off-state to application finalized) with the time that would be required to the operator if production activities were performed manually. These two times are equaled in Equation (1):

$$T_{robot\ setup} = N_{thr} \times t_{robot\ act.} + T_{man\ setup\ for\ robot\ activities} \quad (1)$$

On the left side of the equation, the term $T_{robot\ setup}$ represents the robot setup time, consisting of the sum of:

- Robot positioning in the specific workstation (automatically or manually) and anchorage through a mechanical or magnetic braking system.

- Robot controller startup routine.
- Setup of the hardware components required for a correct execution of the task and product feeding.
- Teaching, correction and low speed test phases for every skill composing the Task.

On the right side of Equation (1), $t_{robot\ act}$ represents the manual cycle time, per unit, of the activities that should be instead allocated to the robot. When multiplied by the number of pieces in the lot (N_{thr}) this gives the time that an operator would need to manually perform all the batch operations. Finally, $T_{man\ setup\ for\ robot\ activities}$ is an additional setup time of those activities that should be made by the workforce in standard manual operation, and that instead can be allocated to the robot in the automatic procedure. The lot threshold value N_{thr} obtained by Equation (1) can finally be multiplied by a safety coefficient to adopt the robot in a more conservative way, only when the saved hours are a relevant percentage of the required workforce time.

The annual working hours saved for the specific process j ($h_{saved\ j}$) are then computed as in Equation (2) only taking into account the batches i bigger than the defined threshold value N_{thr}, which can therefore be carried out through a robotic cell. In the Equation, the saved time is evaluated by summing up, for each batch i, the manual cycle time of the activities to be automatized ($N_{batch_{i,j}} \times t_{robot\ act_j}$) and the setup time that would be necessary to the operator if the activities were performed manually and by finally subtracting the time needed to setup the robot.

$$h_{saved\ j} = \sum_{i}^{suitable\ batches} \left(N_{batch_{i,j}} \times t_{robot\ act_j} + T_{man\ setup\ for\ robot\ act_j} - T_{robot\ setup_j} \right) \quad (2)$$

Dealing with the overall investment, monetary benefit is computed in the easiest scenario by summing up, process by process, the freed person-hours multiplied by the hourly workforce cost. On the other side, the investment costs are the robot, its moveable station, and all the general and process-specific hardware and software costs for developing the required components or software modifications. Comparing these fixed costs and the annual savings, it is possible to compute the payback time and to assess the convenience for the case of interest.

Since the target companies of the proposed work deal with a truly dynamic market request, in order to introduce variability in the study, a Monte Carlo analysis has been used to simulate the demand planning and the consequent outcome on the production process.

Starting from the historical data for every demanded product of the portfolio, the probability density functions of orders' date/size and yearly frequency have been simulated, and random values have been extracted. The constraint of robot production capacity has also been introduced, and the computation of annual savings has been repeated 150 times to get a distribution of the investment benefits, whose average has been used for computing the payback time.

A specific performance indicator on an N-year basis, named *Implementation Efficiency*, has been created for benchmarking different processes in light of robotic implementation. This indicator is computed, as in Equation (3), as the ratio between an equivalent net saving (i.e., savings S_j minus implementation costs in HW and SW) and the equivalent workforce cost (computed by multiplying the hourly cost by the total man cycle time necessary to manually perform all the activities).

$$\eta_N = \frac{S_j - \frac{C_{impl}}{N}}{C_{€/h} \times t_{cycle\ tot/y}} \quad (3)$$

In the equation, N is the year horizon upon which the indicator is computed. The equation thus computes the ratio between the yearly net savings if automation was applied, and the production costs if the operator performed manually all the activities.

5. Industrial Test Case

The proposed methodology has been tested in an industrial environment, in collaboration with the ABB Building Automation department in Vittuone (Milan, Italy). This factory unit, dealing with the manufacturing of personalized electronic components, is aligned with the project target, since its production is highly unstable and customized and it involves small batches and large variety and variability.

The test case is carried out on the production process regarding firmware updates in motherboards for security sensors. This working procedure is rather simple, but it contains crucial ingredients for assessing the feasibility of the proposed method. The board has to be picked out from an input tray, placed on a cable fixture connected to the PC, updated and then picked out and placed on the output tray. The ABB YuMi two-arm robot was used.

For the test case execution, the following required hardware and software was developed:

- *Hardware:* a multi-board 4 × 4 tray module, ad-hoc fixture and specific pliers (3D printed).
- *Skills:* pick and place are the two most important skills developed, completely standard and parametric, allowing for the performing of activities on a large variety of components in all the possible directions. They are developed in the ABB *Robot Studio* Software.
- *Program manager:* the program manager is a no-motion additional software uploaded on the robot controller to coordinate the entire process based on the sequence of skills generated by the engineering office. It allows the management of every phase (i.e., Teaching, Correction or Execution) for every skill, by transferring instructions to each of the two arms and displaying on a screen all the instructions the operator has to follow to define the input of each skill (as an example in the picking operation the operator has to drive the robot in the desired position and set the pose which is saved by the manager software). The communication interface of the developed program is made available on the teach-pendant screen, so that the operator can directly follow the instructions.
- *Correction packages:* lead by demonstration and waypoints addition are two routines useful for solving motion problems, by respectively recording a trajectory directly from the operator dragging the robot, or by adding waypoints interactively to avoid obstacles or force the path to be followed. Two-hands calibration is instead used for improving the precision of a place operation, autonomously calibrating the actual position of the component hold by the robot gripper by using the position of the free arm gripper as a reference (This procedure has to be activated by the operator when a precise insertion is needed). Force calibration and angle rounding routines are instead designed for easing the operator during the teaching phase and for improving the accuracy of teaching-phase parameters.

Figure 4 reports an example of the code through which the program manager code was developed (Figure 4a), and a frame of the developed robotic procedure, in which it is visible that the two arms can operate at the same time (Figure 4b). In the figure the fixture specifically designed to house the electronic boards is visible, for each of the two arms.

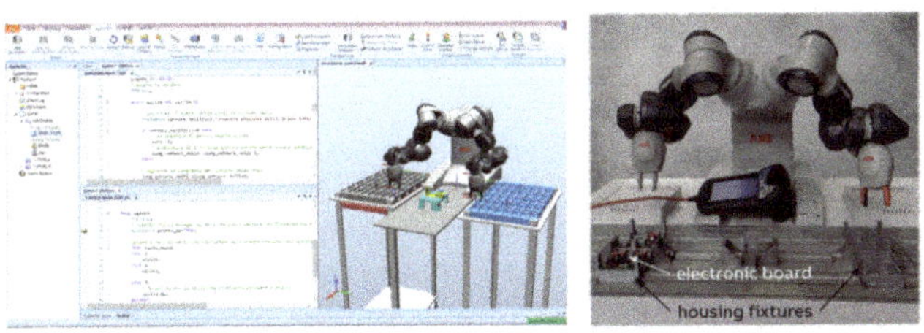

(a) (b)

Figure 4. Implemented Firmware up-date process for security sensor electronic boards. (**a**) Software development for manager program. (**b**) View of the two Yumi arms operating at the same time.

The robotic process has been created and set up by a non-expert user, revealing the great potential of the methodology. Two series of tests have been carried out: a first one with a non-expert user who, however, is up to date on robotic programming. The second series of tests have been carried out by a person who had never used a robot before. In both cases, the operator's learning was rather fast, since the entire process is intuitive and guided by a basic and practical graphical interface.

The results achieved by the two different users were not systematically different in terms of average values and standard deviations, so that the presented results do not differ for the single operator. This fact demonstrates that having skills in robotic does not help to improve usage skills, and the process can be therefore easily set up by every kind of non-expert user.

Table 1 reports in the second column the average teaching and correction times of the two non-expert users after several trials (time of 2nd, 3rd and 4th attempts), together with the maximum and minimum times. In the third column the times of the first trial are reported, which are around 25% higher than the following attempts in which teaching and correction times are stabilized.

Table 1. Times spent on teaching and correction phase by non-expert robotic users.

Skills	**Teaching and Correction Time** (average of 2nd, 3rd and 4th attempts for both the operators)		**1st Attempt Time** (average time on two operators)
Pick a Piece from the multi-board 4 × 4 tray module	1′41″	max 1′48″ min 1′38″	2′06″
Place a piece on the fixture	2′30″	max 2′39″ min 2′21″	3′01″
Connect to PC	-		
Pick a piece from the fixture (after SW updating)	1′49″	max 1′58″ min 1′41″	2′11″
Place a piece on the output tray (after SW updating)	1′54″	max 2′02″ min 1′43″	2′19″

Table 2 reports the time needed for the slow speed tests, which are carried out as a test after the teaching and correction phases for the skills of the entire task are completed, and the final cycle time per piece. By summing up the times of Tables 1 and 2 it can be observed that the complete application has been created in less than 10 min.

Table 2. Slow speed execution time and final cycle time.

Slow-speed test (i.e., test of the complete task executed after defining the input parameter for each skill)	53″
Cycle time per piece	37″

The process ended up not being as optimized in terms of trajectory and cycle time as one created by a specialist in production process engineering, but the flexibility expressed by this approach is by far more relevant than process optimization, considering the peculiar production conditions where robots can be continuously allocated to new processes. The robot could perform the task correctly even without any optimization.

From an economic point of view, a Monte Carlo simulation has been conducted on all the processes which could be feasibly carried out by robots in the considered department, in order to assess if, for the reference factory, a robotic implementation is economically valuable. These processes, in addition to the firmware update which is the considered test case from a technical point of view, are related to the following products: USB Charger (wall charger for USB cable, involving both assemblies and customization activities), Crystal Touch Sensor (CTS electronic boards to be assembled and tested, and frontal motherboard to be tested only), Remote Controller for security sensors (sharing part of the process with other products, as firmware update and laser marking) and product customization. These are only the production processes which involve at least one activity that can be allocated to YuMi, respecting technical feasibility and robot performance. All the processes of interest have been simulated in the Monte Carlo simulation based on probability functions evaluated from historical data and forecast demand.

Table 3 reports a synthesis, for each of the mentioned processes, of the main input data (i.e., time allocated to YuMi, YuMi setup time, average lot size, pieces produced in a year) and output data obtained from the Equations (1)–(3) described previously in Section 4 (i.e., Lot Size threshold, Saved person-hours, implementation efficiency). It can be noted that the assembly of the top and bottom parts of the Crystal Touch Sensor (CTS) accounts for the great majority of saved person-hours, given the high quantities produced per year and the large percentage of activities which can be allocated to the robot.

Table 3. Single process data synthesis.

Process	Firmware Update	Remote Controller	USB Charger	CTS (Bottom)	CTS (Top)	Custom
YuMi allocated time	24 s (39%)	67 s (60%)	34.5 s (81%)	93 s (95%)	165 s (94%)	24 s (75%)
YuMi setup time	57 min	65 min	69 min	89 min	53 min	45 min
average lot size	150 pcs	50 pcs	135 pcs	280 pcs	280 pcs	200 pcs
pieces per year	3899 pcs	576 pcs	2440 pcs	9112 pcs	9880 pcs	2039 pcs
lot size threshold	139 pcs	56 pcs	117 pcs	56 pcs	19 pcs	108 pcs
saved person-hours	10.5 h	5.6 h	10 h	195.6 h	426.9 h	7.7 h
efficiency (N = 2)	8.19%	−1.31%	17.37%	75.16%	86.20%	19.51%

Investment costs are about 50,000 € (i.e., robot 40,000 €, mobile station and general hardware 2000 €, general software bundle 3000 €, operator training 3000 €, process implementation costs 2000 €), whereas saved person-hours are quantified to be in total 656 h/y, which, multiplied for an average the operator hourly cost, leads to an annual savings of 19,688 €. The payback time is therefore around 2.5 years, oscillating with ±0.5 years depending on a positive or negative scenario. In addition to the standard case, indeed, two further scenarios have been analyzed, repeating the simulation with an increased or decreased demand and batch size of 10%.

The mentioned payback scenarios are represented in Figure 5. The two additional cases reported in dashed lines make reference to a specificity of the fiscal regime that the factory department is subjected to, which further reduces the payback time of the investment.

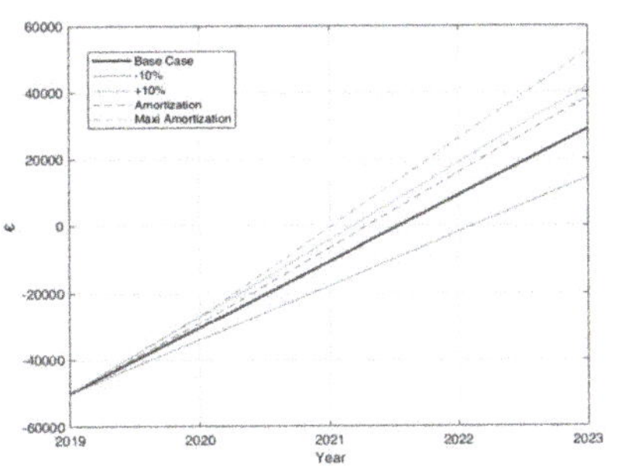

Figure 5. Comparison of Payback time scenarios.

6. Conclusions

The proposed Interactive Refinement Programming (IRP) is an approach aimed at simplifying the creation of a robotic application, allowing non-expert users to actively setup a robot guided by the machine itself, and to correct any occurring error with an iterative procedure. A pyramidal approach is based on primitives and general skills to be developed by expert engineering, which can then be connected in a tree structure to generate a specific task. The only thing that shop floor non-expert users have to do is to provide a set of input parameters for each skill during a Teaching Phase, which specifically relates each skill to the particular process, and a correction phase to fix occurring errors. These teaching and correction phases have been driven by a self-developed program manager software uploaded on the robot controller and able to display instructions on the teach-pendant screen to assist the non-skilled operator.

Ad hoc performance indicators have been created for the economic assessment of process automation benefits and for the daily management of production orders once the robotic application has been set up.

The proposed approach was validated through a real industrial case—in which the base software and hardware needed for the application were developed by expert engineers whereas the actual set-up of the robotic application was established by non-expert users in a very short time—and demonstrated the flexibility of the proposed method. The complexity in creating a robotic application has been drastically reduced in light of easy programming efforts. Collaborative features of cobots are exploited at a higher level, not only during execution but also for correcting and finalizing the application itself.

The main limitation regards the fact that the proposed method is intrinsically based on a trade-off between task optimization and flexibility: skills are designed as generally as possible to obtain the needed flexibility, which implies that the resulting tasks cannot be optimized for the specific process in terms of trajectories and cycle time.

The presented results are a further step toward the implementation of robotics in craft manufacturing, overcoming the limitations of traditional automation and current use of collaborative robots.

Author Contributions: Conceptualization, H.G. and T.A.; methodology, H.G., T.A. and M.C.; software, L.G. and F.C.; validation, T.A., L.G. and F.C.; formal analysis, H.G., T.A., L.G., F.C. and M.C.; investigation, L.G. and F.C.; resources, T.A.; data curation, L.G. and F.C.; writing—original draft preparation, M.C., L.G., F.C. and T.A.; writing—review and editing, H.G. and M.C.; visualization, T.A., L.G. and F.C.; supervision, H.G. and T.A. All authors have read and agreed to the published version of the manuscript.

Funding: This research received no external funding.

Institutional Review Board Statement: Not applicable.

Informed Consent Statement: Not applicable.

Data Availability Statement: Data access is restricted due to the policy of the company involved in the work.

Conflicts of Interest: The authors declare no conflict of interest.

References

1. Cavadias, S.; Ladas, K.; Loch, C. The Transformative Business Model. *Har. Bus. Rev.* 2016. Available online: https://www.apdata.com/upload/file/Edition_October_2016.pdf (accessed on 30 December 2021).
2. Gilmore, J.H.; Pine, B.J., 2nd. The four faces of mass customization. *Harv. Bus Rev.* **1997**, *75*, 91–101. [PubMed]
3. Rosenstrauch, M.J.; Pannen, T.J.; Krüger, J. Human robot collaboration—Using Kinect v2 for ISO/TS 15066 speed and sepa-ration monitoring. *Procedia CIRP* **2018**, *76*, 183–186. [CrossRef]
4. Duque, D.A.; Prieto, F.A.; Hoyos, J.G. Trajectory generation for robotic assembly operations using learning by demonstration. *Robot. Comput. Manuf.* **2019**, *57*, 292–302. [CrossRef]
5. Popov, V.; Ahmed, S.; Shakev, N.; Topalov, A. Gesture-based Interface for Real-time Control of a Mitsubishi SCARA Robot Manipulator. *IFAC-PapersOnLine* **2019**, *52*, 180–185. [CrossRef]

6. Kaczmarek, W.; Panasiuk, J.; Borys, S.; Banach, P. Industrial Robot Control by Means of Gestures and Voice Commands in Off-Line and On-Line Mode. *Sensors* **2020**, *20*, 6358. [CrossRef] [PubMed]
7. Kaczmarek, W.; Lotys, B.; Borys, S.; Laskowski, D.; Lubkowski, P. Controlling an Industrial Robot Using a Graphic Tablet in Offline and Online Mode. *Sensors* **2021**, *21*, 2439. [CrossRef] [PubMed]
8. Benotsmane, R.; Dudás, L.; Kovács, G. Trajectory Optimization of Industrial Robot Arms Using a Newly Elaborated "Whip-Lashing" Method. *Appl. Sci.* **2020**, *10*, 8666. [CrossRef]
9. Urrea, C.; Jara, D. Design, Analysis, and Comparison of Control Strategies for an Industrial Robotic Arm Driven by a Multi-Level Inverter. *Symmetry* **2021**, *13*, 86. [CrossRef]
10. Gašpar, T.; Deniša, M.; Radanovič, P.; Ridge, B.; Savarimuthu, T.R.; Kramberger, A.; Priggemeyer, M.; Roßmann, J.; Wörgötter, F.; Ivanovska, T.; et al. Smart hardware integration with advanced robot programming technologies for efficient reconfiguration of robot workcells. *Robot. Comput. Manuf.* **2020**, *66*, 101979. [CrossRef]
11. Djuric, A.M.; Urbanic, R.; Rickli, J. A Framework for Collaborative Robot (CoBot) Integration in Advanced Manufacturing Systems. *SAE Int. J. Mater. Manuf.* **2016**, *9*, 457–464. [CrossRef]
12. Michalos, G.; Makris, S.; Tsarouchi, P.; Guasch, T.; Kontovrakis, D.; Chryssolouris, G. Design Considerations for Safe Human-robot Collaborative Workplaces. *Procedia CIRP* **2015**, *37*, 248–253. [CrossRef]
13. Lee, J.; Lapira, E.; Bagheri, B.; Kao, H.-A. Recent advances and trends in predictive manufacturing systems in big data environment. *Manuf. Lett.* **2013**, *1*, 38–41. [CrossRef]
14. Wahrburg, A.; Zeiss, S.; Matthias, B.; Peters, J.; Ding, H. Combined pose-wrench and state machine representation for modeling Robotic Assembly Skills. In Proceedings of the 2015 IEEE/RSJ International Conference on Intelligent Robots and Systems (IROS), Hamburg, Germany, 28 September–2 October 2015; pp. 852–857. [CrossRef]
15. Stenmark, M.; Topp, E.A. From Demonstrations to Skills for High-Level Programming of Industrial Robots (2016) AAAI Fall Symposium; Technical Report; ELLIIT: The Linköping-Lund Initiative on IT and Mobile Communication Department of Computer Science Robotics and Semantic Systems FS-16-01–FS-16-05. pp. 75–78. Available online: https://portal.research.lu.se/en/publications/from-demonstrations-to-skills-for-high-level-programming-of-indus (accessed on 30 December 2021).
16. Herrero, H.; Moughlbay, A.A.; Outón, J.L.; Sallé, D.; de Ipiña, K.L. Skill based robot programming: Assembly, vision and Workspace Monitoring skill interaction. *Neurocomputing* **2017**, *255*, 61–70. [CrossRef]
17. Pedersen, M.R.; Nalpantidis, L.; Andersen, R.S.; Schou, C.; Bøgh, S.; Krüger, V.; Madsen, O. Robot skills for manufacturing: From concept to industrial deployment. *Robot. Comput. Manuf.* **2016**, *37*, 282–291. [CrossRef]
18. Sorensen, L.C.; Mathiesen, S.; Waspe, R.; Schlette, C. Towards Digital Twins for Industrial Assembly—Improving Robot Solutions by Intuitive User Guidance and Robot Programming. In Proceedings of the 2020 25th IEEE International Conference on Emerging Technologies and Factory Automation (ETFA2020), Vienna, Austria, 8–11 September 2020; pp. 1480–1484. [CrossRef]
19. Schou, C.; Damgaard, J.; Bogh, S.; Madsen, O. Human-robot interface for instructing industrial tasks using kinesthetic teaching. *IEEE ISR* **2013**, *2013*, 1–6. [CrossRef]
20. Bøgh, S.; Nielsen, S.O.; Pedersen, R.M.; Krüger, V.; Madsen, O. Does your Robot have Skills? In Proceedings of the 43rd International Symposium on Robotics, Taipei, Taiwan, 29–31 August 2012.
21. Pedersen, M.R.; Nalpantidis, L.; Bobick, A.; Krüger, V. On the Integration of Hardware-Abstracted Robot Skills for use in In-dustrial Scenarios. In Proceedings of the 2nd International IROS Workshop on Cognitive Robotics Systems (CRS): Replicating Human Actions and Activities, Tokyo, Japan, 3–7 November 2013.
22. Akkaladevi, S.C.; Pichler, A.; Plasch, M.; Ikeda, M.; Hofmann, M. Skill-based programming of complex robotic assembly tasks for industrial application [Skill-basierte Programmierung von komplexen Roboter-Montageaufgaben für die industrielle Applikation]. *E I Elektrotechnik Und Inf.* **2019**, *136*, 326–333. [CrossRef]
23. Andersen, R.S.; Damgaard, J.S.; Madsen, O.; Moeslund, T.B. Fast calibration of industrial mobile robots to workstations using QR codes. *IEEE ISR* **2013**, *2013*, 1–6. [CrossRef]
24. Hvilshøj, M.; Bøgh, S.; Madsen, O.; Kristiansen, M. Calibration Techniques for Industrial Mobile Manipulators: Theoretical configurations and Best practices (2010) Joint. In Proceedings of the 41st International Symposium on Robotics and 6th German Conference on Robotics 2010 ISR/ROBOTIK, Munich, Germany, 7–9 June 2010; Volume 2, pp. 773–779.
25. Korayem, M.H.; Azargoshasb, S.; Tabibian, S. Design and Implementation of the Voice Command Recognition and the Sound Source Localization System for Human–Robot Interaction. *Robotics* **2021**, *39*, 1779–1790. [CrossRef]
26. Dmytriyev, Y.; Zaki, A.M.A.; Carnevale, M.; Insero, F.; Giberti, H. Brain computer interface for human-cobot interaction in industrial applications. In Proceedings of the 3rd International Congress on Human-Computer Interaction, Optimization and Robotic Applications (HORA), Ankara, Turkey, 11–13 June 2021.

Author Index

MDPI
St. Alban-Anlage 66
4052 Basel
Switzerland
Tel. +41 61 683 77 34
Fax +41 61 302 89 18
www.mdpi.com

Robotics Editorial Office
E-mail: robotics@mdpi.com
www.mdpi.com/journal/robotics